SELECTED SOLUTIONS

for

FUNDAMENTALS OF PHYSICS

SECOND EDITION / SECOND EDITION EXTENDED

DAVID HALLIDAY

ROBERT RESNICK

EDWARD DERRINGH

PREPARED BY

EDWARD DERRINGH
Wentworth Institute of Technology
Boston, Massachusetts

JOHN WILEY & SONS NEW YORK
CHICHESTER BRISBANE TORONTO

ISBN 0 471 06463 7

Printed in the United States of America

10 9 8 7 6 5

Michael DiLeva

PREFACE

This manual contains worked-out solutions to more than one quarter of the problems appearing in the text at the end of each chapter. An effort has been made to select the more difficult of these, but it must be borne in mind that "difficulty" is not a physical quantity for which there exists an absolute standard. The volume is intended to be used with the text: neither the statements of the problems nor any accompanying text figures are reproduced; diagrams are added in the solutions only when these appear to be necessary.

Generally, the equations used are presented first in algebraic form followed by numerical substitution. Often the units are omitted in the latter step: many equations would not fit on one line if this was not done; also, possible confusion in notation (e.g., N for normal force and newton, V for potential and volt) is avoided. However, numerical data substituted into a given equation is always in a consistent set of units; for example, newtons are not combined with centimeters, but the centimeters are first converted to meters. The reader should be alert for this.

In response to comments from student users of previous editions of both FUNDAMENTALS and of the original PHYSICS, the rules with regard to significant digits have been disregarded in this manual.

Selection of the problems and presentation of the solutions was left to the undersigned. Consequently, he alone is responsible for any errors that may be present, and would appreciate being notified of any that are found.

Cohasset, Massachusetts

Edward Derringh

October 26, 1980

CONTENTS

CHAPTER 1

1-6

(a) 1 light year = (186,000 mi/s)(3.156 X 10^7 s) = 5.870 X 10^{12} mi so that

$$1 \text{ mi} = \frac{1}{5.870 \times 10^{12}} = 1.7 \times 10^{-13} \text{ ly}.$$

Therefore,

$$1 \text{ AU} = 92.9 \times 10^6 \text{ mi} = (92.9 \times 10^6 \text{ mi})(1.7 \times 10^{-13} \text{ ly/mi}),$$

$$1 \text{ AU} = 1.58 \times 10^{-5} \text{ ly} \quad \underline{\text{Ans}}.$$

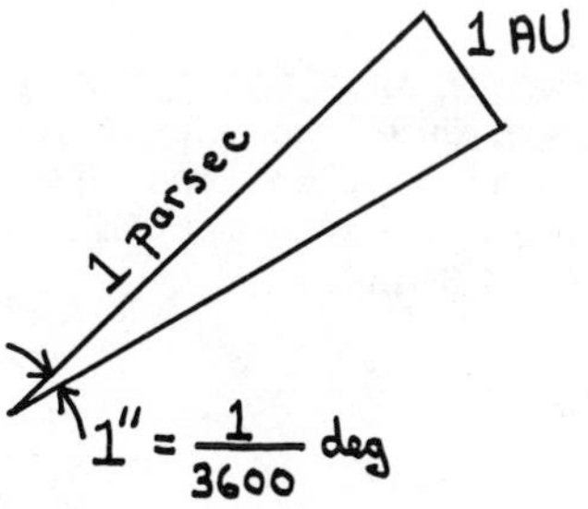

Since 206265″ = 1 rad, then by definition,

$$206265 \text{ AU} = 1 \text{ pc};$$

hence,

$$1 \text{ AU} = \frac{1}{206265} \text{ pc},$$

$$1 \text{ AU} = 4.85 \times 10^{-6} \text{ pc} \quad \underline{\text{Ans}}.$$

(b) From (a), 1 ly = 5.87 X 10^{12} mi <u>Ans</u>.

Since 1 AU = 4.85 X 10^{-6} pc = 15.8 X 10^{-6} ly, it follows that

$$1 \text{ pc} = \frac{15.8}{4.85} = 3.26 \text{ ly}$$

and therefore

$$1 \text{ pc} = (3.26)(5.87 \times 10^{12} \text{ mi}) = 1.91 \times 10^{13} \text{ mi} \quad \underline{\text{Ans}}.$$

1-9

The apparent angular diameters θ of the sun and moon in the sky are virtually identical. Thus the situation is as shown in the sketch. Let the earth-moon distance be r, the earth-sun distance R and call d the diameter of the moon and D the same for the sun.

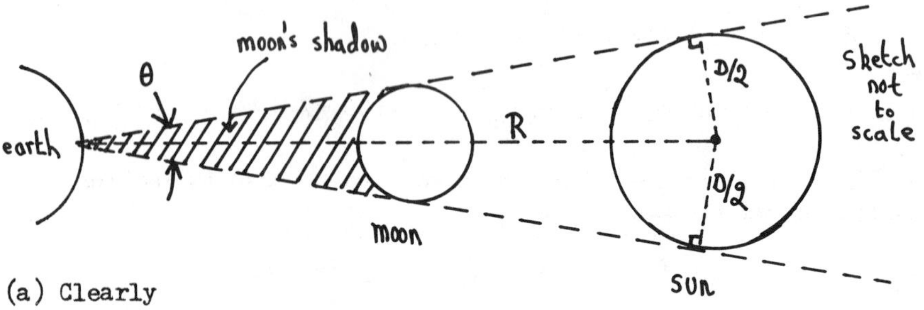

(a) Clearly

$$\sin(\tfrac{1}{2}\theta) = \tfrac{1}{2}d/r = \tfrac{1}{2}D/R,$$

$$D/d = R/r = (400r)/r = 400 \quad \underline{\text{Ans.}}$$

(b) Since the volume of a sphere = $\pi(\text{diameter})^3/6$,

$$\text{volume of sun/volume of moon} = D^3/d^3 = 400^3 = 6.4 \times 10^7 \quad \underline{\text{Ans.}}$$

(c) In the figure, substitute 'dime' for 'moon' and 'moon' for 'sun'. You will find by experiment that θ = 0.0092 rad (i.e., you must hold a dime about 190 cm from your eye to just eclipse the moon; do NOT try this with the sun). The average earth-moon distance r = 380,000 km. Since θ is very small, sinθ ≈ θ and

$$d = r\theta = (3.8 \times 10^5 \text{ km})(0.0092) = 3500 \text{ km} \quad \underline{\text{Ans.}}$$

1-12

(a) Since 1g = 10^{-3} kg and 1 ℓ = 1000 cm^3, it follows that 1 g/cm^3 = 10^{-3} kg/10^{-3} ℓ = 1 kg/ℓ <u>Ans.</u>

(b) 1.0 liter of water contains 1.0 kg of water, from (a). Since 10 h = (10)(3600 s) = 36,000 s, the mass flow rate = mass/time is

$$1.0 \text{ kg}/36{,}000 \text{ s} = 2.78 \times 10^{-5} \text{ kg/s} \quad \underline{\text{Ans.}}$$

1-15

The year used in civil affairs is the tropical year and this contains 365.2422 mean solar days. Using the old definition of the second (1 mean solar day = 86,400 seconds),

$$1 \text{ year} = (365.2422)(86{,}400 \text{ s}) = 3.1557 \times 10^7 \text{ s}.$$

Since $\pi \times 10^7 = 3.1416 \times 10^7$ approximately,

$$\% \text{ error} = \frac{3.1416 - 3.1557}{3.1557} \times 100 = -0.447\%,$$

the sign indicating that the approximate value used is smaller than the actual number.

1-20

Let f = average rotation rate of the earth during the year. From the graph, the deviation in f at midsummer is +60 and the deviation in spring is -90 (the values depend on the exact dates chosen for comparison), leading to a net deviation of 150 parts in 10^{10} of the average rate; that is,

$$\text{net deviation in rate} = 150 \times 10^{-10} f.$$

The "total equivalent" rate is $f + 150 \times 10^{-10} f$, corresponding to a rotation period equal to

$$(f + 150 \times 10^{-10} f)^{-1} \cong f^{-1}(1 - 150 \times 10^{-10})$$

$$= (86{,}400 \text{ s})(1 - 150 \times 10^{-10}).$$

Hence, the difference in period between midsummer and spring is

$$(86{,}400)(150 \times 10^{-10}) \text{ s} = 1.3 \times 10^{-3} \text{ s} \quad \underline{\text{Ans}},$$

the period in midsummer being shorter.

1-22

The last day of the twenty centuries is longer than the first day by

$$(20 \text{ centuries})(10^{-3} \text{ s/century})$$

which is 0.020 s. The average day during the twenty centuries is

(0 + 0.020)/2 = 0.010 s longer than the first day. As the increase occurs uniformly, the total cumulative effect is

$$(\text{average difference})(\text{number of days})$$
$$= (0.010 \text{ s/average day})(365.25 \text{ days} \times 2000),$$
$$= 7305 \text{ s} = 2.0 \text{ h} \quad \underline{\text{Ans.}}$$

1-23

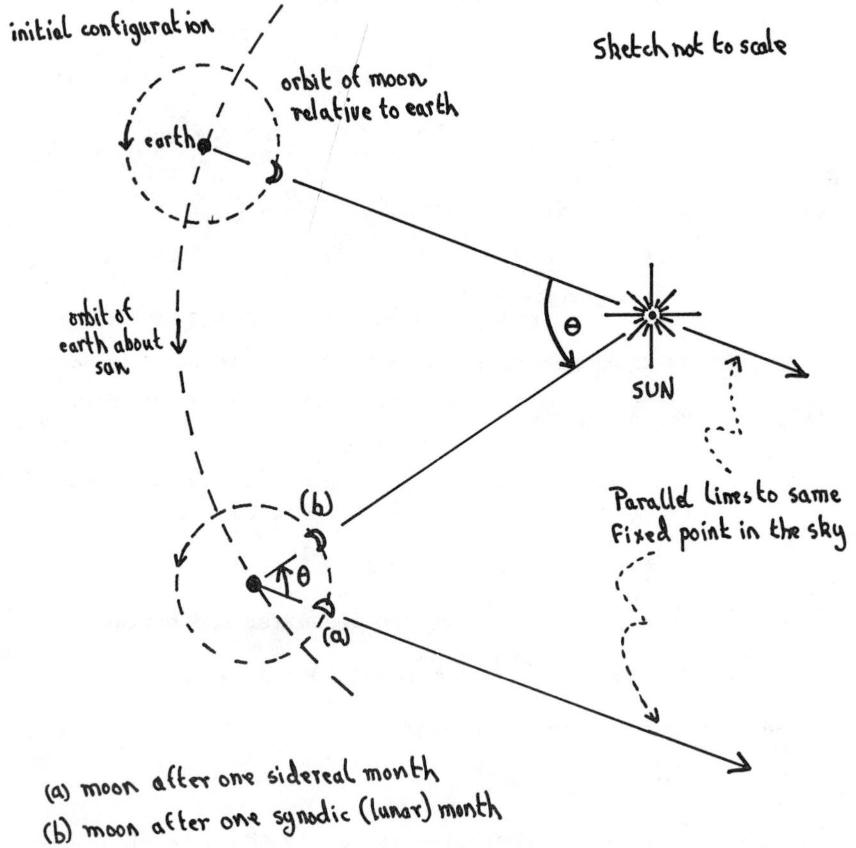

The moon revolves around the earth as the earth revolves around the sun, and the motions are in the same sense, counterclockwise as seen from the north. The synodic period of the moon (lunar month), the period relative to the sun, can be measured as the time required for the moon to pass from new moon to new moon, say, since the phases of the moon depend on the position of the moon relative to the earth-sun line. The sidereal period, the period relative to the stars, can be measured as the time required for the earth-moon line to sweep through 360 degrees with respect to the stars: the earth-moon lines after one sidereal month are parallel, the stars being so far away compared to the size of the earth's orbit about the sun that lines drawn from the earth to any one fixed star are parallel, regardless of where the earth is in its orbit. Thus, the configurations of the earth, moon and sun after one sidereal and one synodic month are as shown in the sketch.

Clearly, one synodic month is longer than one sidereal month, the reason being that the moon revolves in the same direction as the earth revolves. The months differ by the time needed for the moon to cover the angle θ in the diagram. Since the month is roughly 30 days, or about 1/12 year, $\theta = (360 \text{ deg})/12$, approximately; the time sought is about 1/12(30 days) = 2.5 days approximately (more exactly, 2 days 5 hours).

2-5

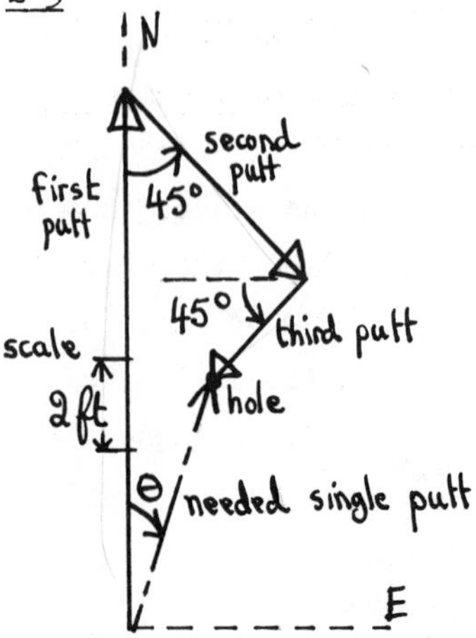

The displacements of the ball due to the successive putts are shown in the sketch, with the scale used shown on the first putt. The putt that should have been made in the first place is drawn from the tail of the first to the head of the last, as required by the rules of vector addition. Use of the scale and a protractor show the desired putt to be about 6 ft (length of vector), at an angle θ of about 20 degrees east of north.

2-12

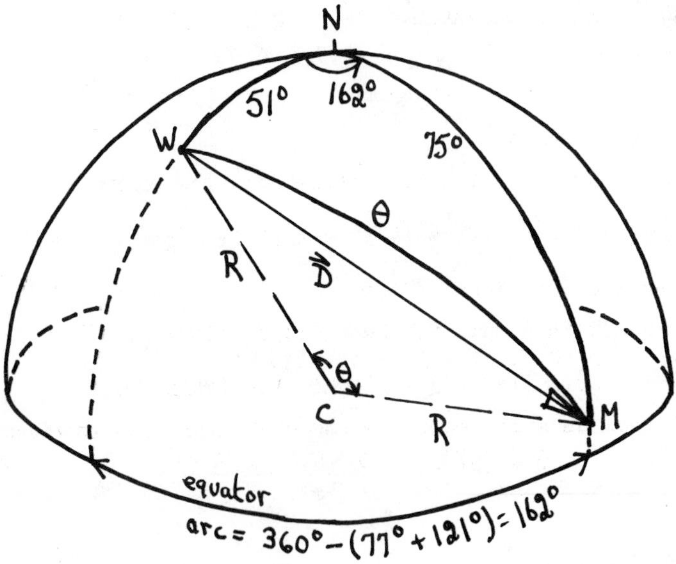

(a) The displacement vector $\vec{D}$ joins Washington and Manila along the straight line connecting them, the line passing through the solid earth, and points towards Manila.

(b) In the spherical triangle NMW, where N is the north pole, the law of cosines gives

$$\cos\theta = \cos 51^\circ \cos 75^\circ + \sin 51^\circ \sin 75^\circ \cos 162^\circ = -0.55105.$$

If R = 6378 km, the radius of the earth, it is clear from the plane triangle WCM, where C is the center of the earth, that

$$D = 2R \sin\tfrac{1}{2}\theta = 11{,}230 \text{ km} \quad \underline{\text{Ans}},$$

using the value of θ from the first equation.

2-16

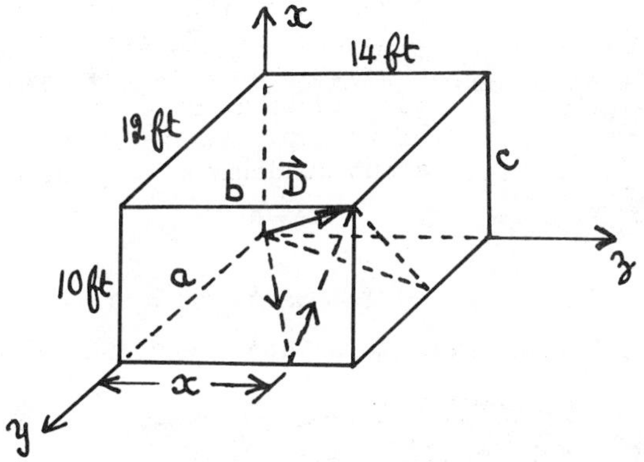

(a) From the figure, $\vec{D} = 10\vec{i} + 12\vec{j} + 14\vec{k}$ and therefore

$$D = (10^2 + 12^2 + 14^2)^{\frac{1}{2}} = 21.0 \text{ ft} \quad \underline{\text{Ans}}.$$

(b) Answering the questions in turn:

no: a straight line is the shortest distance;
yes: fly need not fly in a straight line;
yes: it could fly in a straight line (unlikely).

(c) For the room as oriented above and choice of corners shown,

$$\vec{D} = 10\vec{i} + 12\vec{j} + 14\vec{k} \quad \underline{\text{Ans}}.$$

(d) For this part, call the lengths of the sides of the room a, b, c as indicated (do not set a = 12 ft, b = 14 ft, c = 10 ft), for the fly has a choice of walls to walk on first; two possible paths to the opposite corner are shown; focus on the one with arrows. The total length L of the path is

$$L = \left[a^2 + x^2\right]^{\frac{1}{2}} + \left[(b - x)^2 + c^2\right]^{\frac{1}{2}}.$$

To find the shortest path set dL/dx = 0; this gives

$$(c^2 - a^2)x^2 + 2a^2bx - a^2b^2 = 0.$$

The two solutions follow from the quadratic formula and are

$$x_1 = \frac{ab}{a - c}\ ;\ x_2 = \frac{ab}{a + c}.$$

Now x represents a distance and must be positive. Likewise, b - x is also a distance (see figure to locate it) and must also be positive; these latter distances are

$$b - x_1 = \frac{bc}{c - a}\ ;\ b - x_2 = \frac{bc}{a + c}.$$

Clearly, x_1 and $b - x_1$ both cannot be positive. Therefore, choose $x_2 = x = ab/(a + c)$. Substituting this into the equation for L gives

$$L_{min} = L(x_2) = \left[(a + c)^2 + b^2\right]^{\frac{1}{2}},$$

$$L_{min} = (a^2 + b^2 + c^2 + 2ac)^{\frac{1}{2}}.$$

For the smallest L_{min}, select a, b, c so that 2ac is the smallest possible: a = 10 ft, c = 12 ft or a = 12 ft, c = 10 ft. In either case, b = 14 ft and

$$L_{min} = (12^2 + 14^2 + 10^2 + 240)^{\frac{1}{2}} = 26.1 \text{ ft} \quad \underline{\text{Ans.}}$$

2-18

Orient the coordinate axes so that one of the vectors lies along either the x or y-axis; for example, let $\vec{a}$ lie along the x-axis. Then

$$\vec{a} = a\vec{i}\ ;\ \vec{b} = (b\cos\theta)\vec{i} + (b\sin\theta)\vec{j}$$

and therefore

$$\vec{a} + \vec{b} = (a + b\cos\theta)\vec{i} + (b\sin\theta)\vec{j};$$

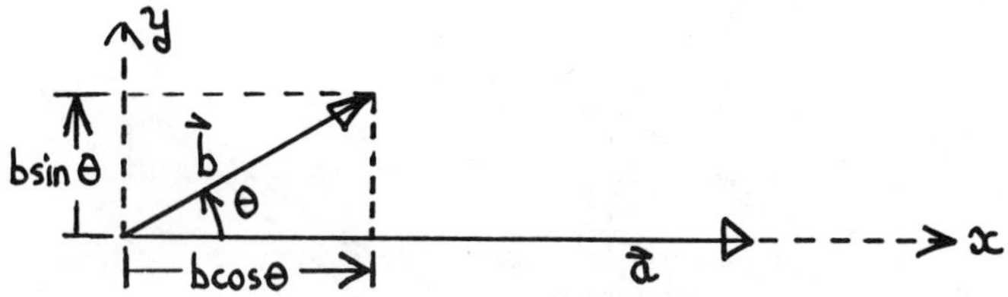

hence,

$$|\vec{a} + \vec{b}| = [(a + b\cos\theta)^2 + (b\sin\theta)^2]^{\frac{1}{2}} = [a^2 + b^2 + 2ab\cos\theta]^{\frac{1}{2}} \quad \underline{\text{Ans}}.$$

2-24

(a) With $\vec{R} = \vec{A} - \vec{B} + \vec{C}$, the rules of vector addition give

$$\begin{aligned} \vec{A} &= 5\vec{i} + 4\vec{j} - 6\vec{k} \\ -\vec{B} &= 2\vec{i} - 2\vec{j} - 3\vec{k} \\ \vec{C} &= 4\vec{i} + 3\vec{j} + 2\vec{k} \\ \hline \vec{R} &= 11\vec{i} + 5\vec{j} - 7\vec{k} \quad \underline{\text{Ans}}. \end{aligned}$$

The magnitude R of $\vec{R}$ is just $R = (11^2 + 5^2 + 7^2)^{\frac{1}{2}} = 14.0$ Ans.

(b) Let the angle sought be θ; then,

$$\cos\theta = R_z/R = (-7)/14 = -0.5;$$
$$\theta = 120^\circ \quad \underline{\text{Ans}}.$$

2-36

Since $\vec{a} = 3\vec{i} + 3\vec{j} - 3\vec{k}$, $a = 3\sqrt{3}$; $\vec{b} = 2\vec{i} + \vec{j} + 3\vec{k}$ so that $b = \sqrt{14}$. Hence,

$$\vec{a}\cdot\vec{b} = ab\cos\theta = (3\sqrt{3})(\sqrt{14})\cos\theta = 3\sqrt{42}\cos\theta.$$

But $a_xb_x + a_yb_y + a_zb_z = (3)(2) + (3)(1) + (-3)(3) = 0$ and since these two expressions for the scalar product are equal, it follows that

$$3\sqrt{42}\cos\theta = 0; \ \cos\theta = 0; \ \theta = 90^\circ \quad \underline{\text{Ans}}.$$

2-37

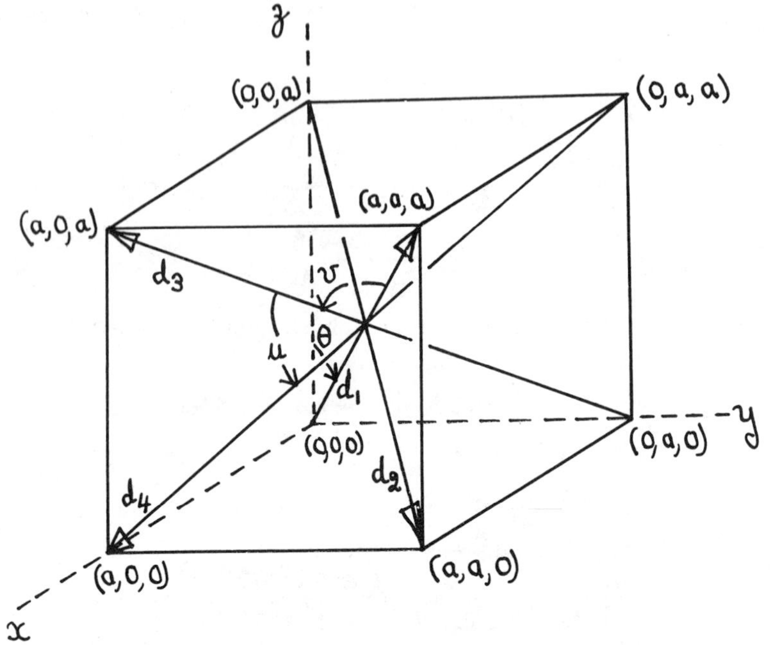

(a) Each diagonal vector can be represented as the difference between vectors drawn from the origin to its head and to its tail; that is,

$$\vec{d}_1 = (a - 0)\vec{i} + (a - 0)\vec{j} + (a - 0)\vec{k} = a\vec{i} + a\vec{j} + a\vec{k};$$
$$\vec{d}_2 = (a - 0)\vec{i} + (a - 0)\vec{j} + (0 - a)\vec{k} = a\vec{i} + a\vec{j} - a\vec{k};$$
$$\vec{d}_3 = (a - 0)\vec{i} + (0 - a)\vec{j} + (a - 0)\vec{k} = a\vec{i} - a\vec{j} + a\vec{k};$$
$$\vec{d}_4 = (a - 0)\vec{i} + (0 - a)\vec{j} + (0 - a)\vec{k} = a\vec{i} - a\vec{j} - a\vec{k},$$

or, since the direction of the diagonal vector is irrelevant, the negatives of any or all of the above would also be correct.

(b) Let $\vec{R}$ lie on the edge through the origin along the +z-axis, so that $\vec{R} = a\vec{k}$. Using the two expressions for scalar product,

$$\vec{d}_1 \cdot \vec{R} = d_1 R \cos\theta = (a\sqrt{3})(a)\cos\theta = a^2$$

$$\cos\theta = 3^{-\frac{1}{2}} \;;\; \theta = 54.7^\circ \quad \underline{\text{Ans}}.$$

(c) Since the components of each diagonal are either +a or -a, the magnitude of each of them is $(a^2 + a^2 + a^2)^{\frac{1}{2}} = 3^{\frac{1}{2}}a$ Ans.

(d) Consider the angle u between diagonals d_3 and d_4. Again, using the two scalar product expressions,

$$\vec{d}_3 \cdot \vec{d}_4 = (3^{\frac{1}{2}}a)(3^{\frac{1}{2}}a)\cos u = a^2 + a^2 - a^2 = a^2;$$

$$3a^2 \cos u = a^2,$$

$$\cos u = \frac{1}{3},$$

$$u = 70.5° \text{ Ans.}$$

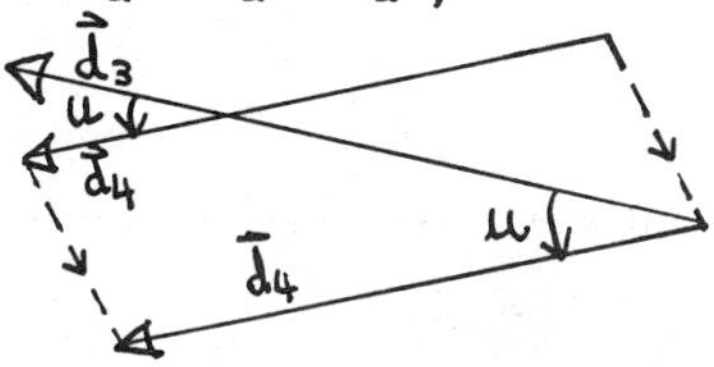

The reader will find the same result for the angle between the diagonals d_1 and d_3.

2-40

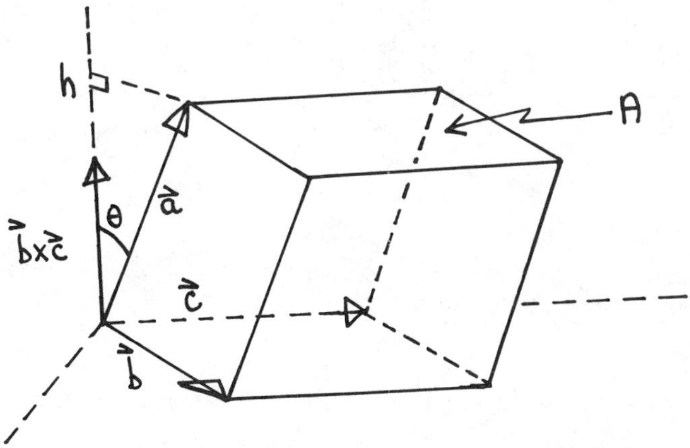

Since the area of a parallelogram is the base times the height,

$$|\vec{b} \times \vec{c}| = \text{area of base} = \text{area of top} = A,$$

and therefore,

$$\vec{a}\cdot(\vec{b} \times \vec{c}) = a|\vec{b} \times \vec{c}|\cos(\vec{a}, \vec{b} \times \vec{c}) = a\ A \cos\theta = (a \cos\theta)A = hA,$$

which is the volume of the parallelopiped.

2-41

Let the vectors be $\vec{a}$ and $\vec{b}$. As the scalar product of perpendicular vectors is zero,

$$(\vec{a} + \vec{b})\cdot(\vec{a} - \vec{b}) = 0,$$

$$\vec{a}\cdot\vec{a} - \vec{a}\cdot\vec{b} + \vec{b}\cdot\vec{a} - \vec{b}\cdot\vec{b} = 0,$$

$$a^2 - b^2 = 0,$$

$$a = b\ .$$

2-44

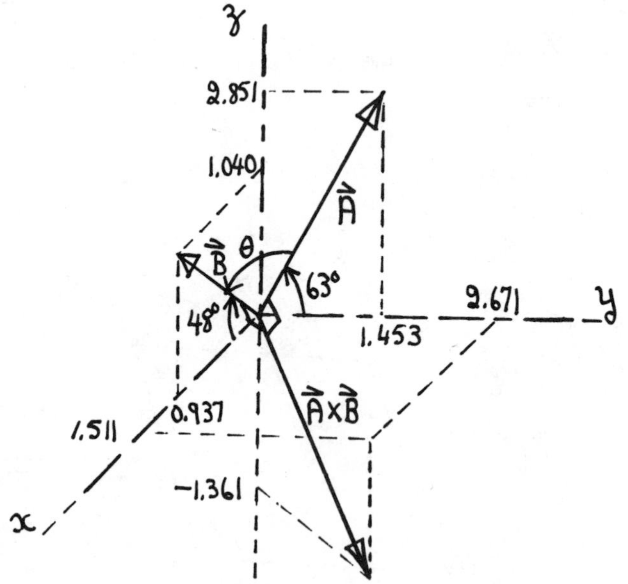

Write the vectors in unit vector (component) form:

$$\vec{A} = 0\vec{i} + (3.2\cos 63^\circ)\vec{j} + (3.2\sin 63^\circ)\vec{k},$$

$$\vec{A} = 0\vec{i} + 1.45277\vec{j} + 2.85122\vec{k};$$

$$\vec{B} = (1.4\cos 48^\circ)\vec{i} + 0\vec{j} + (1.4\sin 48^\circ)\vec{k},$$

$$\vec{B} = 0.93678\vec{i} + 0\vec{j} + 1.04040\vec{k}.$$

Hence,

$$\vec{A}\cdot\vec{B} = (2.85122)(1.04040) = 2.97 \quad \underline{\text{Ans}}.$$

Using the formula for the vector product in unit vector notation,

$$\vec{A} \times \vec{B} = (1.45277)(1.04040)\vec{i}$$
$$+ (2.85122)(0.93678)\vec{j} - (1.45277)(0.93678)\vec{k},$$
$$\vec{A} \times \vec{B} = 1.51\vec{i} + 2.67\vec{j} - 1.36\vec{k} \quad \underline{\text{Ans}}.$$

For the angle θ between $\vec{A}$ and $\vec{B}$,

$$\vec{A}\cdot\vec{B} = (3.2)(1.4)\cos\theta = 2.96641,$$
$$\theta = 48.5° \quad \underline{\text{Ans}}.$$

2-45

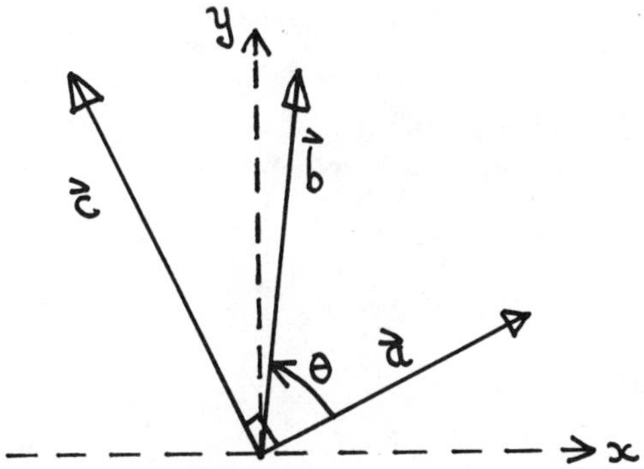

(a) The magnitudes of the vectors are a = 3.5777, b = 4.5277. Using the formulae for scalar product:

$$\vec{a}\cdot\vec{b} = ab\cos\theta = (3.5777)(4.5277)\cos\theta = 16.1988\cos\theta,$$
$$\vec{a}\cdot\vec{b} = (3.2)(0.5) + (1.6)(4.5) = 8.8,$$
$$8.8 = 16.1988\cos\theta,$$
$$\theta = 57° \quad \underline{\text{Ans}}.$$

(b) In this part, the vector $\vec{b}$ has no role. Write $\vec{c} = c_x\vec{i} + c_y\vec{j}$. Since c = 5,

$$25 = c_x^2 + c_y^2.$$

Also, since $\vec{c}$ is perpendicular to $\vec{a}$, then $\vec{c}\cdot\vec{a} = 0$ so that

$$3.2c_x + 1.6c_y = 0.$$

This last equation gives $c_y = -2c_x$. Substituting this into the first equation yields

$$25 = c_x^2 + 4c_x^2 = 5c_x^2,$$

$$c_x = \pm 5^{\frac{1}{2}} = \pm 2.236; \quad c_y = \mp 4.472 \quad \underline{\text{Ans}}.$$

The vector $\vec{c}$ with components given by the lower set of signs is shown in the figure; the other possibility is antiparallel to this.

2-46

(a) Since the vector product of two vectors is perpendicular to each of them (by definition), it follows that $\vec{C} = \vec{B} \times \vec{A}$ is at right angles to $\vec{A}$. But then $\vec{C}\cdot\vec{A} = CA\cos 90° = 0$ and therefore $\vec{A}\cdot(\vec{B} \times \vec{A}) = 0$.

(b) $\vec{B} \times \vec{A}$ is perpendicular to $\vec{A}$ and has a magnitude $BA\sin\theta$. Hence $|\vec{A} \times (\vec{B} \times \vec{A})| = A(BA\sin\theta)\sin 90° = A^2B\sin\theta$. The directions, given by the right-hand rule, are shown below.

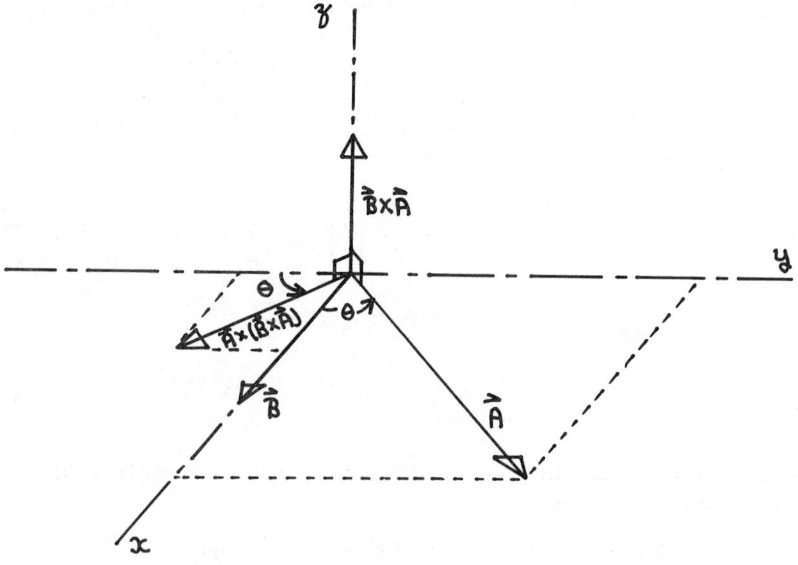

CHAPTER 3

3-4

Let $\vec{D}$ be the train's displacement during the total travel time T. Then, by definition, the average velocity $\vec{\bar{v}}$ is

$$\vec{\bar{v}} = \vec{D}/T .$$

Since 60 km/h = 1 km/min, the successive distances travelled are 40 km, 20 km and 50 km. A diagram showing the displacement vectors is easily constructed and from this the total displacement $\vec{D}$ is found to be

$$\vec{D} = \left[40\vec{i} + (20\sqrt{2}/2\vec{i} + 20\sqrt{2}/2\vec{j}) - 50\vec{i}\right] \text{ km},$$

$$\vec{D} = 4.142\vec{i} + 14.142\vec{j}.$$

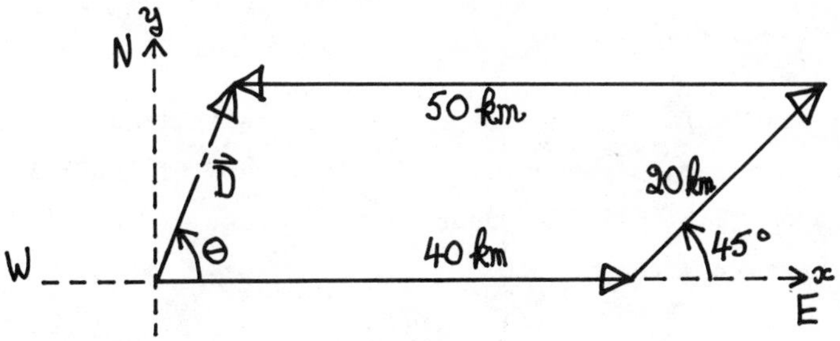

The total time T of travel is, of course, 40 min + 20 min + 50 min = 110 min = 11/6 h. As the magnitude D of $\vec{D}$ is

$$D = (4.142^2 + 14.142^2)^{\frac{1}{2}} = 14.736 \text{ km},$$

then,

$$\bar{v} = \frac{14.736 \text{ km}}{11/6 \text{ h}} = 8.0 \text{ km/h},$$

and the direction of the average velocity is at an angle θ north of east where

$$\theta = \tan^{-1}(14.142/4.142) = 73.7° \quad \underline{\text{Ans.}}$$

3-6

(a) The position at $t = 1$ s is found by replacing t with 1 in the equation given for position; that is,

$$x(1) = 3.0(1) - 4.0(1)^2 + (1)^3 = 0 \quad \underline{\text{Ans.}}$$

Similarly,

$$x(2) = -2;\ x(3) = 0;\ x(4) = 12 \text{ m} \quad \underline{\text{Ans.}}$$

(b) When $t = 0$ the position of the object is $x(0) = 0$; at $t = 4$ s the position is +12 m. Hence, the displacement during this time interval is

$$\text{displacement} = x(4) - x(0) = +12 \text{ m} \quad \underline{\text{Ans}},$$

the positive sign indicating a displacement in the +x-direction.

(c) By definition, the required average velocity is

$$\bar{v} = \frac{x(4) - x(2)}{4 - 2} = \frac{12 - (-2)}{4 - 2} \text{ m/s} = +7 \text{ m/s} \quad \underline{\text{Ans.}}$$

3-22

(a) As the spaceship starts from rest, put the initial velocity $v_0 = 0$; the acceleration $a = 9.8 \text{ m/s}^2$. The speed of light is 3×10^8 m/s so that the final speed desired is $v = 3 \times 10^7$ m/s. The time t required is given by

$$v = at + v_0,$$
$$3 \times 10^7 \text{ m/s} = (9.8 \text{ m/s}^2)t + 0,$$
$$t = 3.06 \times 10^6 \text{ s} \quad \underline{\text{Ans.}}$$

This is about $35\frac{1}{2}$ days.

(b) The distance x travelled in this time is

$$x = \tfrac{1}{2}at^2 + v_0 t + x_0,$$
$$x = \tfrac{1}{2}(9.8 \text{ m/s}^2)(3.06 \times 10^6 \text{ s})^2,$$
$$x = 4.59 \times 10^{13} \text{ m} \quad \underline{\text{Ans}}.$$

3-26

Let $x = 0$ be the position of the car when at rest and x_1 and x_2 be the "first" and "second" points, which the car passes at times t_1 and t_2 with velocities v_1 and v_2. Then, $x_2 - x_1 = 60$ m, $t_2 - t_1 = 6$ s and $v_2 = 15$ m/s.

(a) The average velocity between the two points is

$$\bar{v} = \frac{x_2 - x_1}{t_2 - t_1} = \frac{60 \text{ m}}{6 \text{ s}} = 10 \text{ m/s} .$$

But the acceleration a is constant so that

$$\bar{v} = \tfrac{1}{2}(v_1 + v_2) ,$$
$$10 = \tfrac{1}{2}(v_1 + 15),$$
$$v_1 = 5 \text{ m/s} \quad \underline{\text{Ans}}.$$

(b) The acceleration can be found from the relation involving the two velocities, both now known:

$$v_2 = a(t_2 - t_1) + v_1 ,$$
$$15 = a(6) + 5,$$
$$a = 5/3 = 1.67 \text{ m/s}^2 \quad \underline{\text{Ans}} .$$

(c) To find the distance x_1:

$$v_1^2 = v_0^2 + 2ax_1 ,$$
$$25 = 0 + 2(5/3)x_1,$$
$$x_1 = 7.5 \text{ m} \quad \underline{\text{Ans}}.$$

3-28

If the direction in which the muon is travelling is considered to be positive, then $v_0 = +5.00 \times 10^6$ m/s at $x = 0$ and since the acceleration is in the oppsoite direction, $a = -1.25 \times 10^{14}$ m/s^2. At a distance x from the point where the muon entered the field the velocity $v = 0$; hence:

$$v^2 = v_0^2 + 2ax,$$
$$0^2 = (+5 \times 10^6)^2 + 2(-1.25 \times 10^{14})x,$$
$$x = 0.10 \text{ m} \quad \underline{\text{Ans.}}$$

3-33

Let t = driver reaction time, t' = breaking time at 50 mi/h and t" = breaking time at 30 mi/h. Then, if a is the deceleration, and since 30 mi/h = 44 ft/s and 50 mi/h = 73.33 ft/s,

$$186 = \tfrac{1}{2}at'^2 + 73.33t,$$
$$80 = \tfrac{1}{2}at''^2 + 44t,$$
$$73.33 = at',$$
$$44 = at''.$$

These are four equations for the four unknowns a, t, t', t". Use the third equation to eliminate a from the first equation, and the fourth to eliminate a from the second. Then use the third and the fourth in the form

$$\frac{73.33}{44} = \frac{t'}{t''},$$

to eliminate t', say, from the two equations resulting from the sequence of operations described above. The result is a pair of equations in t, t", which can be solved easily. The results are

$$\text{(a) } t = 0.74 \text{ s} \quad \underline{\text{Ans}}.$$
$$\text{(b) } -a = \text{acceleration} = -20 \text{ ft/s}^2 \quad \underline{\text{Ans}}.$$

3-34

Let the vehicle be moving at a speed v_0 when, at t = 0, the driver slams on the brakes a distance x from the barrier, which it strikes 4.0 s later at a speed of v_f. Then, if a is the acceleration and since 35 mi/h = 51.33 ft/s

(a)

$$x = \tfrac{1}{2}at^2 + v_0t + x_0,$$
$$110 = \tfrac{1}{2}a(4)^2 + (51.33)4 + 0,$$
$$a = -11.9 \text{ ft/s}^2 \quad \underline{\text{Ans}}.$$

(b)

$$v_f = at + v_0,$$
$$v_f = (-11.9)(4) + 51.33,$$
$$v_f = 3.73 \text{ ft/s} \quad \underline{\text{Ans}}.$$

3-36

(a) Since at^2 must have the dimensions of length, $\underline{a}$ must have dimensions of (length)/(time)^2; similarly, $\underline{b}$ must have dimensions of (length)/(time)^3; in the SI system, these are m/s^2 and m/s^3.

(b) With a = 3, b = 1 the given equation for position becomes

$$x = 3t^2 - t^3,$$

which is shown sketched, with x in meters and t in seconds.

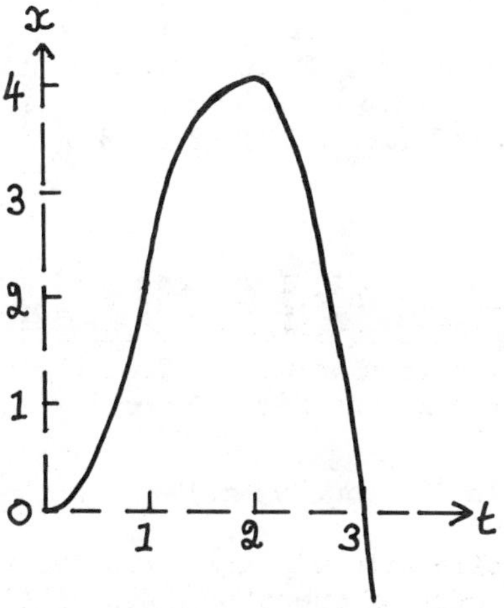

Setting dx/dt = 0 gives

$$0 = 6t - 3t^2;$$

this is satisfied by t = 0 and 2. The value t = 2 gives the maximum position x.

(c) From t = 0 to t = 2 the particle travels x(2) - x(0) = 4 m. From t = 2 to t = 3 the particle travels x(3) - x(2) = 4 m, and from t = 3 to t = 4 it travels x(4) - x(3) = 16 m. Thus, the total distance travelled is 4 + 4 + 16 = 24 m <u>Ans.</u>

(d) Since x(4) = -16, the displacement is x(4) - x(0) = -16 m <u>Ans.</u>

(e) $v = dx/dt = 6t - 3t^2$ and (f) $a = dv/dt = 6(1 - t)$; from these the following table is derived:

t	v(m/s)	a(m/s^2)
1	3	0
2	0	-6
3	-9	-12
4	-24	-18

3-39

Let the upward direction be - and down +; call v the speed of the ball upon striking the floor and u the speed as it leaves; the associated velocities are, then, +v and -u. If the ball is in contact with the floor for a time T, the average acceleration is, by definition,

$$\bar{\vec{a}} = \frac{-u - (+v)}{T} = -\frac{u + v}{T};$$

as this is negative, the average acceleration is directed upward, as expected. Since the speed with which the ball leaves the floor is the same as the speed with which it would strike the floor if dropped from a height of 3.0 ft, both u and v can be found from

$$v^2 = 2gh,$$

using h = 4 ft for v and 3 ft for u; $g = 32\ \text{ft/s}^2$ in each case. Therefore v = 16 ft/s and u = 13.86 ft/s so that

$$\bar{a} = \frac{13.86 + 16}{0.01} = 3000\ \text{ft/s}^2,\ \text{up}\quad \underline{\text{Ans.}}$$

(This is about 93 g.)

3-43

Choose the positive x-axis vertically upward with origin at the ground. At t = 0 the package is released with velocity $\vec{v}_0$ directed upwards (i.e. $v_0 = +12$ m/s) from height $x_0 = +80$ m. At t = T it strikes the ground at x = 0. Hence,

$$x = \tfrac{1}{2}(-9.8)t^2 + 12t + 80,$$

gives the position x of the package as a function of time t; a is negative since the acceleration is down, in the negative direction. At t = T, x = 0 (package strikes ground):

$$0 = \tfrac{1}{2}(-9.8)T^2 + 12T + 80,$$

$$T = 5.4\ \text{s}\quad \underline{\text{Ans.}}$$

3-46

(a) The initial speed v of the ball relative to the ground is 10 + 20 = 30 m/s. The height x reached by the ball above the elevator is found from

$$v^2 = 2gx,$$
$$(30)^2 = 2(9.8)x,$$
$$x = 46 \text{ m}.$$

Since the ball was thrown from a point 30 m above the ground, the highest point reached by the ball is 30 + 46 = 76 m above the ground.

(b) Let t = time for the ball to reach maximum height. Then,

$$t = \frac{v}{g} = \frac{30 \text{ m/s}}{9.8 \text{ m/s}^2} = 3.06 \text{ s}.$$

During this time the elevator has moved a distance (10 m/s)(3.06 s) = 30.6 m up the shaft, so that ball and elevator are separated by 46 - 30.6 = 15.4 m at the moment the ball is at its highest point. Call T the time needed for the ball to fall back to the elevator. Relative to the elevator, the ball is projected downward, with a speed of 10 m/s, from a height of 15.4 m. Choosing down as the positive direction, origin at the maximum height,

$$15.4 = \tfrac{1}{2}(9.8)T^2 + 10T,$$
$$T = 1.03 \text{ s},$$

and therefore the total elapsed time is 3.06 + 1.03 = 4.1 s Ans.

3-50

Let h be the height of fall and t the time to fall this distance. Since the object falls from rest,

$$h = \tfrac{1}{2}gt^2,$$
$$\frac{h}{2} = \tfrac{1}{2}g(t - 1)^2.$$

(a) Eliminating h between these equations gives

$$\frac{t}{t - 1} = 2^{\frac{1}{2}} = 1.4142,$$
$$t = 3.41 \text{ s} \quad \text{Ans.}$$

(b) The height of fall is

$$h = \tfrac{1}{2}(9.8)(3.41)^2 = 57 \text{ m} \quad \text{Ans.}$$

3-51

Let H be the height of the building and h the distance above the ground of the bottom of the window; also, let the ball bearing be released at time $t = 0$ from $x = 0$, the downward direction being taken as positive. If T is the time taken for the ball bearing to fall to the top of the window, then at the instants when the ball bearing passes the top of the window, the bottom of the window and strikes the ground, the following equations are successively satisfied:

$$H - h - 1.20 = \tfrac{1}{2}gT^2,$$
$$H - h = \tfrac{1}{2}g(T + 0.125)^2,$$
$$H = \tfrac{1}{2}g(T + 0.125 + 1)^2.$$

From the first two equations, eliminating $(H - h)$,

$$\tfrac{1}{2}g(T + 0.125)^2 - 1.20 = \tfrac{1}{2}gT^2,$$
$$0.125gT + 0.0078125g - 1.20 = 0.$$

Putting $g = 9.8\ \text{m/s}^2$ into this last gives $T = 0.917$ s. Finally, substituting this value of T into the third equation above yields

$$H = 20.4\ \text{m} \quad \underline{\text{Ans}}.$$

3-53

Let the positive x-axis fixed on the elevator shaft be vertical, pointing upward, with origin at that point of the shaft level with the elevator floor at the moment, $t = 0$, the bolt left the ceiling of the elevator (see figure). Also, let x = position of bolt and X = position of the elevator floor.

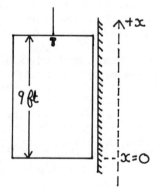

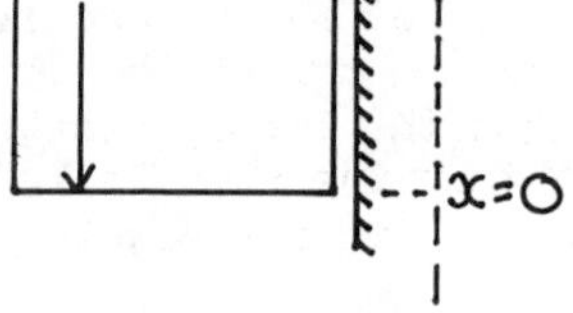

(a) At any time while the bolt is falling,

$$x = \tfrac{1}{2}(-32)t^2 + 8t + 9,$$
$$X = \tfrac{1}{2}(4)t^2 + 8t.$$

If T is the time of flight of the bolt, then at $t = T$, $x = X$:

$$-16T^2 + 8T + 9 = 2T^2 + 8T,$$

$$T = 2^{-\frac{1}{2}} = 0.71 \text{ s} \quad \underline{\text{Ans}}.$$

(b) Since the bolt started from the ceiling 9 ft above the floor, the desired distance is

$$D = 9 - X = 9 - 2T^2 - 8T = 9 - 1 - 5.68,$$
$$D = 2.3 \text{ ft} \quad \underline{\text{Ans}}.$$

3-54

Let y, Y be the distances of the first and second bodies, measured from the common point of release, and let the second body be released at $t = 0$. Then,

$$y = \tfrac{1}{2}g(t + 1)^2; \; Y = \tfrac{1}{2}gt^2.$$

For $y - Y = 10$,

$$\tfrac{1}{2}g(t + 1)^2 - \tfrac{1}{2}gt^2 = 10,$$
$$g(t + \tfrac{1}{2}) = 10.$$

Using $g = 9.8 \text{ m/s}^2$, this gives $t = 0.52$ s, so that the elapsed time since the first body was released is $0.52 + 1 = 1.52$ s Ans.

3-55

As indicated in the sketch, let distances be measured downwards from the highest point reached by the pot in its flight. Also, let the pot be at the highest point at $t = 0$ and, in its descent pass the top of the window and the bottom of the window at the times t_1 and t_2. If the total time that the pot is visible is 1.0 s, then the time it is visible in its descent is $\frac{1}{2}$ s. Therefore,

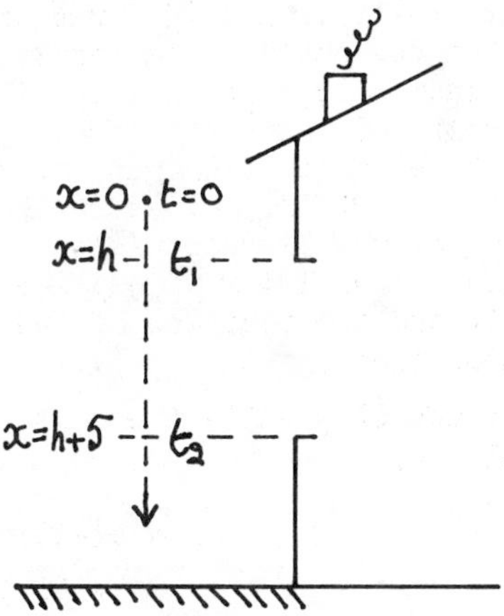

$$h = \tfrac{1}{2}gt_1^2,$$

$$h + 5 = \tfrac{1}{2}gt_2^2 = \tfrac{1}{2}g(t_1 + \tfrac{1}{2})^2.$$

Eliminating h between these equations gives

$$\tfrac{1}{2}gt_1^2 + 5 = \tfrac{1}{2}g(t_1^2 + t_1 + \tfrac{1}{4});$$

since $g = 32$ ft/s^2, $t_1 = \frac{1}{16}$ s and

$$h = \tfrac{1}{2}(32)(\tfrac{1}{16})^2 = \frac{1}{16}\text{ ft} \quad \underline{\text{Ans}}.$$

CHAPTER 4

4-3

(a) For the coordinates at $t = 3$ s, substitute $t = 3$ into the equations given for $x(t)$ and $y(t)$:

$$x(3) = 3(3) - 4(3)^2 = -27 \text{ m},$$
$$y(3) = -6(3)^2 + (3)^2 = -27 \text{ m}.$$

Since $x(0) = y(0) = 0$, the displacement at $t = 3$ s is

$$\vec{D}(3) = -27\vec{i} - 27\vec{j},$$

which has a magnitude

$$D = (27^2 + 27^2)^{\frac{1}{2}} = 38.2 \text{ m},$$

and is at an angle

$$\theta = \tan^{-1}\left(\frac{-27}{-27}\right) = 45°$$

below the negative x-axis, or 180° + 45° = 225° counterclockwise from the positive x-axis.

(b) By definition, the average velocity is, since $\vec{D}(0) = 0$,

$$\bar{\vec{v}} = \frac{\vec{D}(3) - \vec{D}(0)}{3 - 0} = 12.7 \text{ m/s},$$

in the direction of $\vec{D}(3)$ given in (a).

(c) The instantaneous velocity is

$$\vec{v} = \frac{dx}{dt}\vec{i} + \frac{dy}{dt}\vec{j},$$
$$\vec{v} = (3 - 8t)\vec{i} + (-12t + 3t^2)\vec{j}.$$

For $\vec{v}(3)$ substitute $t = 3$ to obtain

$$\vec{v}(3) = -21\vec{i} - 9\vec{j},$$

in m/s. The magnitude is

$$v(3) = (21^2 + 9^2)^{\frac{1}{2}} = 22.8 \text{ m/s},$$

at an angle

$$\theta = \tan^{-1}\left(\frac{-9}{-21}\right) = 23^\circ$$

below the negative x-axis, or $180^\circ + 23^\circ = 203^\circ$ counterclockwise from the positive x-axis.

(d) By definition, the desired average acceleration is

$$\bar{\vec{a}} = \frac{\vec{v}(3) - \vec{v}(0)}{3 - 0}.$$

From the equation for $\vec{v}$ above, $\vec{v}(0) = 3\vec{i}$; hence,

$$\bar{\vec{a}} = \frac{-21\vec{i} - 9\vec{j} - 3\vec{i}}{3} = -8\vec{i} - 3\vec{j}.$$

This has a magnitude

$$a = (8^2 + 3^2)^{\frac{1}{2}} = 8.54 \text{ m/s}^2$$

at an angle

$$\theta = \tan^{-1}\left(\frac{-3}{-8}\right) = 21^\circ$$

below the negative x-axis, or $180^\circ + 21^\circ = 201^\circ$ counterclockwise from the positive x-axis.

(e) The instantaneous acceleration is

$$\vec{a} = \frac{d^2x}{dt^2}\vec{i} + \frac{d^2y}{dt^2}\vec{j} = -8\vec{i} + (-12 + 6t)\vec{j}.$$

At $t = 3$ this has a value

$$\vec{a} = -8\vec{i} + 6\vec{j},$$

with a magnitude

$$a = (8^2 + 6^2)^{\frac{1}{2}} = 10 \text{ m/s}^2$$

at an angle

$$\theta = \tan^{-1}\left(\frac{+6}{-8}\right) = 143.1^\circ$$

counterclockwise from the positive x-axis.

4-6

By definition of velocity and acceleration,
(a)

$$\vec{v} = \frac{d\vec{r}}{dt} = 8t\vec{j} + \vec{k} \quad \underline{\text{Ans.}}$$

$$\vec{a} = \frac{d\vec{v}}{dt} = 8\vec{j} \quad \underline{\text{Ans.}}$$

(b) Since the acceleration is independent of the time, the acceleration is constant and, therefore, the path is a parabola. From the given equation for the position, it is seen that the x-coordinate does not vary with time, so the motion is in a plane parallel to the y,z-plane and cutting the x-axis at x = 1.

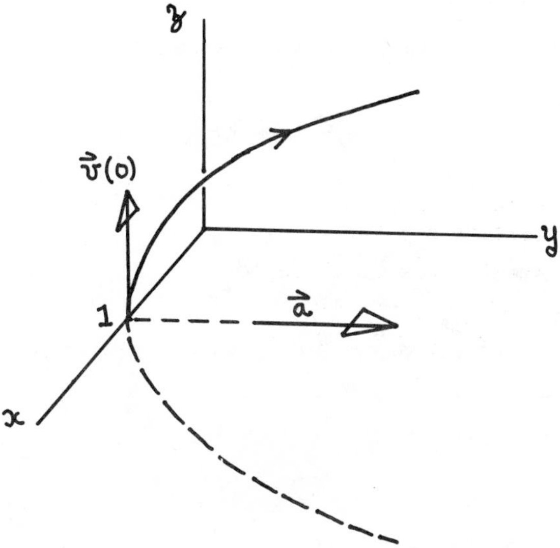

4-8

(a) The velocity at any time subsequent to leaving the origin is

$$\vec{v} = (v_{x0} + a_x t)\vec{i} + (v_{y0} + a_y t)\vec{j},$$
$$\vec{v} = (3 - t)\vec{i} + (0 - 0.5t)\vec{j},$$
$$\vec{v} = (3 - t)\vec{i} - 0.5t\vec{j}.$$

At the maximum x-position, $v_x = 0$. From the equation above, this is seen to occur at t = 3 s, at which time $v_y = -0.5(3) = -1.5$ m/s. Hence, at this position,

$$\vec{v} = -1.5\vec{j}, \text{ m/s} \quad \underline{\text{Ans}}.$$

(b) Equations 4-4c and 4-4c', with $x_0 = y_0 = 0$, $v_{x0} = 3$, $v_{y0} = 0$, $a_x = -1$, $a_y = -0.5$, become

$$x = 3t - 0.5t^2,$$
$$y = -0.25t^2,$$

so that at t = 3 the position of the particle is

$$x = 4.5 \text{ m}, \; y = -2.25 \text{ m} \quad \underline{\text{Ans}}.$$

4-13

The motions of the bullet in the x- and y-directions are given by

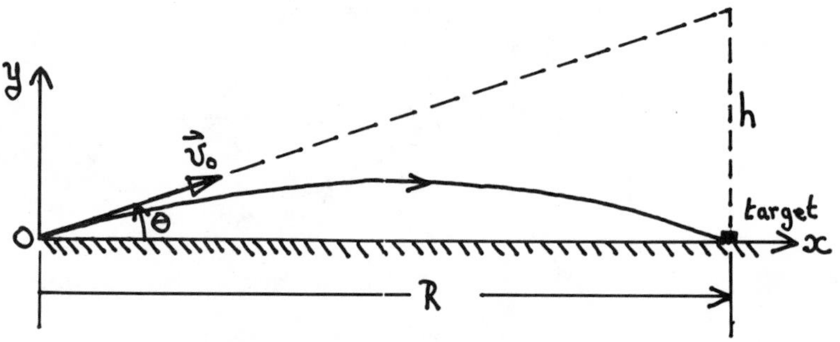

$$x = (v_0\cos\theta)t,$$
$$y = -\tfrac{1}{2}gt^2 + (v_0\sin\theta)t,$$

when the x and y axes are oriented as in the sketch above. If T is the time of flight, then x(T) = R and y(T) = 0. Hence

$$R = v_0T\cos\theta,$$
$$0 = -\tfrac{1}{2}gT^2 + v_0T\sin\theta.$$

The first equation above gives $T = R/v_0\cos\theta$; substituting this into the second equation yields

$$\sin 2\theta = gR/v_0^2,$$

since $\sin 2\theta = 2\sin\theta\cos\theta$. Putting $g = 32\ \text{ft/s}^2$, $R = 150$ ft and $v_0 =$ 1500 ft/s gives $\sin 2\theta = 0.002133$, or $\theta = 0.0611°$. Finally,

$$h = R\tan\theta = (150\ \text{ft})(0.0010667) = 0.16\ \text{ft} = 1.92\ \text{in.}\quad \underline{\text{Ans}}.$$

4-22

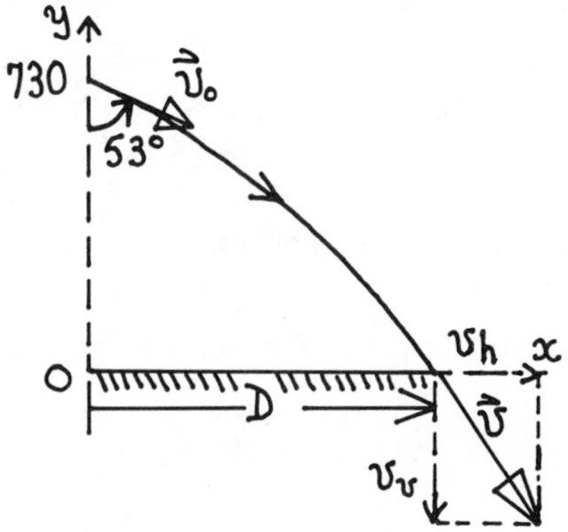

(a) The coordinate system used here is shown on the sketch. The projectile is released at $t = 0$ and its time of flight is T, so that when $t = T$, $y = 0$ (projectile strikes the ground). Hence, $y_0 = +730$ m and $v_{y0} = -v_0\cos 53°$ and therefore

$$0 = -\tfrac{1}{2}gT^2 - (v_0\cos 53°)T + 730,$$

$$v_0 = 202\ \text{m/s}\quad \underline{\text{Ans}},$$

since $T = 5$ s and $g = 9.8\ \text{m/s}^2$.

(b) The horizontal velocity remains unchanged during flight; thus

$$D = (v_0\sin 53°)T = 806\ \text{m}\quad \text{Ans}.$$

(c) As the projectile strikes the ground,

$$v_h = v_x = v_{0x} = (202\ \text{m/s})(\sin 53°) = 161\ \text{m/s}\quad \underline{\text{Ans}},$$

$$v_v = v_y = (v_0\cos 53°) + gT,$$

$$v_v = (202)(\cos 53°) + (9.8)(5) = 171\ \text{m/s}\quad \underline{\text{Ans}}.$$

4-25

(a) The positive x-axis is chosen to be horizontal and in the direction in which the electrons enter the region between the plates. The y-axis is taken to be vertical, positive downwards. The origin of the coordinates is at the point at which the electrons enter the field. The acceleration due to gravity, in view of the acceleration imparted by the field, is disregarded. The time T required by the electrons to pass through the region is

$$T = \frac{2.0 \text{ cm}}{1.0 \times 10^9 \text{ cm/s}} = 2 \times 10^{-9} \text{ s},$$

since the horizontal component of the velocity persists throughout the motion. Since the vertical velocity at injection into the field is zero, the vertical displacement is

$$y = \tfrac{1}{2}aT^2 = \tfrac{1}{2}(1.0 \times 10^{17} \text{ cm/s}^2)(2 \times 10^{-9} \text{ s})^2,$$

$$y = 0.2 \text{ cm} = 2 \text{ mm} \quad \underline{\text{Ans}}.$$

(b) The horizontal velocity upon emerging is the same as upon entering:

$$v_x = 1.0 \times 10^9 \text{ cm/s} \quad \underline{\text{Ans}}.$$

The vertical velocity upon emerging is

$$v_y = aT = (1.0 \times 10^{17} \text{ cm/s}^2)(2 \times 10^{-9} \text{ s}),$$

$$v_y = 2 \times 10^{-8} \text{ cm/s} \quad \underline{\text{Ans}},$$

directed downward.

4-27

If x(t), y(t) are the coordinates of the ball in flight then, with axes arranged as shown,

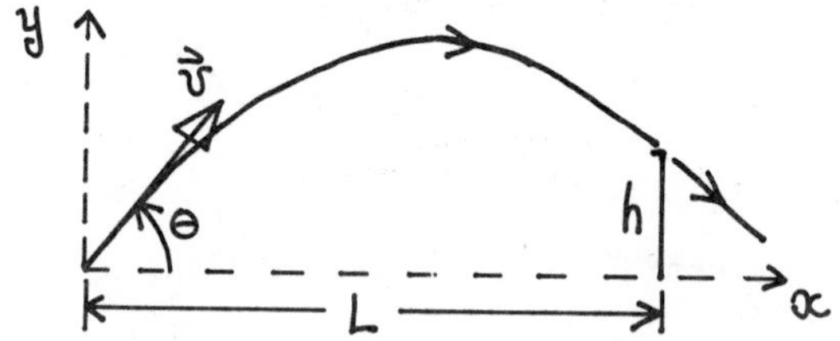

$$y = -\tfrac{1}{2}gt^2 + (v\sin\theta)t,$$

$$x = (v\cos\theta)t.$$

Let T = time interval from the kick until the ball reaches a horizontal distance L, where L = distance from kick to goal posts. Then, $T = L/v\cos\theta$ from the x-equation. When t = T, y = h (ball just clears bar):

$$h = -\tfrac{1}{2}g(L/v\cos\theta)^2 + (v\sin\theta)(L/v\cos\theta),$$

$$h = -\tfrac{1}{2}gL^2/(v^2\cos^2\theta) + L\tan\theta.$$

This can be written as

$$ax^2 - Lx + (h + a) = 0$$

if

$$a = \tfrac{1}{2}gL^2/v^2, \quad x = \tan\theta.$$

Putting in the numbers gives a = 19.60 m and, since h = 3.44 m,

$$19.6x^2 - 50x + 23.04 = 0;$$

$$x = 1.9474,\ 0.6036$$

as the two solutions to the quadratic equation. With $x = \tan\theta$, these yield angles of 63° and 31° as the required limits.

4-30

The situation is shown in the sketch. The ball is 'launched' from a point 4 ft above the ground, so that, if the axes are arranged as in the figure, the position of the ball as a function of time is given by

$$x = (v\cos 45^\circ)t,$$

$$y = -\tfrac{1}{2}(32)t^2 + (v\sin 45^\circ)t + 4.$$

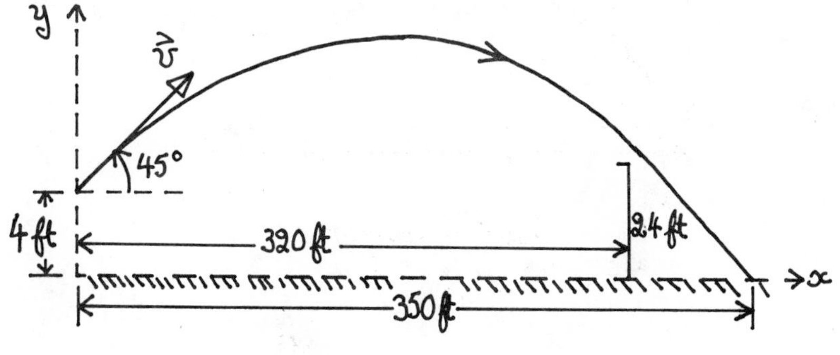

The initial speed v of the ball is unknown and must be obtained from the range. Let T be the time of flight of the ball:

$$350 = (v\cos 45°)T.$$

Solve for T in terms of v and substitute into the equation for y to obtain

$$0 = -(3.92 \times 10^6)v^{-2} + 354,$$
$$v = 105.23 \text{ ft/s}.$$

Return to the original equations for x and y, but now v is known. This time set x = 320 ft and solve for t, the time to reach x = 320 ft from home plate:

$$320 \text{ ft} = (105.23 \text{ ft/s})(\cos 45°)t,$$
$$t = 4.30 \text{ s}.$$

Finally, substitute this together with v into the equation for y: the result is y = 28 ft. Since the fence is 24 ft high, the ball clears the fence by about 4 ft.

4-39

(a) Let R be the radius of the earth and T its period of rotation. Then, if v is the speed of an object due to the earth's rotation,

$$vT = 2\pi R$$

and the acceleration of the object is

$$a = \frac{v^2}{R} = 4\pi^2 R/T^2 = 0.0337 \text{ m/s}^2 \quad \underline{\text{Ans}},$$

since $R = 6.378 \times 10^6$ m and T = 86,400 s.

(b) If T' = rotation period of the earth required for the acceleration at the equator to equal the current value of g, then

$$g = 4\pi^2 R/T'^2.$$

In actuality,

$$a = 4\pi^2 R/T^2.$$

Therefore,

$$a/g = 0.0337/9.8 = T'^2/T^2,$$

$$T' = 0.0586\ T = T/17,$$

which means that the earth would have to be spinning 17 times faster than it is actually doing.

4-40

The acceleration in uniform circular motion is v^2/R. To find v, note that when the string breaks, the stone is projected in a horizontal direction, i.e., with no initial vertical velocity. Hence, $2 = \frac{1}{2}gt^2$ and, since the horizontal velocity is preserved, $10 = vt$. Combining these equations gives

$$2 = \tfrac{1}{2}g(10/v)^2,$$
$$v^2 = 25g = 245\ m^2/s^2,$$
$$a = v^2/R = 245/1.5 = 163\ m/s^2 \quad \underline{\text{Ans.}}$$

4-47

Let: $\vec{v}_{sc}$ = velocity of snow relative to the car,
$\vec{v}_{se}$ = velocity of snow relative to the earth,
$\vec{v}_{ce}$ = velocity of car relative to the earth.

The snow falls vertically relative to the earth and the car moves horizontally; thus, the vectors $\vec{v}_{se}$ and $\vec{v}_{ce}$ are perpendicular. Hence, the relation between the vectors,

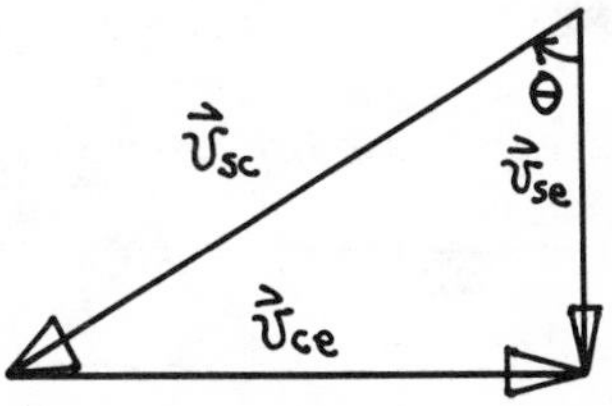

$$\vec{v}_{sc} + \vec{v}_{ce} = \vec{v}_{se},$$

is as shown in the diagram. Since 50 km/h = 50,000/3600 = 13.89 m/s, the angle θ sought is

$$\theta = \tan^{-1}(13.89/8) = 60^\circ \quad \underline{\text{Ans}}.$$

4-49

Let t = time required to get up the escalator and w = person's walking speed, s = escalator's speed, m = person's speed relative to the ground when walking on the escalator and L = length of the escalator. Then,

$$L/w = t_w = 90\ s;\ L/s = t_s = 60\ s;\ L/m = t_m,$$

where t_m is the time needed to walk up the moving escalator. But

$$m = s + w,$$

so that

$$t_m = \frac{L}{s + w} = \frac{L}{L/t_s + L/t_w} = \frac{1}{1/60 + 1/90} = 36\ s \quad \underline{\text{Ans}}.$$

4-52

Let: $\vec{v}_{rt}$ = velocity of rain relative to train,
$\vec{v}_{rg}$ = velocity of rain relative to ground,
$\vec{v}_{tg}$ = velocity of train relative to ground.

Then,

$$\vec{v}_{rt} + \vec{v}_{tg} = \vec{v}_{rg},$$

and this is illustrated in the figure. Since v_{tg} = 30 m/s,

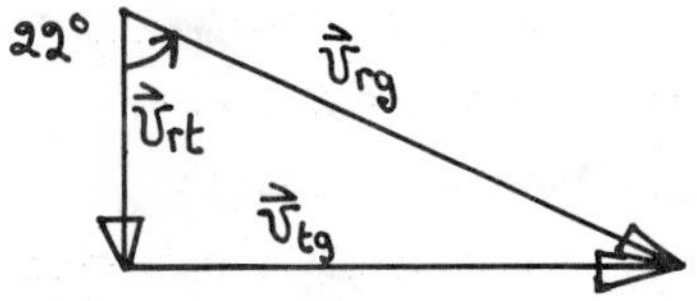

$$v_{rg} = \frac{v_{tg}}{\sin 22°} = \frac{30\ m/s}{\sin 22°},$$

$$v_{rg} = 80.1\ m/s \quad \underline{\text{Ans}}.$$

4-53

Let: $\vec{v}_{pg}$ = velocity of plane relative to the ground,
$\vec{v}_{pa}$ = velocity of plane relative to the air,
$\vec{v}_{ag}$ = velocity of the air relative to the ground.

By the law of relative velocities,

$$\vec{v}_{pa} + \vec{v}_{ag} = \vec{v}_{pg}.$$

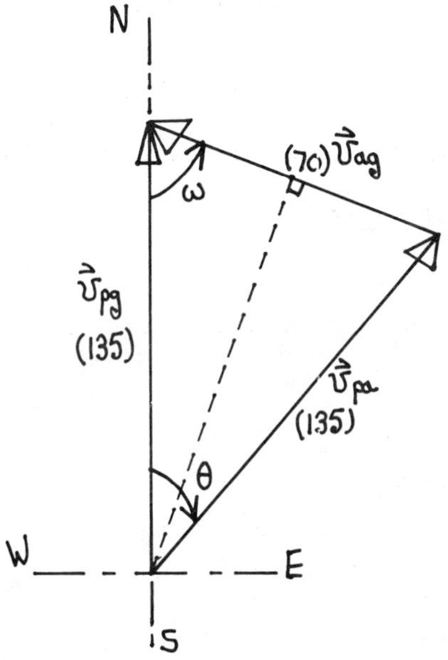

Now, the speed $v_{pa} = 135$ mi/h; also $v_{pg} = 135$ mi/h and the wind speed $v_{ag} = 70$ mi/h so that the lengths of all the vectors are known. As far as directions are concerned, it is given that $\vec{v}_{pg}$ points north. This is drawn first, and then the other two are arranged to obey the law for adding vectors by the geometric method. The triangle obtained contains no right angle. (A solution with θ counterclockwise from north is also possible.)

(b) Since the triangle is isosceles,

$$\sin\tfrac{1}{2}\theta = \frac{35}{135}\,,$$

$$\theta = 30^\circ \text{ E of N} \quad \underline{\text{Ans}}\,.$$

(a) The angles of the triangle must add up to 180° and as the other base angle is w also,

$$w = \tfrac{1}{2}(180^\circ - \theta) = \tfrac{1}{2}(180^\circ - 30^\circ) = 75^\circ \text{ E of S} \quad \underline{\text{Ans}}.$$

4-55

Let: $\vec{v}_{rg}$ = velocity of river relative to ground,

$\vec{v}_{br}$ = velocity of boat relative to river (v_{br} = 4 mi/h),

$\vec{v}_{bg}$ = velocity of boat relative to ground.

Then,

$$\vec{v}_{br} + \vec{v}_{rg} = \vec{v}_{bg}.$$

(a) Here it is desired that $\vec{v}_{bg}$ be perpendicular to the river.

$$\sin\theta = \frac{v_{rg}}{v_{br}} = \frac{2\text{ mi/h}}{4\text{ mi/h}}; \quad \theta = 30^\circ \;;$$

that is, she should head the boat 30° upstream.

(b) If T is the crossing time,

$$T = \frac{D}{v_{bg}} = \frac{D}{v_{br}\cos\theta} = \frac{4\text{ mi}}{(4\text{ mi/h})\cos 30^\circ} = 69\text{ min} \quad \underline{\text{Ans}}.$$

(c) Clearly, $v_{bg} = v_{br} + v_{rg}$ = 4 mi/h + 2 mi/h = 6 mi/h when moving downstream, and $v_{bg} = v_{br} - v_{rg}$ = 4 mi/h - 2 mi/h = 2 mi/h when rowing upstream. Therefore,

$$T = \frac{2\text{ mi}}{6\text{ mi/h}} + \frac{2\text{ mi}}{2\text{ mi/h}} = 1\tfrac{1}{3}\text{ h} = 80\text{ min} \quad \underline{\text{Ans}}.$$

(d) Since this is just (c) in reverse, T = 80 min also Ans.

(e) To cross in the shortest time, $\vec{v}_{bg}$ must have as large a component perpendicular to the banks as possible. As $\vec{v}_{rg}$ is parallel to the banks, it cannot affect the velocity of the boat perpendicular to the banks. Thus, to make this last as large as possible, head the boat directly across the river. Then,

$$T = \frac{D}{v_{br}} = \frac{4\text{ mi}}{4\text{ mi/h}} = 1\text{ h} = 60\text{ min} \quad \underline{\text{Ans}}.$$

Of course, the boat will end up somewhere downstream.

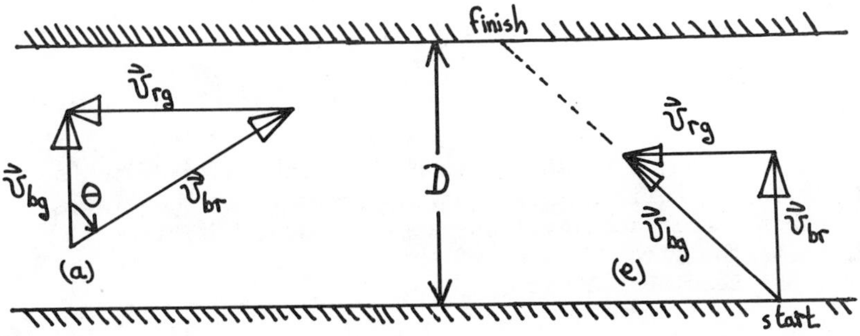

Note that in each case $\vec{v}_{br} + \vec{v}_{rg} = \vec{v}_{bg}$.

5-13

(a) Draw a free-body diagram for each block and apply Newton's second law to the horizontal forces, noting that as the blocks move together, their accelerations are the same. The blocks accelerate to the right so that if this direction is chosen as positive, the resulting equations are

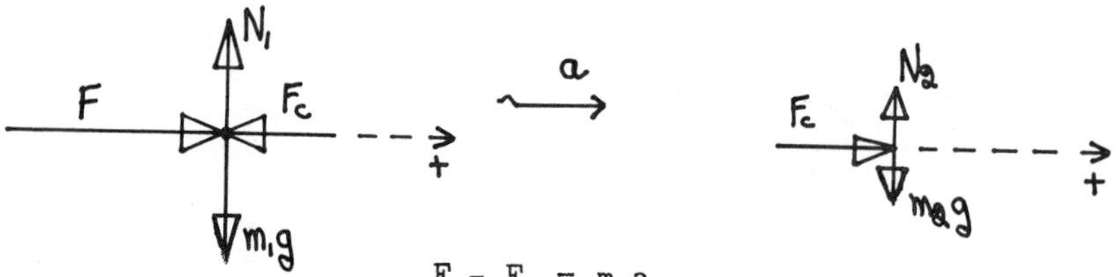

$$F - F_c = m_1 a,$$
$$F_c = m_2 a,$$

where F_c is the force of contact. Eliminating the acceleration a gives

$$F_c = \frac{m_2}{m_2 + m_1} F = \frac{1}{1 + 2}(3) = 1.0 \text{ N} \quad \underline{\text{Ans.}}$$

(b) Here it is necessary only to interchange the blocks:

$$F_c = \frac{m_1}{m_1 + m_2} F = \frac{2}{2 + 1}(3) = 2.0 \text{ N} \quad \underline{\text{Ans.}}$$

The acceleration is the same in parts (a) and (b) since the net external force and total mass are the same in each case. On one of the blocks, the internal force of contact is the sole force providing the acceleration and, by Newton's second law, this force is proportional to the mass it acts upon.

5-16

The spring scale reads the tension in the string. In both cases this tension is 10 lb, which is most easily seen by isolating either of the 10-lb weights in (a) and the single 10-lb weight in (b). The only forces acting on one of these weights are its weight

$W = 10$ lb acting down and the tension T in the string acting up. Since the system is at rest, $a = 0$ and

$$W - T = 0,$$
$$T = W = 10 \text{ lb},$$

both for (a) and (b). Hence the reading of the scale is 10 lb in each case.

5-23

The force $\vec{P}$ the fireman exerts on the pole is equal and opposite to the force the pole exerts on him. This latter force must act upward on the fireman in order for his acceleration to be less than $g = 32 \text{ ft/s}^2$. Choosing up as positive the free-body diagram gives

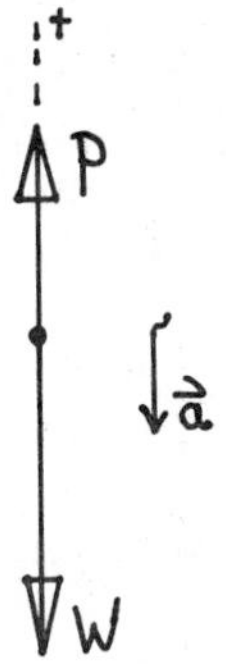

$$P - W = m(-a),$$

since his acceleration $\vec{a}$ points downward, the negative direction by choice. His mass $m = W/g = (160 \text{ lb})/(32 \text{ ft/s}^2) = 5$ slugs, and since $a = 10 \text{ ft/s}^2$,

$$P - 160 = -5(10),$$
$$P = 110 \text{ lb} \quad \underline{\text{Ans.}}$$

5-30

Let the frictional force be $\vec{f}$; this acts opposite to the velocity of the body, that is upwards. The weight W of the body is $W = mg$ and acts, as always, vertically down. Choosing down as positive, Newton's second law gives

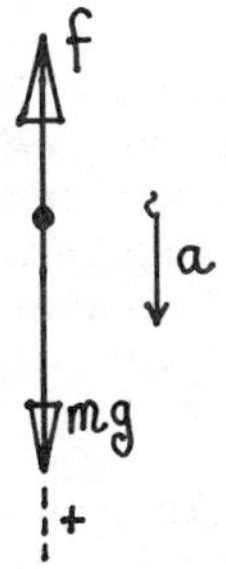

$$mg - f = m(+a),$$

since the downward acceleration $\vec{a}$ points in the chosen positive direction. Hence,

$$f = m(g - a) = (0.25)(9.8 - 9.2),$$
$$f = 0.15 \text{ N} \quad \underline{\text{Ans}}.$$

5-31

On a free-body diagram of the sphere are shown the forces acting on the sphere: F due to the electric field, T the tension in the string and W = mg the weight of the sphere. Newton's second law with the acceleration a = 0 applied to the horizontal and vertical directions gives

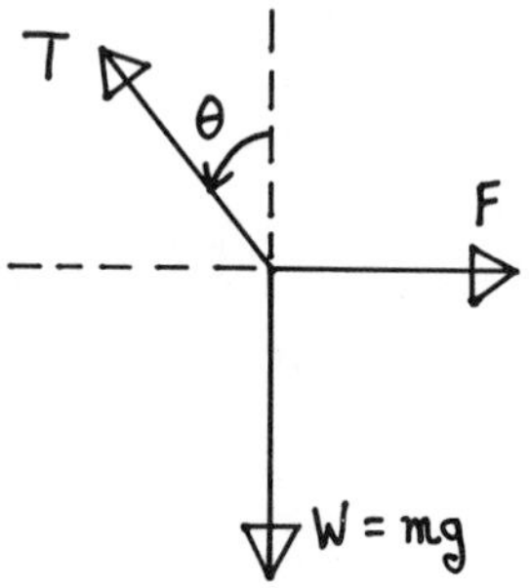

$$T\sin\theta - F = 0,$$
$$T\cos\theta - W = 0.$$

(a) Upon dividing one by the other, these equations imply that

$$\frac{\sin\theta}{\cos\theta} = \tan\theta = \frac{F}{W} = \frac{F}{mg},$$

and since $\theta = 37°$ this gives

$$F = mg\tan\theta = 2.22 \times 10^{-3} \text{ N} \quad \underline{\text{Ans}}.$$

(b) With this result, the tension T is, from the first equation,

$$T = \frac{F}{\sin\theta} = 3.68 \times 10^{-3} \text{ N} \quad \underline{\text{Ans}}.$$

5-32

(a) Examine the forces acting on a small portion of the string: these forces are the weight W of the portion acting vertically down and the tensions T and T' exerted by particles of the string immediately to the left and right of the portion considered. If the string is in equilibrium, the downward acting W must be balanced by upward components of T and T'. If the string is horizontal, however, the tensions have no vertical components. Thus, the string must sag to some degree, although this sag may not be noticeable.

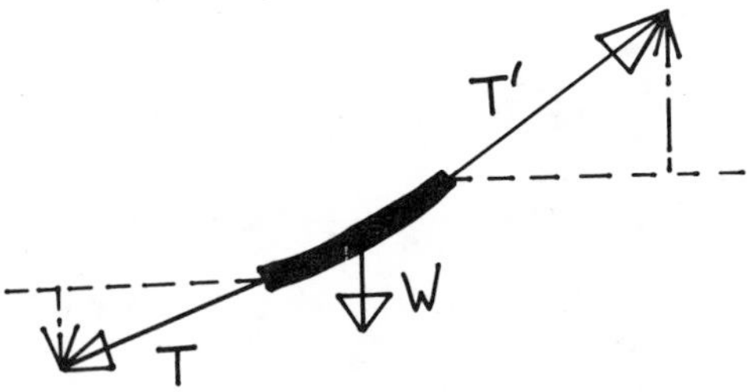

(b)

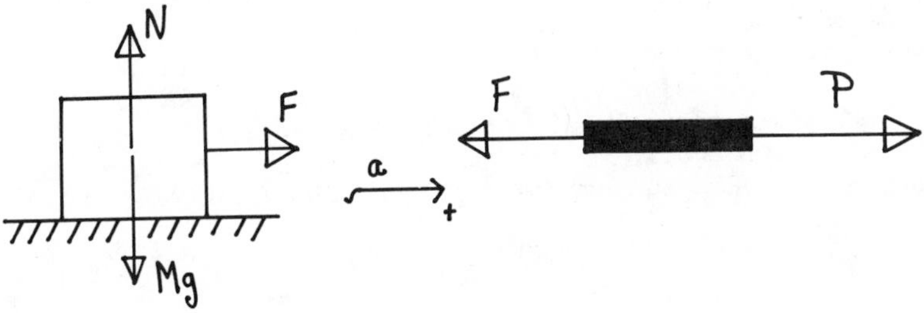

For the rope, mass m, and the block, mass M, the equations of motion are

$$P - F = ma\,,$$

$$F = Ma\,,$$

F the force exerted by the rope on the block; by Newton's third law the block exerts a force of equal magnitude on the rope. Adding these equations gives

$$P = (m + M)a\,,$$

$$a = \frac{P}{m + M} \quad \underline{\text{Ans}}\,.$$

(c) Directly from (b),

$$F = Ma = \frac{M}{m + M}\,P \quad \underline{\text{Ans}}.$$

(d) Let T be the tension in the midpoint of the rope. Draw a free-body diagram of the leading half of the rope. From this, it is clear that

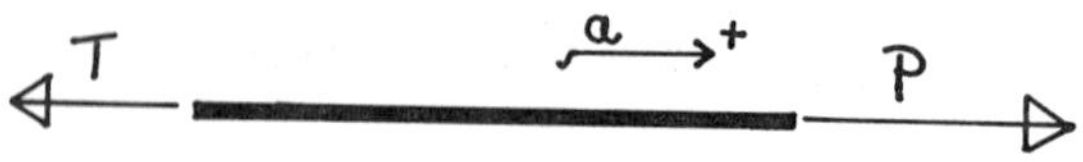

$$P - T = \frac{m}{2}a = \frac{m}{2}\frac{P}{m + M},$$

$$T = \frac{P}{2}\frac{m + 2M}{m + M} \quad \underline{\text{Ans}}.$$

5-34

The pebble is falling vertically with constant speed: i.e., with constant velocity. Thus, the acceleration of the pebble is zero. By Newton's second law, this indicates that the total force on the pebble is zero also. The two forces acting are the weight of the pebble, vertically downwards, and the force due to the water acting opposite to the velocity, that is, vertically upwards. The only way that two oppositely directed forces can sum to zero is if their magnitudes are equal, so that the force F due to the water, in magnitude, equals the weight W of the pebble:

$$F = W = mg = (0.150 \text{ kg})(9.8 \text{ m/s}^2) = 1.47 \text{ N} \quad \underline{\text{Ans}}.$$

5-36

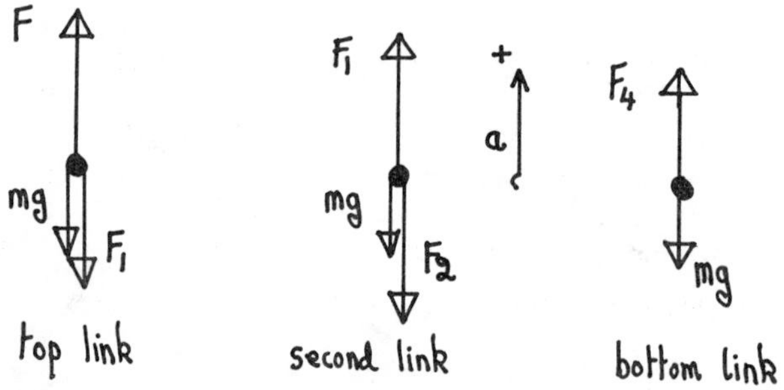

In applying Newton's second law to each link, the following quantities are needed:

$$mg = (0.10 \text{ kg})(9.8 \text{ m/s}^2) = 0.98 \text{ N},$$
$$ma = (0.10 \text{ kg})(2.5 \text{ m/s}^2) = 0.25 \text{ N}.$$

(c) From the latter equation, the net force F_{net} on each link is, by Newton's second law,

$$F_{net} = ma = 0.25 \text{ N} \quad \underline{\text{Ans}}.$$

(a) Now let F_i = forces between adjacent links, counting from the top pair. Then, using Newton's third law and since F is the external force lifting the chain (note that F acts only on the top link), the equations for each link, starting at the top, are

$$F - F_1 - mg = ma ,$$
$$F_1 - F_2 - mg = ma,$$
$$F_2 - F_3 - mg = ma,$$
$$F_3 - F_4 - mg = ma,$$
$$F_4 - mg = ma .$$

The last equation gives

$$F_4 = mg + ma = 0.98 + 0.25 = 1.23 \text{ N} \quad \underline{\text{Ans}} .$$

Substituting this into the fourth equation gives F_3, which can then be put into the third equation to give F_2, and so on. The results obtained are,

$$F_3 = 2.46 \text{ N}, \; F_2 = 3.69 \text{ N}, \; F_1 = 4.92 \text{ N} \quad \underline{\text{Ans}},$$

and (b),

$$F = 6.15 \text{ N} \quad \underline{\text{Ans}}.$$

This result for F can be obtained also by considering the chain as a single object of mass 5m; then,

$$F - W = (5m)a,$$
$$F - (5m)g = (5m)a,$$
$$F = (5m)(g + a) = 6.15 \text{ N}.$$

<u>5-39</u>

Let T be the tension in the cord. A choice must be made for the direction of the acceleration, as it is not given; the assumed direction is shown on the diagram. (The direction of acceleration must be consistent with the conditions of the problem: one cannot

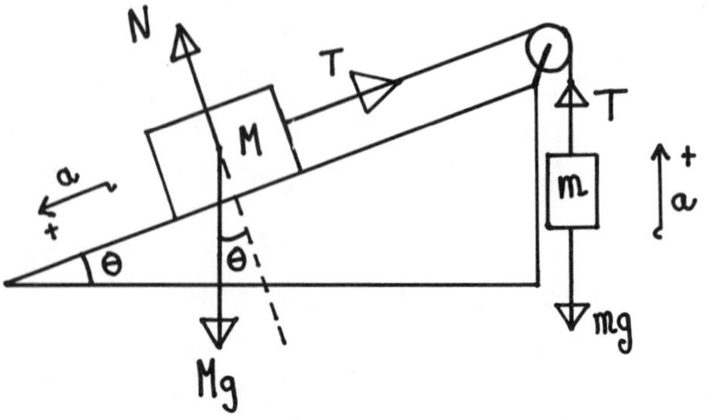

assume, for example, that M accelerates down the plane and that m falls, for this would require the string to stretch.) Since the pulley is implied to be frictionless and massless, the tensions on the two sides of the pulley are equal. The equations of motion for M, m respectively are

$$Mg\sin\theta - T = Ma,$$
$$T - mg = ma.$$

(a) Adding the equations gives

$$a = \frac{M\sin\theta - m}{M + m} g = \frac{(3)(\frac{1}{2}) - 2}{3 + 2} (32) = -3.2 \text{ ft/s}^2.$$

As this is negative, the acceleration actually is in the direction opposite to that which was assumed; i.e., M accelerates up the incline.

(b) From the second equation above, and using the result for a:

$$T = m(g + a) = (2)(32 - 3.2) = 57.6 \text{ lb} \quad \underline{\text{Ans.}}$$

5-40

(a) The only force parallel to the incline acting on the block is the component of its weight W parallel to the incline, and this is $W\sin\theta = mg\sin\theta$. Thus, the acceleration of the block (whether it is moving up or down the incline) is

$$a = \frac{mg\sin\theta}{m} = g\sin\theta,$$

acting down the plane. Let L be the distance the block moves up the incline. At its highest point on the incline, the speed of the block is zero. The speed at the bottom x = 0 is v_0; if x is positive up the incline, a = -gsinθ and

$$v^2 = v_0^2 + 2a(x - x_0),$$
$$0^2 = v_0^2 + 2(-g\sin\theta)L,$$
$$L = v_0^2/(2g\sin\theta),$$
$$L = (8)^2/(2)(32)(\tfrac{1}{2}) = 2 \text{ ft} \quad \underline{\text{Ans}}.$$

(b) Let t = 0 as the block is projected up the incline and let T be the time required for the block to reach its greatest distance up the incline. Then,

$$v = at + v_0,$$
$$0 = (-g\sin\theta)(T) + v_0,$$
$$T = v_0/g\sin\theta\ ,$$
$$T = (8)/(32)(\tfrac{1}{2}) = 0.5 \text{ s} \quad \underline{\text{Ans}}.$$

(c) Since there is no friction, the speed of the block when it gets back to the bottom must be the speed with which it was projected up the incline, i.e.,

$$v_0 = 8 \text{ ft/s} \quad \underline{\text{Ans}}.$$

<u>5-43</u>

The weight of the man is (80 kg)(9.8 m/s^2) = 784 N and the weight of the parachute is (5 kg)(9.8 m/s^2) = 49 N.

(a) Draw a free-body diagram of the man+parachute system. The total weight is 784 + 49 = 833 N. Let F be the force exerted by the air on the parachute. Then,

$$W - F = Ma,$$
$$833 - F = (85)(2.5),$$
$$F = 620.5 \text{ N} \quad \underline{\text{Ans}}.$$

(b) Now draw a free-body diagram of the parachute (or man) alone. Let f be the force exerted by the man on the parachute. Hence,

$$w + f - F = ma,$$
$$49 + f - 620.5 = (5)(2.5),$$
$$f = 584 \text{ N} \quad \textbf{Ans}.$$

5-44

Let F be the upward lift of the balloon (assumed constant) and m the mass of ballast discarded. If the upward direction is taken as positive, the equations of motion before and after dropping the ballast are

$$F - Mg = -Ma,$$
$$F - (M - m)g = (M - m)a.$$

Subtracting and solving for m,

$$m = 2M \frac{a}{a + g} \quad \underline{\text{Ans}}.$$

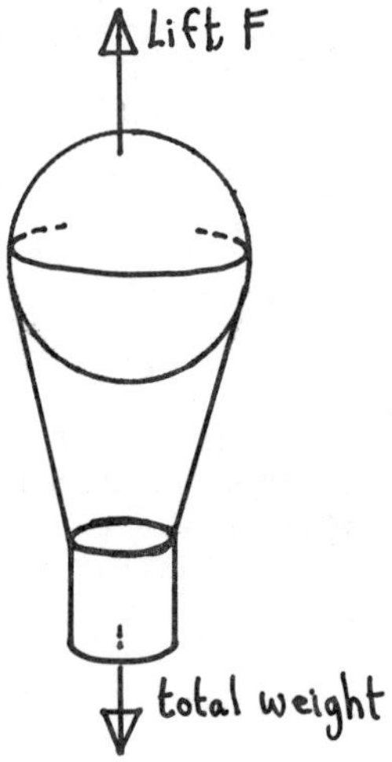

5-45

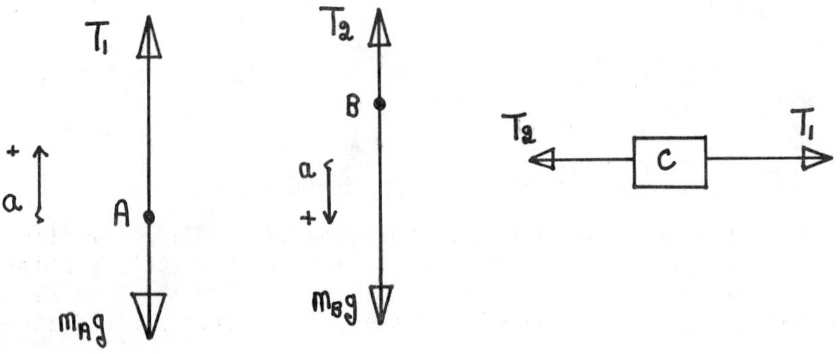

Draw free-body diagrams of the cage (A) and counterweight (B); also shown is the mechanism (C) and the forces exerted on it by the cables.

(a) For the cage,

$$T_1 - m_A g = m_A a,$$

$$T_1 = m_A(g + a) = (1100)(9.8 + 2) = 12980 \text{ N} \quad \underline{\text{Ans.}}$$

(b) For the counterweight,

$$m_B g - T_2 = m_B a,$$

$$T_2 = m_B(g - a) = (1000)(9.8 - 2) = 7800 \text{ N} \quad \underline{\text{Ans}}.$$

(c) The net force $\vec{F} = \vec{T}_1 + \vec{T}_2$ exerted by the cable on the mechanism is $F = T_1 - T_2 = 12980 - 7800 = 5180$ N, to the right as pictured above. Hence, by Newton's third law, the force exerted by the mechanism on the cable is 5180 N to the left, that is, toward the mechanism.

<u>5-46</u>

(a) In order to lift the mass, the tension T in the rope must be at least the weight of the object: that is

$$T = W = (15 \text{ kg})(9.8 \text{ m/s}^2) = 147 \text{ N}$$

as a minimum requirement. As for the monkey, his equation of motion is

$$T - w = ma,$$

w the monkey's weight = 98 N. Putting T = 147 N as found above gives

$$a = \frac{147 - 98}{10} = 4.9 \text{ m/s}^2,$$

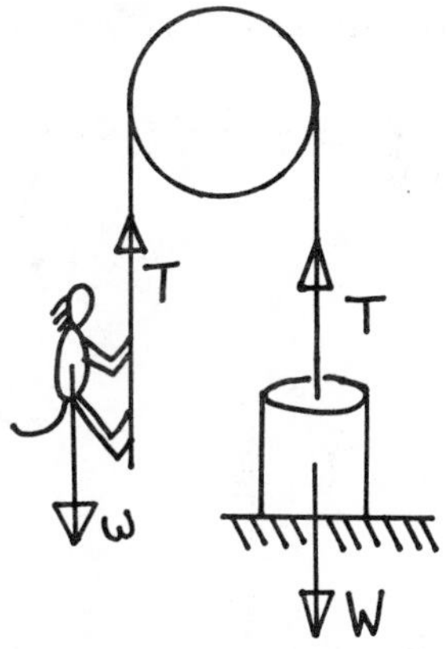

so that the monkey must climb up the rope with an acceleration greater than this to lift the weight. Note that in this problem, with the length of rope between the two 'objects' not held fixed, they may have different accelerations. It has been assumed above, in order to find the minimum requirements, that the acceleration of the weight to be lifted is virtually zero.

(b) When the monkey stops climbing, the system forms an Atwood machine so that, by Example 8,

$$a = \frac{15 - 10}{15 + 10}(9.8) = 1.97 \text{ m/s}^2 \quad \underline{\text{Ans;}}$$

(c)

$$T = \frac{(2)(10)(15)}{15 + 10}(9.8) = 117.6 \text{ N} \quad \underline{\text{Ans.}}$$

CHAPTER 6

6-1

(a) Let v be the initial speed and x the distance required to come to a stop. The only force acting in the horizontal direction is the force f_k of kinetic friction. Therefore,

$$f_k = ma.$$

But $v^2 = 2ax$, giving

$$a = \frac{v^2}{2x} = \frac{6^2}{2(15)} = 1.2 \text{ m/s}^2.$$

Therefore,

$$f_k = (0.110 \text{ kg})(1.2 \text{ m/s}^2) = 0.132 \text{ N} \quad \underline{\text{Ans}}.$$

(b) Since $f_k = \mu_k N$ and $N = mg$ in this problem,

$$f_k = \mu_k N = \mu_k mg = ma,$$
$$\mu_k = \frac{a}{g} = \frac{1.2}{9.8} = 0.122 \quad \underline{\text{Ans}}.$$

6-5

(a) The tension T in the rope is just the force exerted by the man in pulling the crate; f is the force of friction. With the crate on the verge of moving, $f = \mu_s N$. Note that $N \neq W$ in this case since part of the weight W is balanced by the vertical component of T. Applying the second law:

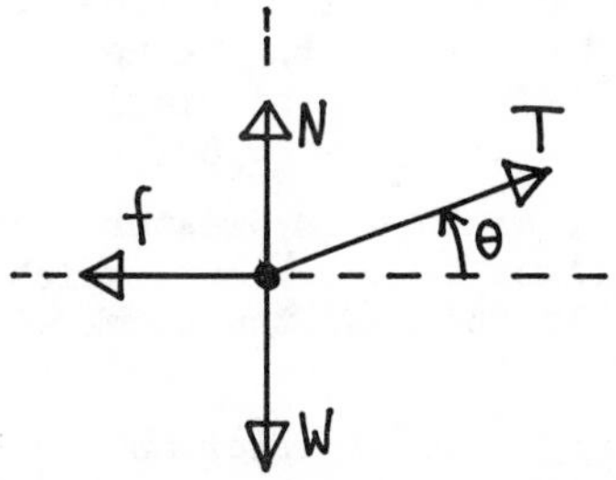

$$T\cos\theta - \mu_s N = 0,$$
$$N + T\sin\theta - W = 0.$$

Eliminate the normal force N and solve for T:

$$T = \frac{\mu_s W}{\cos\theta + \mu_s \sin\theta},$$

$$T = \frac{(0.5)\ (150\ \text{lb})}{\cos 15° + (0.5)\sin 15°} = 68.47\ \text{lb} \quad \underline{\text{Ans.}}$$

(b) With the crate moving the force of friction becomes $\mu_k N$ and the acceleration is no longer zero:

$$T\cos\theta - \mu_k N = ma,$$
$$N + T\sin\theta - W = 0.$$

Set $m = W/g$ and solve for a by eliminating N, as before; the result can be written

$$\frac{a}{g} = \frac{T}{W}(\cos\theta + \mu_k \sin\theta) - \mu_k,$$
$$a = 4.23\ \text{ft/s}^2 \quad \underline{\text{Ans}},$$

using the value of T found in part (a).

6-13

(a) Clearly, if the block slips it will slip downward; hence, the force $\vec{f}$ of static friction points upward. The block will not move in the horizontal direction, so that N = F = 12 lb. The maximum "available" force of static friction is $\mu_s N$ = (0.6)(12 lb) = 7.2 lb. As this is greater than the weight of the block, which force it opposes, the block will not move.

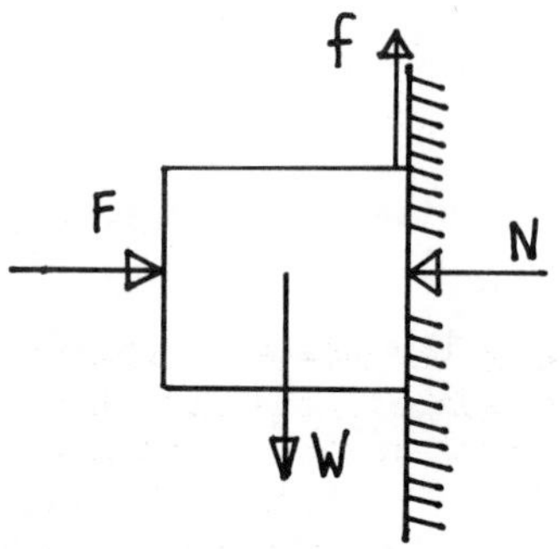

(b) The force of friction can be no greater than the weight of the block, for if it was, the block would accelerate up the wall. Thus, the forces exerted by the wall on the block are a force N = 12 lb to the left and a force f of static friction equal to 5 lb and directed upward.

6-19

Assume, for the moment, that the block slips on the slab. It is essential to note that, by Newton's third law, forces of friction of equal magnitudes act on each object; indeed, it is solely a force of friction that causes the slab to accelerate. Let F be the 100 N force acting on the block. As usual, the equations of motion follow from the free-body diagrams; since $N_b = m_b g$, the equations are,

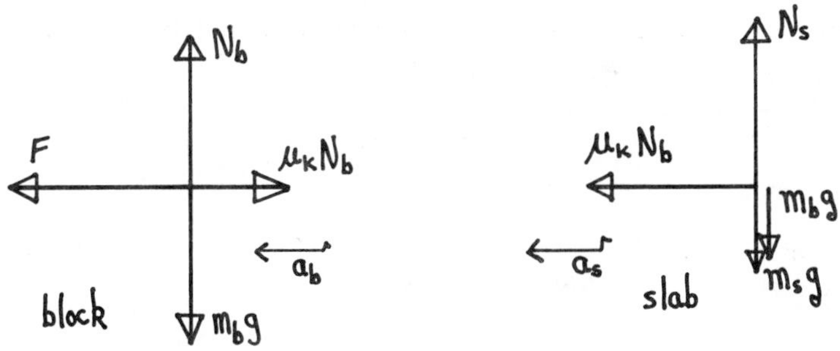

(a) for the block,

$$F - \mu_k m_b g = m_b a_b,$$
$$100 - (0.4)(10)(9.8) = (10)a_b,$$
$$a_b = 6.08\ \text{m/s}^2 \quad \underline{\text{Ans.}}$$

(b) For the slab,

$$\mu_k m_b g = m_s a_s,$$
$$a_s = \mu_k (m_b/m_s) g = 0.98\ \text{m/s}^2 \quad \underline{\text{Ans.}}$$

It remains to determine whether the block does, in fact, slip on the slab. (If it does not, block and slab move together with a common acceleration.) The maximum force of static friction is

$$f_{s,max} = \mu_s N_b = \mu_s m_b g = (0.6)(10\ \text{kg})(9.8\ \text{m/s}^2),$$
$$f_{s,max} = 58.8\ \text{N} < 100\ \text{N},$$

indicating that static friction cannot hold the block and the motion of the system is, indeed, as was outlined in the solution above.

6-20

With block C placed on A, the two objects form a single object of weight $W_T = W_A + W_C$. With C removed, only object A with weight W_A remains on the horizontal surface. Let f be the force of friction and draw the free-body diagrams.

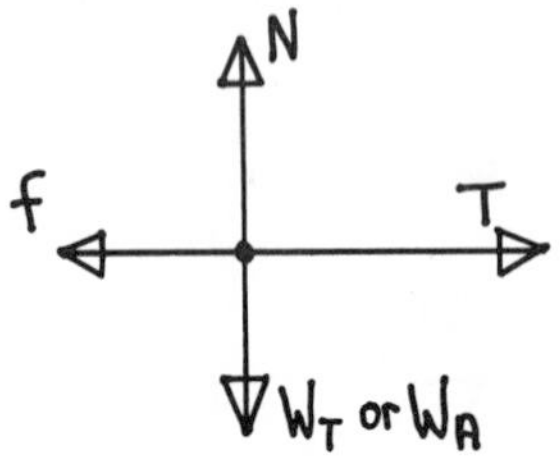

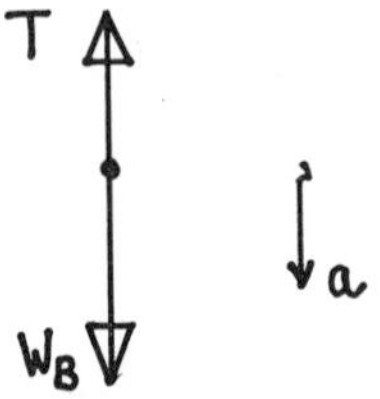

(a) Use W_T, $f = f_{s,max} = \mu_s N$ and $a = 0$. Then,

$$T - \mu_s W_T = 0,$$
$$W_B - T = 0.$$

Eliminate the tension T to obtain

$$W_B - \mu_s(W_A + W_C) = 0,$$
$$W_C = (W_B/\mu_s) - W_A \quad \underline{\text{Ans}}.$$

This gives either 15 lb or 66 N depending on the units selected.
(b) With block C removed use W_A in place of W_T, $f = f_k = \mu_k N$;

$$T - \mu_k W_A = m_A a,$$
$$W_B - T = m_B a.$$

Adding these to remove T,

$$a = \frac{W_B - \mu_k W_A}{m_B + m_A} = \frac{W_B - \mu_k W_A}{W_B + W_A} g \quad \underline{\text{Ans}}.$$

Numerically, this is 6.4 ft/s^2 or 1.96 m/s^2.

6-21

(a) Draw a diagram showing the forces acting, and with the 8-lb block leading.

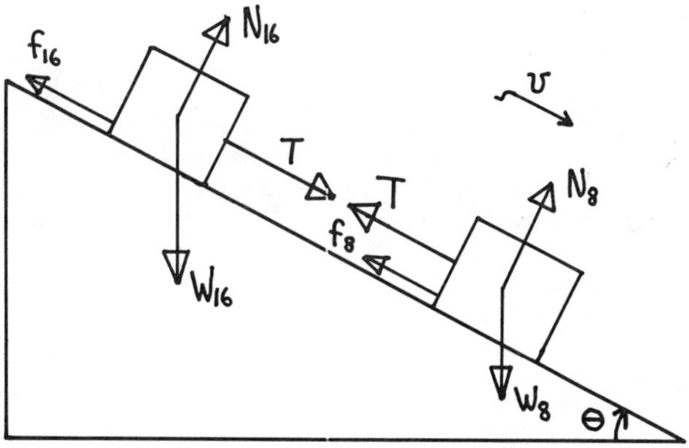

Since

$$f_{16} = \mu_{16} m_{16} g\cos\theta,$$

$$f_8 = \mu_8 m_8 g\cos\theta,$$

the equations of motion are

$$m_8 a_8 = m_8 g\sin\theta - T - \mu_8 m_8 g\cos\theta\ ,$$

$$m_{16} a_{16} = m_{16} g\sin\theta + T - \mu_{16} m_{16} g\cos\theta.$$

Now $\mu_{16} = 2\mu_8$ so that if $T = 0$ initially, then $a_8 > a_{16}$ and the leading block eventually will pull on the following, with the blocks subsequently moving with a common acceleration, a tension present in the string. Set $a_{16} = a_8 = a$ and add the equations;

$$(m_8 + m_{16})a = (m_8 + m_{16})g\sin\theta - (m_8\mu_8 + m_{16}\mu_{16})g\cos\theta,$$

$$a = 11.38 \text{ ft/s}^2 \quad \underline{\text{Ans}},$$

since $m_8 = \frac{1}{4}$ slug, $m_{16} = \frac{1}{2}$ slug, etc.

(b) From the equation of motion for the heavier block,

$$T = m_{16}a - m_{16}g\sin\theta + \mu_{16}m_{16}g\cos\theta,$$

$$T = 0.46 \text{ lb} \quad \underline{\text{Ans.}}$$

(c) If the blocks are reversed, interchange the subscripts '8' and '16' in the equations of motion. If T = 0 initially, $a_8 > a_{16}$ as before. But as the 16-lb block leads to start with, the string remains slack and the blocks move independently, unless the plane is long enough for the 8-lb block to catch up with the other. The separate accelerations can be found from the equations of motion above with T = 0.

6-22

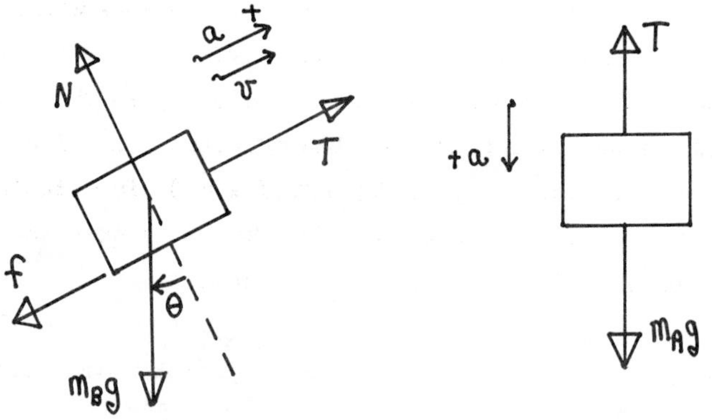

The free-body diagrams are drawn for block B moving up the plane, and also accelerating up the plane so that A accelerates downward. For the opposite direction of acceleration, substitute -a for a in equations that follow. Note that there is no correlation between the directions of velocity and acceleration. When friction is present, the direction of the velocity must be specified, for the force of friction is opposite to the velocity and not necessarily to the acceleration. In general, reversing the direction of the velocity leads to a different acceleration, so the two directions of velocity must be treated separately. In writing the equations of motion below, acceleration up the plane is considered positive. The equations of motion are,

$$m_Ag - T = m_Aa,$$

$$T \mp f - m_Bg\sin\theta = m_Ba,$$

$$f = \mu N = \mu m_Bg\cos\theta.$$

In the middle equation the upper sign applies if B is moving up the plane, as pictured, since then f, opposing v, points downward in the negative direction, up being chosen as positive. Substitute for f as indicated in the last equation and then add the first two equations together to obtain

$$a = \frac{m_Ag - m_Bg\sin\theta \mp \mu m_Bg\cos\theta}{(m_A + m_B)g}\,g.$$

(a) For $v = 0$ use $\mu = \mu_s$, the static coefficient. Numerically, the terms in the numerator are $m_Ag = 32$ lb, $m_Bg\sin\theta = 70.71$ lb since $\theta = 45°$, and $\mu_s m_Bg\cos\theta = 39.60$ lb. For $f = 0$, $a < 0$ indicating that if the system is released from rest block B will try to slide down the plane. But if the maximum possible value of f_s is used (with the lower sign in the expression for the acceleration), i.e., μ_sN, then $a > 0$. Thus, in actuality, $f < \mu_sN$ and $a = 0$, for static friction will not provide a force so large as to reverse the "natural" tendency of B to move down the plane.

(b) For $v \neq 0$, $\mu = \mu_k = 0.25$ so that $f = \mu_k m_Bg\cos\theta = 17.68$ lb. Substituting the numbers into the expression for the acceleration gives

$$a = \frac{32 - 70.71 \mp 17.68}{132\text{ lb}}(32\text{ ft/s}^2),$$

$$a = -13.7\text{ ft/s}^2 \quad \text{(B moving up the plane)},$$

$$a = -5.10\text{ ft/s}^2 \quad \text{(B moving down the plane)}.$$

(The terms in the numerator of the equation for a also have units of lb.) The negative signs indicate that the acceleration is directed down the plane in both cases.

6-25

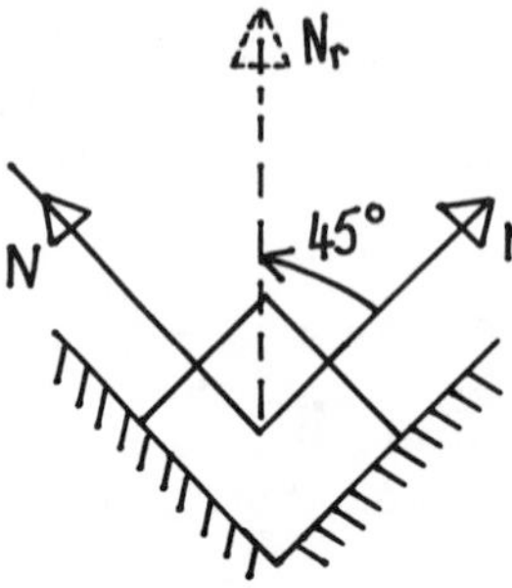

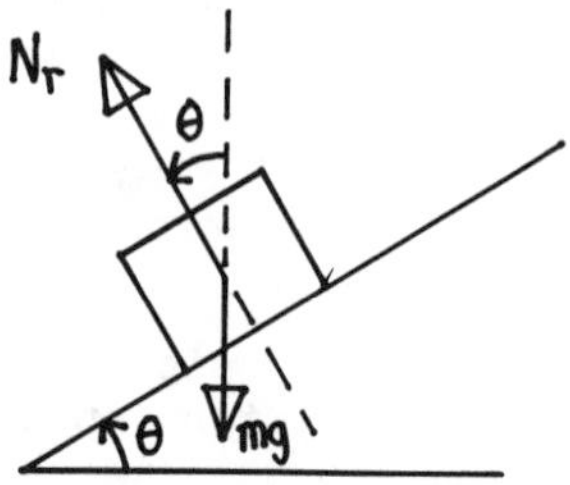

The resultant of the two equal, in magnitude, normal forces N is $N_r = 2N\cos 45° = 2^{\frac{1}{2}}N$. This acts at θ to the vertical and therefore

$$2^{\frac{1}{2}}N = mg\cos\theta.$$

The total friction force is $2(\mu_k N)$, so that

$$mg\sin\theta - 2\mu_k N = ma.$$

Substituting for N from the first equation gives

$$mg\sin\theta - 2\mu_k(2^{-\frac{1}{2}}mg\cos\theta) = ma,$$

$$a = (\sin\theta - 2^{\frac{1}{2}}\mu_k\cos\theta)g.$$

6-38

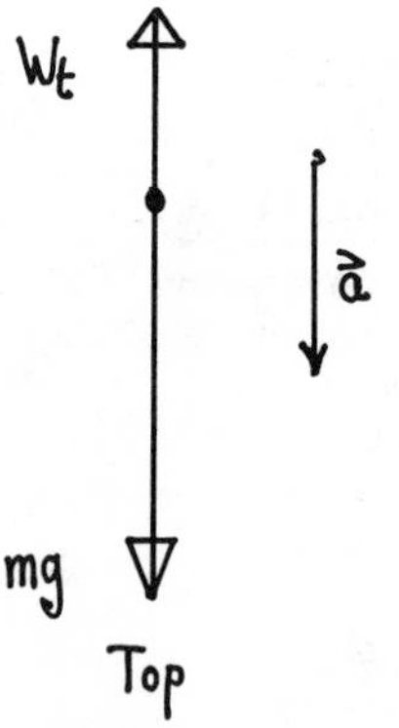

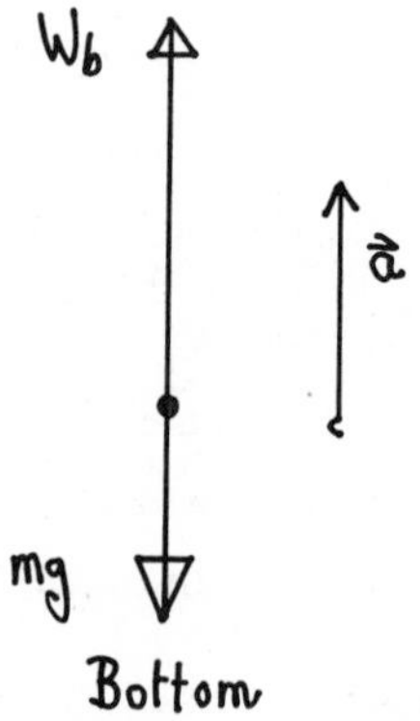

(a) The student's apparent weight W is the force exerted by him on the Ferris wheel. By Newton's third law this, in magnitude, equals the force exerted on him by the wheel. Note that $W \neq mg$ in this problem, **for** W is taken to represent his apparent weight. Let b refer to the bottom of the wheel and t to the top. The student's acceleration is always directed to the center of the wheel. Thus, at the top,

$$mg - W_t = mv^2/R,$$
$$150 - 125 = mv^2/R,$$
$$mv^2/R = 25 \text{ lb.}$$

At the bottom,

$$W_b - mg = mv^2/R,$$
$$W_b - 150 = 25,$$
$$W_b = 175 \text{ lb} \quad \underline{\text{Ans}}.$$

(b) If the new speed $u = 2v$, then $mu^2/R = 4mv^2/R = 100$ lb. Under these conditions, $W_t = mg - mu^2/R = 150 - 100 = 50$ lb Ans.

6-40

(a)

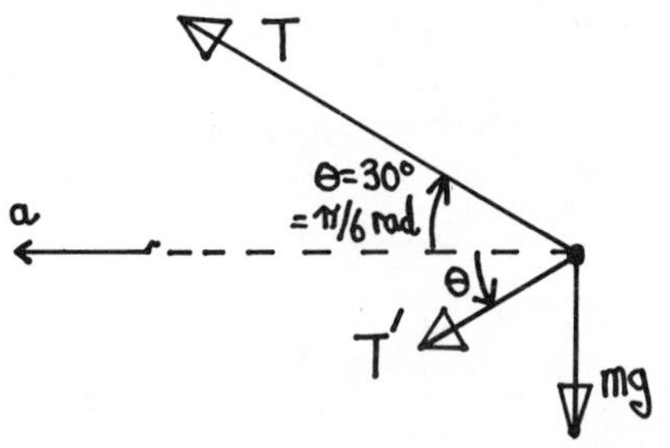

(b) The acceleration of the ball is directed horizontally toward the rod, that is, to the center of its circular path. Hence, the vertical force components must sum to zero; since the angle between the cords and the horizontal is π/6 rad, this gives

$$T\sin(\tfrac{\pi}{6}) = T'\sin(\tfrac{\pi}{6}) + mg,$$

$$(25)(\tfrac{1}{2}) = T'(\tfrac{1}{2}) + 9.8,$$

$$T' = 5.4 \text{ N} \quad \underline{\text{Ans}}.$$

(c) The net force F acts in the direction of the acceleration, therefore toward the rod. Thus,

$$F = T\cos(\frac{\pi}{6}) + T'\cos(\frac{\pi}{6}),$$

$$F = (30.4)3^{\frac{1}{2}}/2 = 26.3 \text{ N} \quad \underline{\text{Ans.}}$$

(d) $F = ma = mv^2/r$, where r is the perpendicular distance of the ball to the rod (i.e., the radius of its circular path). This distance is 0.866 m and therefore,

$$F = m\frac{v^2}{r} = (1)\frac{v^2}{0.866},$$

$$v = 4.8 \text{ m/s} \quad \underline{\text{Ans.}}$$

6-41

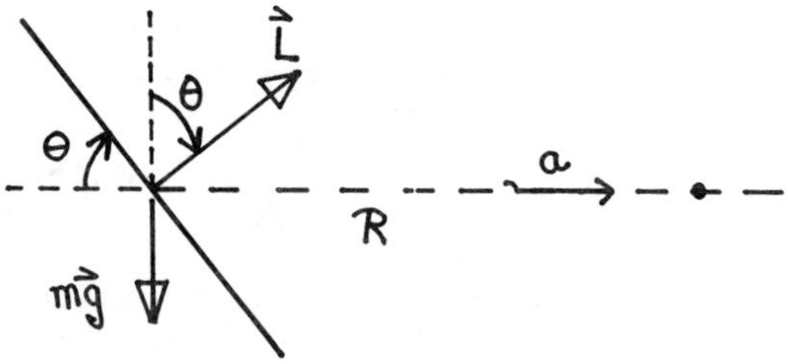

Let R be the radius of the circle. The forces acting on the plane are its weight mg vertically down and the lift L exerted by the air perpendicular to the wings. The acceleration is directed to the center of the circle. Therefore

$$L\cos\theta = mg,$$

$$L\sin\theta = mv^2/R.$$

Dividing gives

$$\cot\theta = gR/v^2,$$

$$R = \frac{v^2}{g}\cot\theta = 1.8 \text{ km} \quad \underline{\text{Ans,}}$$

since $v = 400/3$ m/s and $\cot\theta = 1$.

6-42

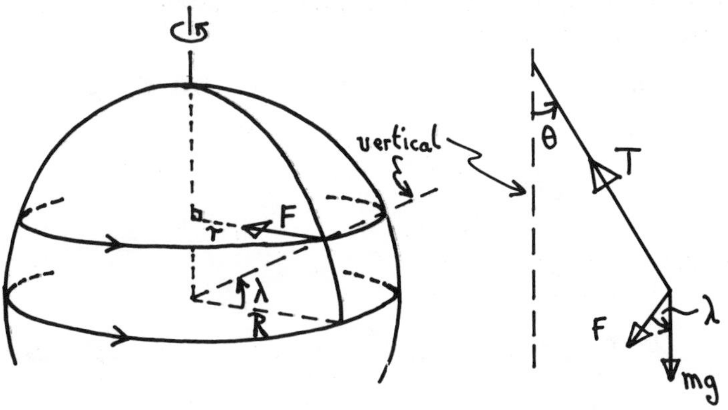

Due to the rotation of the earth, the plumb bob is moving at constant speed in a circle of radius $r = R\cos\lambda$, where R is the radius of the earth and λ is the latitude. Consequently, there must be a resultant force F on the bob directed at the axis of rotation. F makes an angle λ with the local vertical and has a magnitude

$$F = m\frac{v^2}{r} = m\frac{(2\pi R\cos\lambda/T)^2}{R\cos\lambda},$$

where m is the mass of the bob and T = 86,400 seconds is the rotation period of the earth. The local situation is shown in the figure on the right. Resolving into vertical and horizontal components,

$$mg - T\cos\theta = F\cos\lambda,$$

$$T\sin\theta = F\sin\lambda.$$

Transpose mg in the first equation and then divide the second equation by the first. The result is,

$$\tan\theta = \frac{F\sin\lambda}{mg - F\cos\lambda}.$$

Now θ is a small angle so that $\tan\theta = \theta$ approximately. Also, mg is much larger than $F\cos\lambda$. To a good degree of accuracy, then,

$$\theta = \frac{F\sin\lambda}{mg} = \frac{2\pi^2 R\sin 2\lambda}{gT^2}.$$

(a) With $R = 6.4 \times 10^6$ m, $g = 9.8$ m/s^2, $T = 86{,}400$ s and $\lambda = 40°$, $\theta = 0.0017$ rad $= 5.8'$ Ans.

(b), (c) At either $\lambda = 0°$, or $90°$, it is clear that the formula gives $\theta = 0$ Ans.

CHAPTER 7

7-2

(a) Since the speed of the block does not change, neither does its kinetic energy; thus, the total work W_t done by all the forces acting on the block is zero since, with $v_i = v_f$,

$$W_t = \tfrac{1}{2}mv_f^2 - \tfrac{1}{2}mv_i^2 = 0 \quad \underline{\text{Ans}}.$$

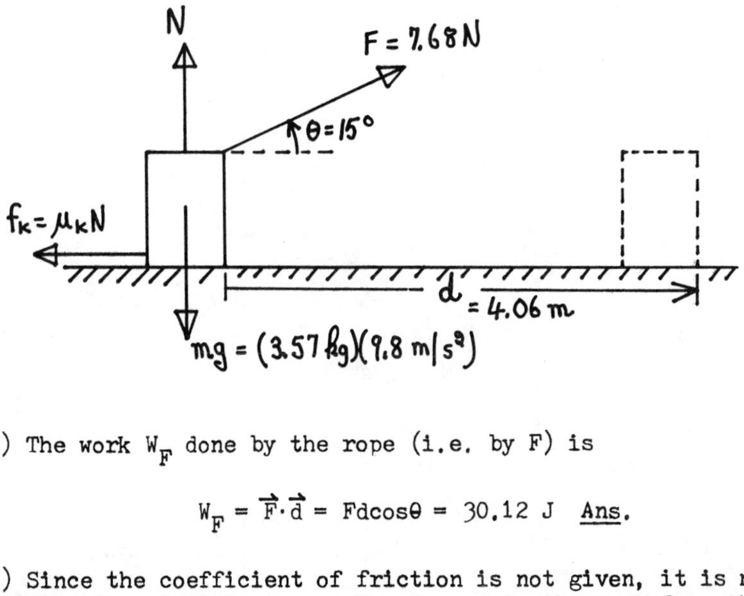

(b) The work W_F done by the rope (i.e. by F) is

$$W_F = \vec{F}\cdot\vec{d} = Fd\cos\theta = 30.12 \text{ J} \quad \underline{\text{Ans}}.$$

(c) Since the coefficient of friction is not given, it is not possible to calculate $f_k = \mu_k N$ and compute the work from the definition. But, the result from (a) can be used as follows.

$$W_t = W_F + W_N + W_{mg} + W_f = 0.$$

Now, the weight $m\vec{g}$ and the normal force $\vec{N}$ are perpendicular to the displacement $\vec{d}$ of the block and therefore these forces do no work on the block, for $\cos(\pi/2) = 0$. Hence

$$W_f = -W_F = -30.12 \text{ J} \quad \underline{\text{Ans.}}$$

(d) By the definition of work

$$W_f = f_k d\cos\pi = -\mu_k Nd,$$

the force of friction being directed opposite to the displacement. The normal force $N \neq mg$ since $\vec{F}$ also has a vertical component:

$$N + F\sin\theta - mg = 0 .$$

Solve for N and substitute into the expression for W_f to obtain

$$W_f = -\mu_k(mg - F\sin\theta)d,$$

$$\mu_k = \frac{W_f/d}{F\sin\theta - mg} = 0.225 \quad \underline{\text{Ans}},$$

using the result from (c).

7-4

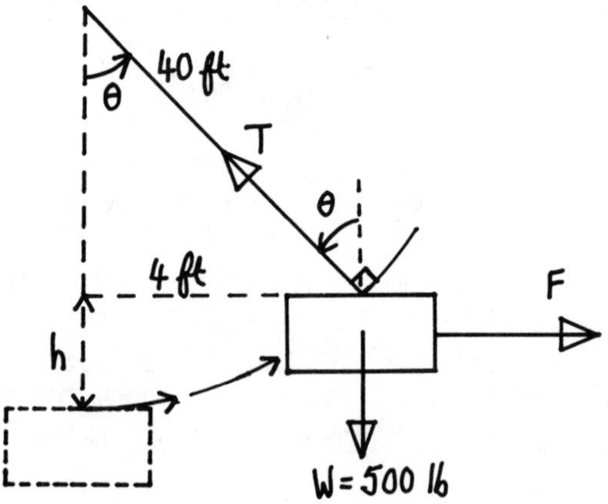

(a) In the final position, the sums of the vertical and horizontal force components acting on the crate must each be zero:

$$T\cos\theta - W = 0,$$

$$T\sin\theta - F = 0.$$

Upon dividing, these give

$$F = W\tan\theta.$$

But,

$$\tan\theta = \frac{4}{(40^2 - 4^2)^{\frac{1}{2}}} = 0.1005,$$

so that

$$F = 50.25 \text{ lb} \quad \underline{\text{Ans}}.$$

(b) Once the block has reached its final position, the work done in holding it there is zero, since there is no additional displacement.

(c) The work done by gravity as the block is moved aside depends only on the displacement in the vertical **direction.** Since the force of gravity acts down, and the crate was lifted upward, the angle between these vectors is 180° and the cosine of this is -1; therefore

$$W_g = -mgh.$$

The distance h is

$$h = 40 - (40^2 - 4^2)^{\frac{1}{2}} = 0.2005 \text{ ft}$$

giving

$$W_g = -(500 \text{ lb})(0.2005 \text{ ft}) = -100.25 \text{ ft}\cdot\text{lb} \quad \underline{\text{Ans}}.$$

(d) The crate is at rest at the beginning and end of the displacement: $v_i = v_f = 0$. Hence, the total work W_t done is

$$W_t = \tfrac{1}{2}mv_f^2 - \tfrac{1}{2}mv_i^2 = 0.$$

But

$$W_t = W_T + W_F + W_g.$$

The forces acting on the crate are the tension $\vec{T}$ in the rope, $\vec{F}$ and the weight $m\vec{g}$. During the displacement, the crate moves on the arc of a circle of which the rope forms a radius. The tension, therefore, is at right angles to the displacement and the work done by the tension is zero. The work done by the force pushing the crate aside is, then,

$$W_F = -W_g = +100.25 \text{ ft·lb} \quad \underline{\text{Ans}}.$$

(e) Clearly, this last result is not (50.25 lb)(4 ft), the result anticipated if $\vec{F}$ and the displacement vector are constant in magnitude and direction. But, if $\vec{F}$ does not change in direction, it almost certainly changes in magnitude in order to bring the crate to rest at the end of the displacement and then to assume the value needed to hold it in place there. Also, the angle between $\vec{F}$ and the displacement continually changes: $\vec{F}$ is (assumed) horizontal and the displacement is tangent to the circular path, so that the angle between the two increases from 0 to θ from the initial to the final position. To calculate the work done by F from the definition of work, the general relation

$$W_F = \int \vec{F}\cdot d\vec{r}$$

must be used. However, regardless of how $\vec{F}$ varies during the displacement, the value of the integral is +100.25 ft·lb provided that the crate is brought to rest at the end. Since the net displacement is about 4 ft, the average value of F during the displacement is about 25 lb.

(f), (g) Both answers are zero; see (d).

<u>7-6</u>

First, obtain the tension T. As $a = g/4$,

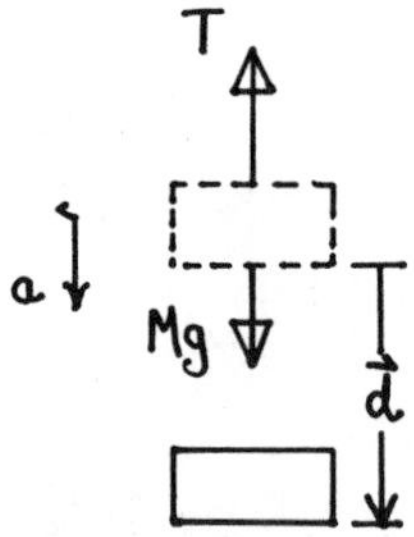

$$Mg - T = M\left(\frac{g}{4}\right),$$

$$T = \frac{3}{4}\,Mg\,.$$

The work done by the tension is

$$W = \vec{T}\cdot\vec{d} = Td\cos\pi = \left(\frac{3}{4}\,Mg\right)(d)(-1),$$

$$W = -\frac{3}{4}\,Mgd \quad \underline{\text{Ans}}.$$

7-10

Taking x_0 positive, the graph of F vs. x is as shown below: this is a straight line with slope F_0/x_0.

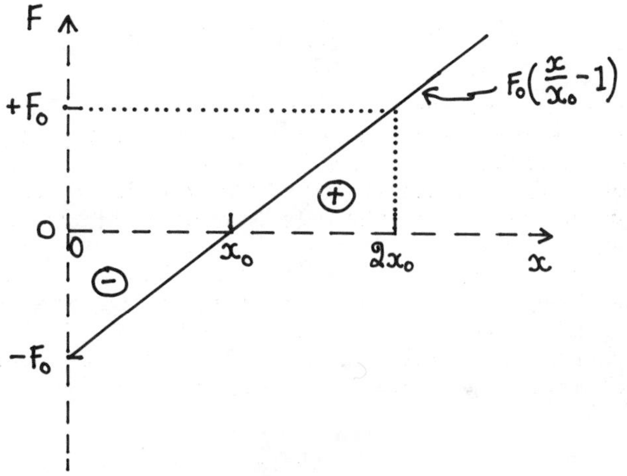

The work done from x = 0 to $2x_0$ is the area "under" the straight line between these limits. From the sketch, this area is zero, since the positive area (that above the F = 0 axis) of $\frac{1}{2}F_0x_0$ equals the negative area (that below the F = 0 axis) of $-\frac{1}{2}F_0x_0$, giving a sum of zero. Analytically, the work is

$$W = \int F\,dx = F_0\int_0^{2x_0}(x/x_0 - 1)dx,$$

$$W = F_0\left(\frac{x^2}{2x_0} - x\right)\Big|_0^{2x_0} = 0,$$

the same result as obtained from the graph.

7-11

The total work W is obtained by integrating dW from the initial point i to the final point f.

$$W = \int_i^f \vec{F}\cdot d\vec{r} = m\int_i^f \left(\frac{d^2x}{dt^2}\,\vec{i} + \frac{d^2y}{dt^2}\,\vec{j} + \frac{d^2z}{dt^2}\,\vec{k}\right)\cdot(dx\,\vec{i} + dy\,\vec{j} + dz\,\vec{k}),$$

$$W = m\int_i^f \left(\frac{d^2x}{dt^2}\,dx + \frac{d^2y}{dt^2}\,dy + \frac{d^2z}{dt^2}\,dz\right).$$

Consider a typical term.

$$\frac{d^2x}{dt^2}\,dx = \frac{d}{dt}\left(\frac{dx}{dt}\right)dx = \frac{d}{dx}\left(\frac{dx}{dt}\right)\frac{dx}{dt}\,dx = \frac{dx}{dt}\,d\left(\frac{dx}{dt}\right),$$

the last expression integrating to $\frac{1}{2}v_x^2 - \frac{1}{2}v_{x0}^2$. Therefore,

$$W = m\left(\tfrac{1}{2}v_x^2 - \tfrac{1}{2}v_{x0}^2 + \tfrac{1}{2}v_y^2 - \tfrac{1}{2}v_{y0}^2 + \tfrac{1}{2}v_z^2 - \tfrac{1}{2}v_{z0}^2\right),$$

$$W = \tfrac{1}{2}m(v_x^2 + v_y^2 + v_z^2) - \tfrac{1}{2}m(v_{x0}^2 + v_{y0}^2 + v_{z0}^2),$$

$$W = \tfrac{1}{2}mv^2 - \tfrac{1}{2}mv_0^2.$$

7-21

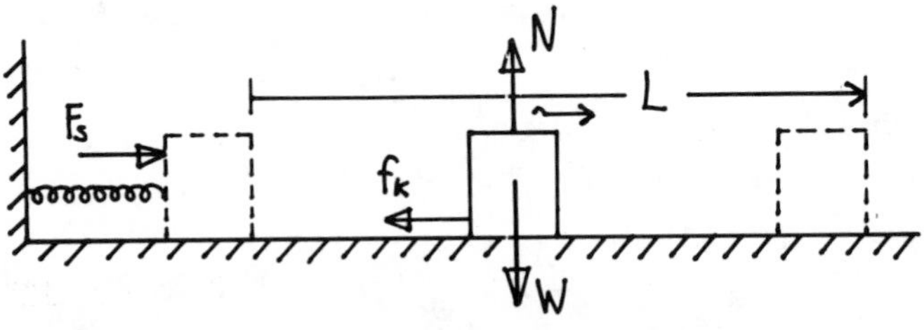

As the initial and final speeds are zero, so are the initial and final kinetic energies, and therefore the total work done on the block during its motion is zero, by the work-energy theorem. The forces acting on the block during all or part of its motion are the spring force F_s, kinetic friction f_k, the weight W and the normal force N. The last two forces do no work as they act in the vertical direction and the block moves horizontally (force and displacement at right angles). Thus,

$$W_t = W_s + W_f.$$

The force exerted by the spring is in the direction of the block's motion, so that W_s is positive; if x is the distance the spring was compressed,

$$W_s = +\tfrac{1}{2}kx^2.$$

The work done by friction is negative, for friction acts opposite to the displacement:

$$W_f = -f_k L = -\mu_k NL = -\mu_k mgL.$$

Thus,

$$0 = \tfrac{1}{2}kx^2 - \mu_k mgL,$$

$$\mu_k = \tfrac{1}{2}kx^2/mgL,$$

$$\mu_k = (\tfrac{1}{2})(200)(0.15)^2/(2)(9.8)(0.6) = 0.19 \quad \underline{\text{Ans.}}$$

7-22

(a) It is necessary first to find the tension T in the cable. From the free-body diagram of the astronaut,

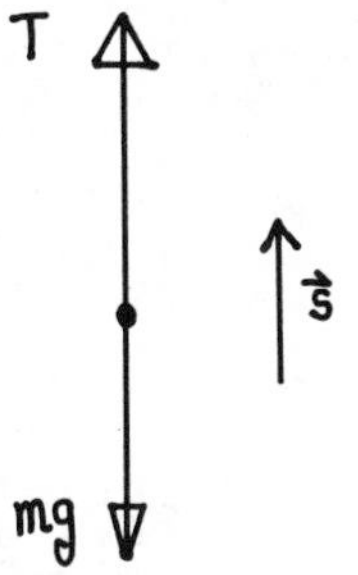

$$T - mg = ma = m(\tfrac{g}{10}),$$

$$T = \frac{11}{10}\,mg.$$

Let the total displacement of the astronaut be $\vec{s}$. The work W done by the helicopter (i.e. by the cable) is

$$W = \vec{T}\cdot\vec{s} = (\tfrac{11}{10}\,mg)s = \tfrac{11}{10}(160\text{ lb})(50\text{ ft}) = 8800\text{ ft}\cdot\text{lb} \quad \underline{\text{Ans.}}$$

(b) The work W_g done by gravity is

$$W_g = m\vec{g}\cdot\vec{s} = mgs\cos\pi = -mgs,$$

$$W_g = -(160\text{ lb})(50\text{ ft}) = -8000\text{ ft}\cdot\text{lb} \quad \underline{\text{Ans.}}$$

(c) The net work done on the astronaut is the sum of (a) and (b), to wit, 800 ft·lb. But this is equal to the change $K_f - K_i$ in the astronaut's kinetic energy. Presumably $K_i = 0$ so that, if v is the speed with which he reaches the helicopter,

$$800 \text{ ft}\cdot\text{lb} = \tfrac{1}{2}mv^2 = \tfrac{1}{2}\left(\frac{160}{32}\text{ slug}\right)v^2,$$

$$v = 17.9 \text{ ft/s} \quad \underline{\text{Ans.}}$$

7-26

The force F applied to the tool must be directed at some angle to the radial direction in order to provide a tangential component to counteract friction: $F_t = f_k$. The radial component F_r of F is the force actually exerted against the wheel, and this equals 180 N. The wheel must do work against F_t and therefore,

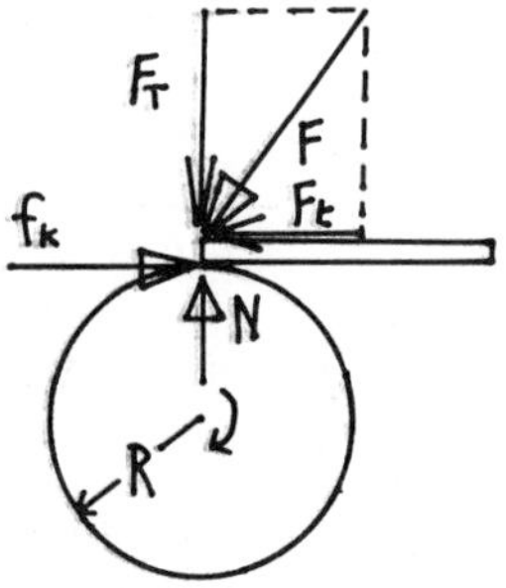

$$P = F_t v = f_k v = \mu_k N v = \mu_k F_r v.$$

In one second, a point on the rim of the wheel moves a distance

$$(2.5)(2\pi R) = (2.5)(2\pi)(0.2 \text{ m}) = \pi \text{ m}.$$

Hence, the speed v of a point on the rim is π m/s. Therefore,

$$P = (0.32)(180)(\pi) = 180.956 \text{ W},$$

$$P = \frac{180.956}{745.7} = 0.24 \text{ hp} \quad \underline{\text{Ans.}}$$

7-34

Let F, F' be the forces pulling the truck up the hill with speed v and then down with speed v' (these are friction forces exerted by the road). The resisting force in each direction is $F_R = mg/25$. In moving up and down the hill, the total work done on the truck is zero, for with the speed constant the kinetic energy does not change and $W = \delta K$. Thus, the rate at which work is done is zero also. The forces that do work are F (or F'), the weight mg and the resisting force F_R. Taking note of the direction of these forces,

for motion up the hill,

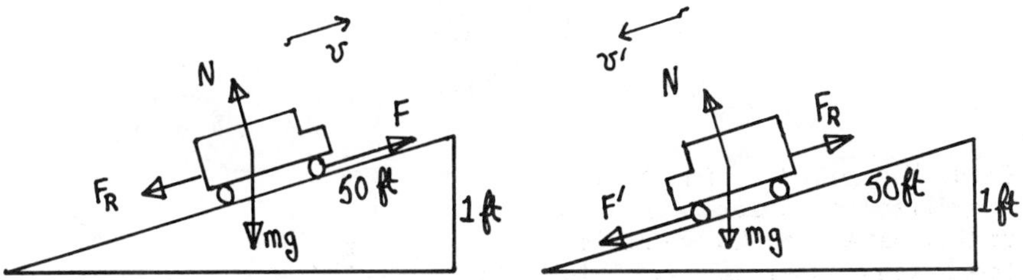

$$P - mg(v/50) - F_R v = 0,$$

where $P = Fv$ and $v/50$ is the vertical component of the truck's velocity. Since $F_R = mg/25$,

$$P = \frac{3}{50}\,mgv.$$

Moving down the hill, with $P' = F'v'$,

$$P' + mg(v'/50) - F_R v' = 0,$$

$$P' = \frac{1}{50}\,mgv'.$$

But $P = P'$ by assumption, so that

$$\frac{3}{50}\,mgv = \frac{1}{50}\,mgv',$$

$$v' = 3v = 3(15) = 45 \text{ mph} \quad \underline{\text{Ans.}}$$

7-37

(a) By the work-energy theorem,

$$W = P\,\delta t = (1.5 \times 10^6 \text{ W})(360 \text{ s}) = 5.4 \times 10^8 \text{ J} = \delta K,$$

$$\delta K = \tfrac{1}{2}mv_f^2 - \tfrac{1}{2}mv_i^2 = \tfrac{1}{2}m(625 - 100),$$

$$m = 2.057 \times 10^6 \text{ kg} \quad \underline{\text{Ans.}}$$

(b) If v is the speed of the train a time t (in seconds) after the train starts to accelerate,

$$Pt = \tfrac{1}{2}mv^2 - \tfrac{1}{2}mv_i^2,$$

$$v^2 = v_i^2 + \frac{2P}{m}t,$$

$$v = (100 + 1.458t)^{\frac{1}{2}} \quad \underline{\text{Ans}},$$

v in m/s and t in s.

(c) From (b),

$$\tfrac{1}{2}mv^2 - \tfrac{1}{2}mv_i^2 = Pt,$$

$$mv\frac{dv}{dt} = mva = P,$$

$$a = \frac{P}{mv},$$

since v_i and P are constants. The force is F = ma, giving for F,

$$F = \frac{P}{v} = \frac{1.5 \times 10^6}{(100 + 1.458t)^{\frac{1}{2}}} \quad \underline{\text{Ans}},$$

F in newtons.

(d) The total distance x moved is

$$x = \int v(t)dt = \int_0^{360} (100 + 1.458t)^{\frac{1}{2}}dt,$$

$$x = 6.7 \text{ km} \quad \underline{\text{Ans}}.$$

7-38

(a) Since the body started from rest, the work done after time t is

$$W = \tfrac{1}{2}mv^2,$$

where v is the speed after time t. As the acceleration a is constant,

$$v = at,$$

$$v_f = at_f,$$

so that

$$v = \frac{t}{t_f} v_f,$$

giving

$$W = \tfrac{1}{2}m \frac{v_f^2}{t_f^2} t^2 .$$

(b) The power is

$$P = \frac{dW}{dt} = m \frac{v_f^2}{t_f^2} t \quad \underline{\text{Ans}} .$$

(c) Since v_f = 90 km/h = 25 m/s, t = t_f = 10 s and m = 1500 kg, the power is 93750 W.

CHAPTER 8

8-1

The minimum work W required is that work performed by exerting a force F on the chain just equal to the weight of chain still hanging over the table's edge. If at some instant a length x is hanging over, then $F(x) = (m/L)gx$. Since the chain is being moved in a way to decrease x,

$$W = -\int F\,dx = -\int_{L/5}^{0} (m/L)gx\,dx = mgL/50 \quad \underline{\text{Ans.}}$$

8-9

The initial and final positions of the block are shown in the diagram, with the zero-level of gravitational potential energy marked. In both of these positions the speed and hence kinetic energy of the block is zero. By conservation of energy, with U_s the spring's potential energy,

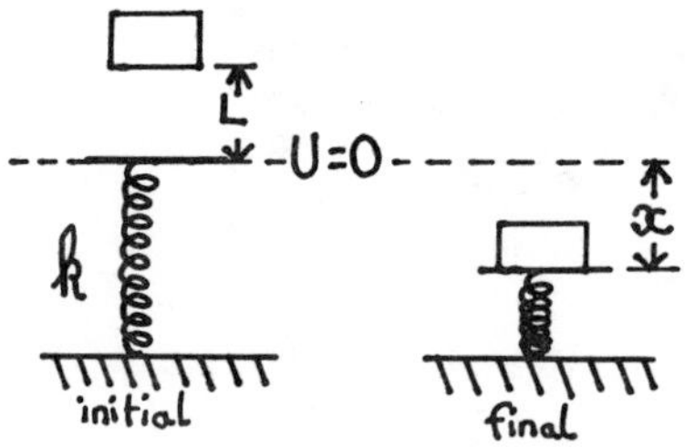

$$U_i + U_{si} + K_i = U_f + U_{sf} + K_f,$$

$$mg(+L) + \tfrac{1}{2}k(0)^2 + 0 = mg(-x) + \tfrac{1}{2}kx^2 + 0.$$

This can be rearranged to give

$$x^2 - (2mg/k)x - (2mgL/k) = 0\,.$$

But $mg/k = (2\text{ kg})(9.8\text{ m/s}^2)/(1960\text{ N/m}) = 0.01\text{ m}$; $mgL/k = 0.004\text{ m}^2$. With these substitutions the equation becomes,

$$x^2 - (0.02)x - 0.008 = 0\,,$$

$$x = 0.10\text{ m} \quad \underline{\text{Ans}},$$

taking the positive root.

8-16

(a) The force F is given by

$$F = -\frac{dU}{dx} = -(\text{slope of U vs x graph}).$$

Near x = 0 the slope is about -4, so $F(0) \cong 4$ N. U has zero slope at x = 2 m and therefore F(2) = 0. Between x = 2 and x = 6 m, the graph of U is almost a straight line with slope 2/4, giving F roughly constant at $F = -\frac{1}{2}$ N in this region. The negative sign indicates that the force points in the -x-direction.

(b) The kinetic energy is K = 4 - U, so the graph of K vs x is just a reflection of U vs x about the line 4 J.

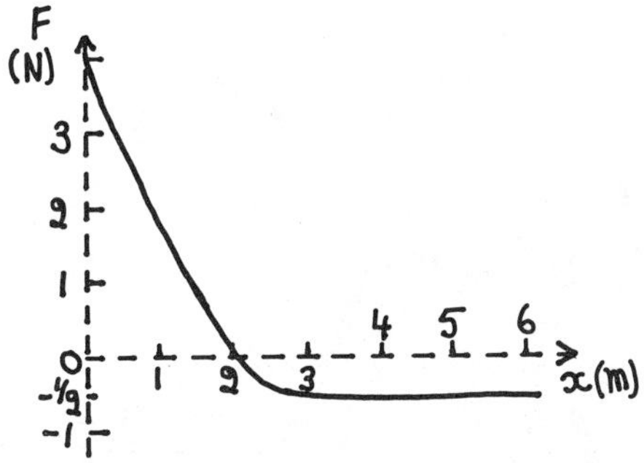

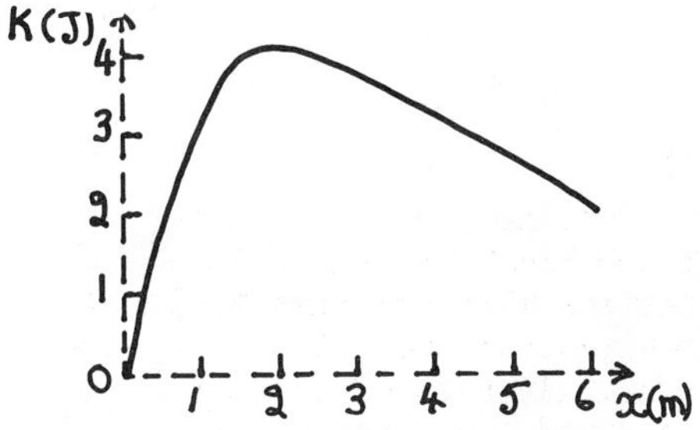

8-19

If the ball just swings around the nail, the tension in the string vanishes as the ball passes the top of its circular path with speed v, leaving its weight the only force acting. By Newton's second law and energy conservation,

$$mv^2/(L - d) = mg,$$

$$mgL + 0 = mg \cdot 2(L - d) + \tfrac{1}{2}mv^2.$$

Eliminating v between these and solving for d gives

$$d = \frac{3}{5}L \quad \underline{\text{Ans}}.$$

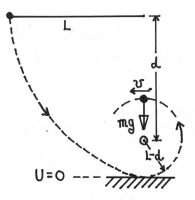

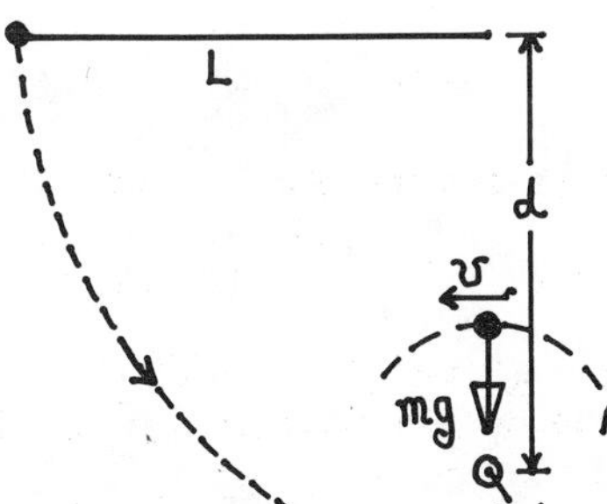

8-21

Call h the height of the table, m the mass of each marble and L the horizontal distance to impact. The horizontal speed v needed to carry the marble a distance L can be found from

$$L = vt,$$
$$h = \tfrac{1}{2}gt^2,$$

giving

$$v^2 = L^2g/2h.$$

$$t = \frac{L}{v} \qquad \frac{L}{v} = \sqrt{\frac{h}{\frac{1}{2}g}}$$
$$t = \sqrt{\frac{h}{\frac{1}{2}g}} \qquad v^2 = \frac{g}{2h}L^2$$

If the spring is compressed a distance x, the speed v with which the marble leaves the spring-gun is given by

$$\tfrac{1}{2}kx^2 = \tfrac{1}{2}mv^2.$$

Substituting for v^2 from the previous equation leads to

$$kx^2 = \tfrac{1}{2}mL^2g/h.$$

Call x_1, L_1 and x_2, L_2 the values of x, L for the two attempts. The equation above implies that

$$x_1/x_2 = L_1/L_2,$$

$$\frac{1 \text{ cm}}{x_2} = \frac{1.8 \text{ m}}{2.0 \text{ m}},$$

$$x_2 = 1.11 \text{ cm} \quad \underline{\text{Ans.}}$$

8-26

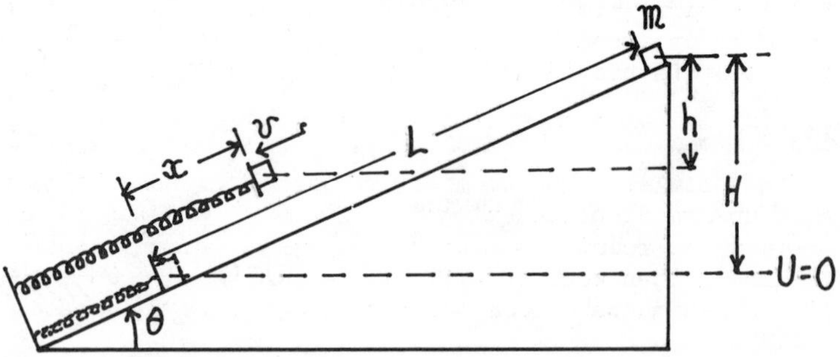

(a) The force exerted by an ideal spring is $F_s = kx$; with $x = 0.02$ meters, $F_s = 100$ N, it follows that $k = (100\text{ N})/(0.02\text{ m}) = 5000$ N/m By conservation of energy,

$$U_i + U_{si} + K_i = U_f + U_{sf} + K_f,$$

$$MgH + 0 + 0 = 0 + \tfrac{1}{2}kx^2 + 0,$$

$$H = kx^2/2Mg.$$

Using $k = 5000$ N/m found above, $x = 0.05$ m, $M = 10$ kg, $g = 9.8$ m/s^2 yields $H = 6.3776$ cm. The distance L through which the mass slides is

$$L = H/\sin\theta = 12.8 \text{ cm} \quad \underline{\text{Ans.}}$$

(b) Clearly $h = H - x\sin\theta = 6.3776 - (5)(\tfrac{1}{2}) = 3.8776$ cm. Again, by conservation of energy,

$$Mgh = \tfrac{1}{2}Mv^2,$$

$$v = 87 \text{ cm/s} \quad \underline{\text{Ans.}}$$

8-27

(a) The work done in one minute = mgh, where m is the mass carried in one minute and h is the vertical displacement; $m = 7500$ kg and $h = 8$ m. Therefore, the power "consumed" is

$$P = \frac{mgh}{t} = \frac{(7500)(9.8)(8)}{60} = 9800 \text{ W} \quad \underline{\text{Ans.}}$$

(b) The work that must be done against gravity is mgh = (80 kg)$\cdot$(g)(8 m) = 6272 J. The escalator would require (12 m)/(0.6 m/s) = 20 s to carry him up if he stood still. Since he reaches the top in 10 s, apparently he is supplying half the required work and the escalator is providing the other half, or 3136 J.

(c), (d) In walking to stay at the same level in space, at least part of the man (his feet) are alternatively being carried upward by the escalator and then allowed to fall back. Thus, there is some upward displacement and work is done by the motor. If at any instant, at least one foot is in contact with the step (normal walking), then work is being done continually. On the other hand, if the man jumps down after each short upward movement, then he is in contact with the escalator for only part of the time, affecting the power "consumed" from the motor. Evidently, the power "consumed" from an escalator motor depends on precisely "how" one walks.

8-28

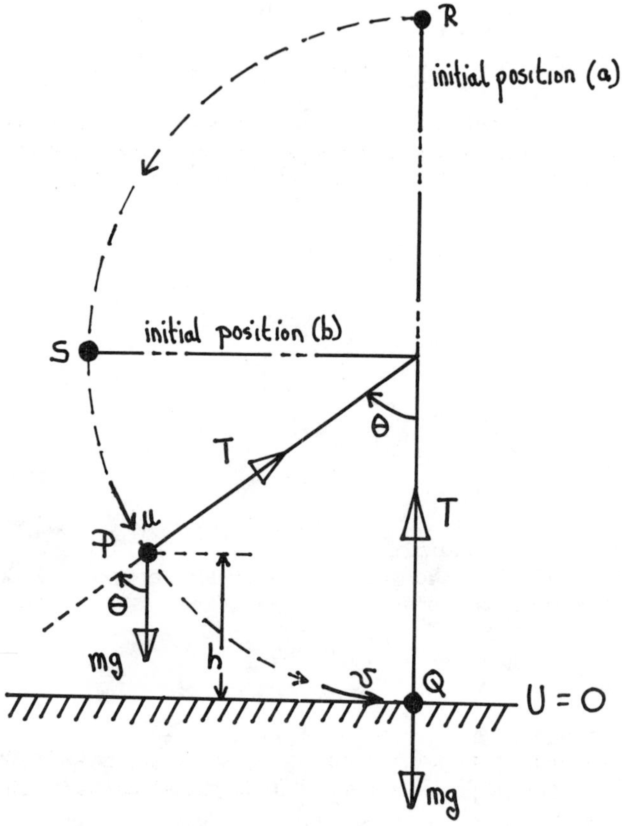

(a) By conservation of energy,

$$U_R + K_R = U_Q + K_Q,$$

$$mg(2L) + 0 = 0 + \tfrac{1}{2}mv^2,$$

$$v = 2(gL)^{\frac{1}{2}} \quad \underline{\text{Ans.}}$$

From the free-body diagram of the mass when at Q,

$$T - mg = mv^2/L,$$

since L is the radius of the circular path. But the second equation above gives

$$mv^2/L = 4mg,$$

so that

$$T = mg + 4mg = 5mg \quad \underline{\text{Ans}}.$$

(b) Applying Newton's second law along the direction of the suspension (i.e., along the direction to the center of the circle) to the mass at P, where the angle with the vertical has the desired value,

$$T - mg\cos\theta = mu^2/L,$$

u the speed at P. Energy conservation gives

$$U_S + K_S = U_P + K_P,$$

$$mgL + 0 = mgh + \tfrac{1}{2}mu^2,$$

$$u^2 = 2gL\cos\theta$$

since $h = L - L\cos\theta$. The tension T is, therefore

$$T = 2mg\cos\theta + mg\cos\theta = 3mg\cos\theta.$$

If $T = mg$, $\cos\theta = \frac{1}{3}$ or $\theta = 71°$.

<u>8-29</u>

Let v be the speed with which the boy leaves the ice; as he is given a very small push, let the initial speed be zero. Energy conservation gives

$$0 = \tfrac{1}{2}mv^2 - mg(R - h).$$

From Newton's second law,

$$mg\cos\theta - N = mv^2/R.$$

Combining these equations yields

$$mg\cos\theta - N = 2mg\,\frac{R - h}{R}.$$

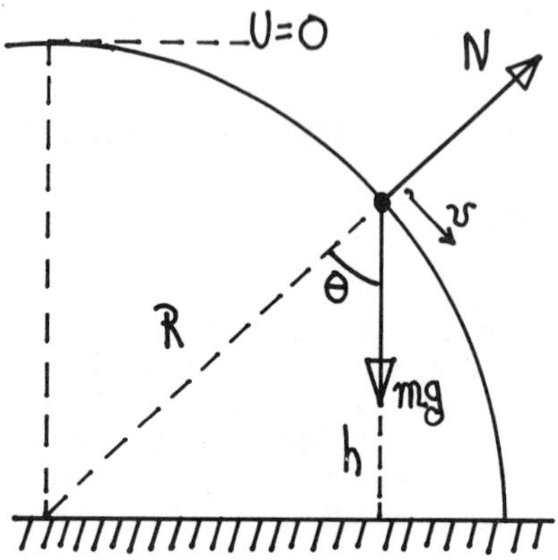

As the boy leaves the ice mound, the normal force N vanishes; putting N = 0 into the previous equation gives

$$\cos\theta = 2 - 2\frac{h}{R},$$

$$h = \frac{2}{3}R,$$

since $\cos\theta = h/R$.

8-30

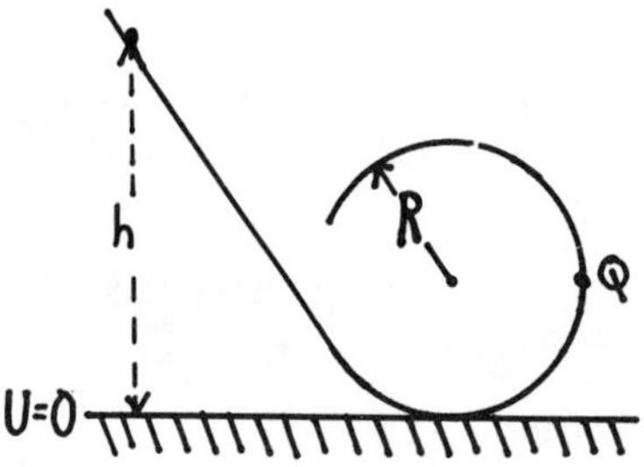

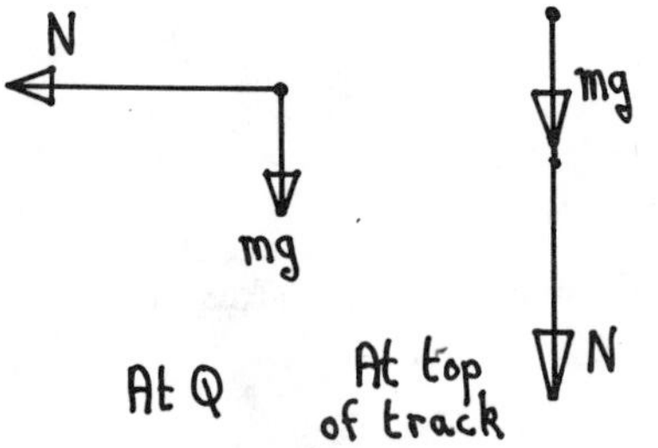

(a) Conservation of energy and Newton's second law give

$$mg(5R) + 0 = mgR + \tfrac{1}{2}mv_Q^2,$$

$$N = mv_Q^2/R = 8mg,$$

so that the forces acting at Q are N = 8mg to the left, normal to the track, and the weight mg directed vertically down as always.

(b) At the top of the track

$$N + mg = mv_t^2/R,$$

since N, normal to the track surface, here is in the same direction as the weight. By energy conservation,

$$mgh + 0 = mg(2R) + \tfrac{1}{2}mv_t^2.$$

Set N = mg as requested; the former equation gives $\frac{1}{2}mv_t^2 = mgR$; then the latter equation yields h = 3R.

8-33

(a) The vertical distance of the stone above the lowest point of swing when the string makes an angle θ with the vertical is L - Lcosθ = L(1 - cosθ). Using the lowest point as the zero level of potential energy, and letting v = speed at the lowest point, u = speed at angle θ, conservation of energy gives

$$\tfrac{1}{2}mv^2 = \tfrac{1}{2}mu^2 + mgL(1 - \cos\theta).$$

At the moment of release cosθ = ½; also v = 8 m/s and L = 4 m. Numerically, then, the equation, after dividing out the mass, yields

$$32 = \tfrac{1}{2}u^2 + 19.6,$$

$$u = 4.98\ \text{m/s}\quad \text{Ans.}$$

(b) Now let θ refer to the highest point of the swing; here u = 0 so that

$$32 = 39.2(1 - \cos\theta),$$

$$\theta = \cos^{-1}(0.18367) = 79.4^\circ\quad \underline{\text{Ans}}.$$

(c) The total energy is most easily calculated at the bottom of the swing, where the potential energy is zero; this energy is only kinetic at this point and is $\frac{1}{2}mv^2 = \frac{1}{2}(2)(8)^2 = 64$ J.

8-42

(a) Let the line AC serve as zero level for potential energy. The conservation of energy applied at points A and D yields

$$\tfrac{1}{2}mv_0^2 = mgl,$$

$$v_0 = (2gl)^{\frac{1}{2}}\quad \underline{\text{Ans}}.$$

(b) At B, the tension $\vec{T}$ acts up and the weight $m\vec{g}$ acts down:

$$T - mg = mv^2/l,$$

v the speed at B. Applying conservation of energy at A and B gives

$$\tfrac{1}{2}mv_0^2 = \tfrac{1}{2}mv^2 - mgl.$$

Combining these equations to eliminate v yields

$$T = mg + (2mg + mv_0^2/l),$$

$$T = 3mg + mv_0^2/l,$$

$$T = 5mg \quad \underline{\text{Ans}},$$

using (a)

(c) The total work done by all the forces = change in kinetic energy. The tension $\vec{T}$ does no work, being at right angles to the instantaneous velocity of the ball. Hence, if W_f is the work done by friction,

$$W_f + W_g = 0 - \tfrac{1}{2}mv_0^2.$$

Since C is on the same horizontal level as A, the work W_g done by gravity is zero, and therefore

$$W_f = -\tfrac{1}{2}mv_0^2 = -mgl \quad \underline{\text{Ans}},$$

since v_0 is the same as in (a).

(d) Apply the former equation, but now $W_g = +mgl$, B being at a vertical distance l below A; thus,

$$W_f = -\tfrac{1}{2}mv_0^2 - mgl,$$

$$W_f = -2mgl \quad \text{Ans},$$

again invoking (a).

8-43

(a) Let f be the friction force. The mass of the elevator is m = (4000)/(32) = 125 slug. If L is the distance traveled before hitting the spring, the work-energy theorem says that

$$\tfrac{1}{2}mv^2 - 0 = mgL - fL,$$
$$\tfrac{1}{2}(125)v^2 = (4000 - 1000)(12),$$
$$v = 24 \text{ ft/s} \quad \underline{\text{Ans.}}$$

(b) At the moment of maximum compression of the spring, the elevator is at rest momentarily. Thus

$$0 - \tfrac{1}{2}mv^2 = mgx - \tfrac{1}{2}kx^2 - fx,$$

since the spring does negative work in trying to push the elevator back up the shaft; the friction force f always does negative work. Numerically,

$$-36{,}000 = (4000)x - \tfrac{1}{2}(10{,}000)x^2 - (1000)x,$$
$$5x^2 - 3x - 36 = 0,$$
$$x = 3 \text{ ft} \quad \underline{\text{Ans.}}$$

(c) Let the distance be s. The elevator is at rest momentarily at the highest point reached and also at the point of maximum compression of the spring. Hence,

$$0 = -mgs + \tfrac{1}{2}kx^2 - fs,$$

the spring force now acting in the same direction the elevator moves; it is assumed that $s > x$. Using the result from (b),

$$0 = -(4000)s + \tfrac{1}{2}(10{,}000)(3)^2 - (1000)s,$$
$$s = 9 \text{ ft} \quad \underline{\text{Ans.}}$$

(d) Gravity and the spring force are conservative forces; all of the energy dissipated is dissipated by friction. If static friction can be neglected (f represents kinetic friction), the spring will be compressed a distance z when the elevator finally comes to rest, where kz = mg, or z = (4000 lb)/(10,000 lb/ft) = 0.4 ft. The total energy of the system the instant the cable snapped is

$$E_i = (4000 \text{ lb})(12.4 \text{ ft}) = 49{,}600 \text{ ft}\cdot\text{lb},$$

taking the final position as the zero level of gravitational potential energy. The final energy is

$$E_f = \tfrac{1}{2}kz^2 = \tfrac{1}{2}(10,000)(0.4)^2 = 800 \text{ ft·lb.}$$

The difference was removed by friction doing work on the elevator. If the elevator moved a total distance y,

$$(49,600 - 800) = fy = 1000y,$$

$$y = 48.8 \text{ ft} \quad \underline{\text{Ans.}}$$

The answer is exact to the extent that static friction can be neglected in determining the elevator's final resting place.

8-49

If M is the mass of the spaceship and m the mass that must be converted to energy, then using the classical expression for the kinetic energy as a simple approximation,

$$\tfrac{1}{2}Mv^2 = mc^2,$$

$$\tfrac{1}{2}M\left(\frac{c}{10}\right)^2 = mc^2,$$

$$m = \frac{1}{200} M,$$

$$m = \frac{1}{200}(2000 \text{ X } 10^3 \text{ lb})/(32 \text{ ft/s}^2),$$

$$m = 312.5 \text{ slug} \quad \underline{\text{Ans.}}$$

9-6

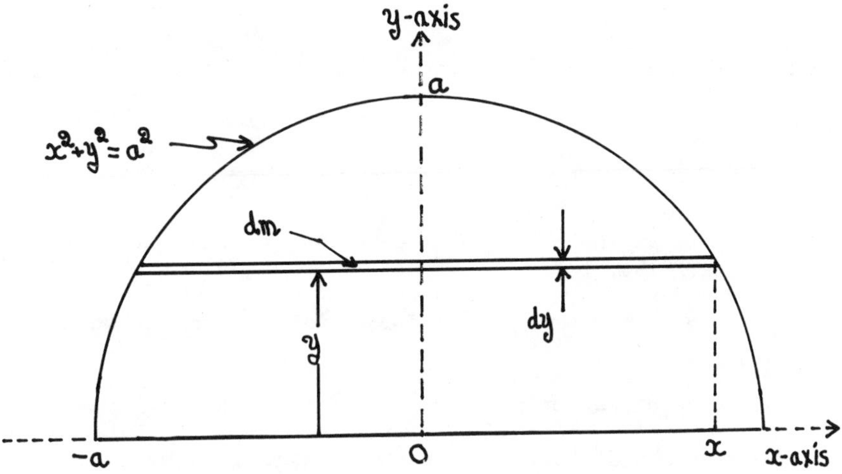

If the plate is arranged as shown with respect to the x,y-axes, then clearly the x-coordinate of the center of mass is zero, that is, the center of mass lies along the y-axis (the axis of symmetry of the plate). For the y-coordinate of the center of mass,

$$My_{cm} = \int y \, dm,$$

by definition; M is the mass of the plate. Consider a thin strip of the plate parallel to the x axis; its mass dm is

$$dm = \left(\frac{M}{A}\right)(2x \, dy),$$

where A is the area of the plate and 2xdy the area of the strip. The lower boundary of the plate is y = 0 and the upper is the equation

$$x^2 + y^2 = a^2.$$

Hence,

$$My_{cm} = \frac{2M}{A}\int_0^a y(a^2 - y^2)^{\frac{1}{2}}dy.$$

The integral can be put into the form $\int u^{\frac{1}{2}}du$, giving

$$y_{cm} = \frac{4a}{3\pi} \quad \underline{\text{Ans.}}$$

since $A = \frac{1}{2}\pi a^2$.

9-11

Let the shore be at the origin of the x-axis, x increasing outward from the shore. Also, let

x = distance of center of mass of dog+boat system from shore,

x_d = distance of dog from shore,

x_b = distance of center of mass of boat from shore.

Then, if W = weight of boat, w = weight of dog, and primes denote quantities after the dog takes his walk,

$$(W + w)x = Wx_b + wx_d,$$
$$(W + w)x' = Wx_b' + wx_d'.$$

But $x = x'$ so that

$$W(x_b' - x_b) = w(x_d - x_d').$$

From Chapter 4, (displacement of dog relative to boat) = (the displacement of dog relative to shore) + (displacement of shore relative to boat). Recalling that x increases away from the shore and that the dog walks toward the shore,

$$-8 = (x_d' - 20) - (x_b' - x_b),$$

$$x_b' - x_b = x_d' - 12.$$

Substitute this into the third equation to get

$$W(x_d' - 12) = w(20 - x_d'),$$

since $x_d = 20$ ft. With $W = 40$ lb, $w = 10$ lb this gives

$$x'_d = 13.6 \text{ ft} \quad \underline{\text{Ans.}}$$

9-14

With the ice being frictionless, the center of mass of (iceboat + man) must move with constant velocity even though the man changes his position relative to the iceboat. The center of mass of

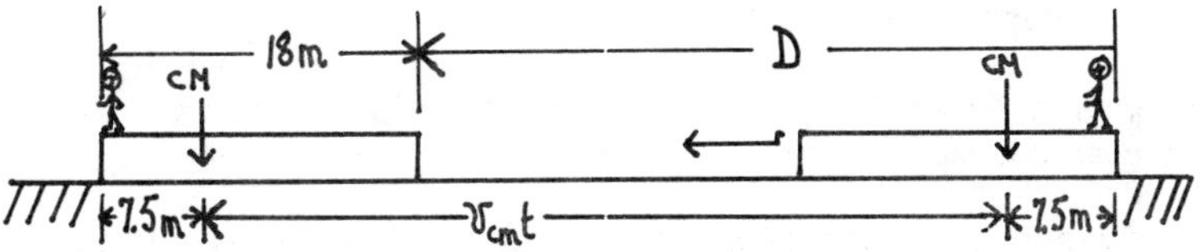

(iceboat + man) is at a distance

$$\frac{(80)(0) + (400)(9)}{400 + 80} = 7.5 \text{ m}$$

from the end on which the man is standing, before and after his walk. The man takes a time $t = (18 \text{ m})/(2 \text{ m/s}) = 9$ s to walk from one end of the boat to the other. In that time, the center of mass of the system moves across the ice a distance

$$d = v_{cm}t = (4 \text{ m/s})(9 \text{ s}) = 36 \text{ m}.$$

The distance D the boat moved in this time measured, say, by the motion of the end of the boat, is

$$D = d + 2(7.5) - 18 = 36 + 15 - 18 = 33 \text{ m} \quad \underline{\text{Ans.}}$$

9-16

(a) The external force F_{ext} exerted by the ground during the leap acts over a distance $s_{cm} = 90 - 40 = 50$ cm $= 0.5$ m. If v is the speed with which she leaves the ground,

$$F_{ext}s_{cm} = \tfrac{1}{2}Mv^2,$$

where M = 55 kg. Subsequently, she rises to a height h = 120 - 90 = 30 cm = 0.3 m above her "launching" position; hence,

$$v^2 = 2gh.$$

Combining these gives

$$F_{ext}s_{cm} = Mgh,$$

$$F_{ext} = (55)(9.8)(0.3)/(0.5) = 323.4\ \text{N} \quad \underline{\text{Ans.}}$$

(b) From the second equation,

$$v = (2gh)^{\frac{1}{2}} = 2.425\ \text{m/s} \quad \underline{\text{Ans.}}$$

9-22

(a) Clearly, the center of mass lies midway between the bodies, 2.5 cm from each.

(b) Let x, y be the distances of the center of mass from the 520 g, 480 g bodies respectively. Then,

$$x + y = 5,$$

$$520x = 480y.$$

These give x = 2.4 cm, or the center of mass has moved 1 mm closer to the more massive body.

(c) In terms of the accelerations of the two bodies, the acceleration a_{cm} of the center of mass is given by

$$a_{cm} = \frac{m_1a_1 + m_2a_2}{m_1 + m_2} = \frac{m_1a + m_2(-a)}{m_1 + m_2},$$

$$a_{cm} = \frac{m_1 - m_2}{m_1 + m_2}\,a,$$

where m_1 = 520 g and a is the acceleration of m_1 (directed down, as this is the heavier body). From Example 8, Chapter 5,

$$a = \frac{m_1 - m_2}{m_1 + m_2}\,g,$$

and therefore

$$a_{cm} = \left[\frac{m_1 - m_2}{m_1 + m_2}\right]^2 g = (0.04)^2 g,$$

$$a_{cm} = 0.0016\ g \quad \underline{\text{Ans}},$$

directed down.

9-26

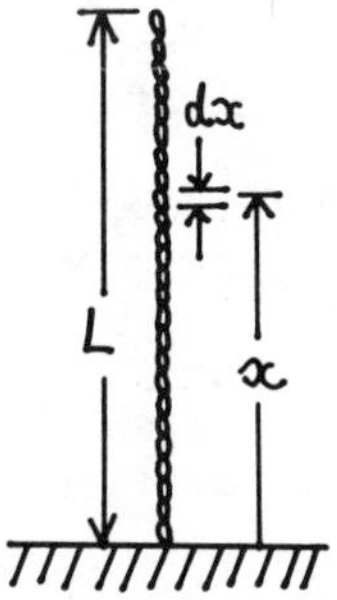

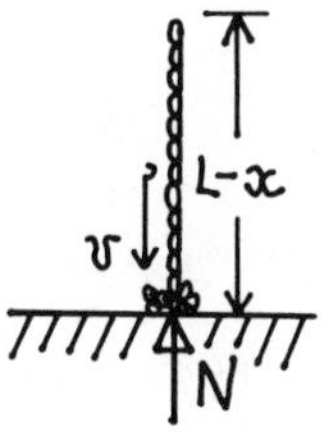

In magnitude, the force N exerted by the table on the chain is equal to the force exerted on the table by the chain. This is

$$N = (\frac{M}{L}\ x)g + \frac{dp}{dt}.$$

M is the mass of the chain and L is its length. The first term is the weight of that part of the chain already on the table and the second term is the force exerted by the link just landing and being brought to rest. If dM, dx are the mass and length of a link and v the speed with which it strikes the table,

$$\frac{dp}{dt} = \frac{(dM)v}{dt} = \frac{(\frac{M}{L}\ dx)v}{dt} = \frac{M}{L}\ v^2 = \frac{M}{L}(2gx),$$

x being the original height above the table of the link just now landing. With this, N becomes

$$N = \frac{M}{L}\ xg + 2\ \frac{M}{L}\ xg = 3(\frac{M}{L}\ x)g \quad \underline{\text{Ans}},$$

or three times the weight of chain already on the table.

9-30

(a) Let V, v be the velocities of the rocket case, mass M and payload, mass m, respectively after separation. The velocity of the center of mass of the rocket is unchanged by the separation; hence, if u = 7600 m/s, the speed before separation, is

$$u = \frac{MV + mv}{M + m},$$

initial direction taken as positive. The relative speed after separation is 910 = v - V (the less massive component expected to have the greater 'absolute' velocity), and therefore

$$(440)(7600) = 290(v - 910) + 150v,$$

$$v = 8200 \text{ m/s} \quad \underline{\text{Ans}},$$

$$V = 8200 - 910 = 7290 \text{ m/s} \quad \underline{\text{Ans}}.$$

(b) The kinetic energies before and after separation are

$$K_b = \tfrac{1}{2}(M + m)u^2 = 1.271 \times 10^{10} \text{ J} \quad \underline{\text{Ans}};$$

$$K_a = \tfrac{1}{2}mv^2 + \tfrac{1}{2}MV^2 = (0.5043 + 0.7706) \times 10^{10},$$

$$K_a = 1.275 \times 10^{10} \text{ J} \quad \underline{\text{Ans}}.$$

Evidently, 4×10^7 J $(= \tfrac{1}{2}kx^2)$ was stored in the spring prior to separation.

9-34

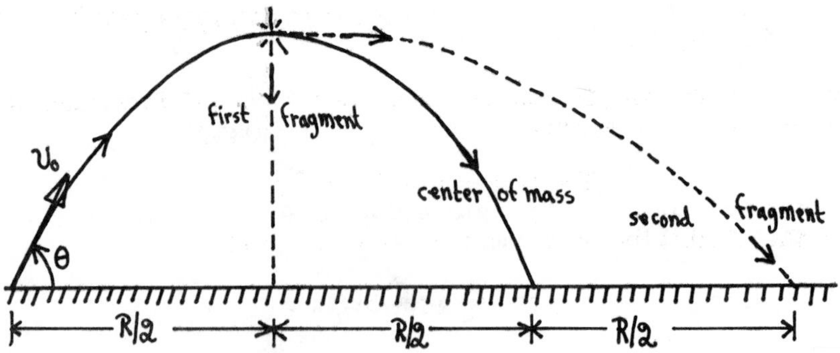

The time t required for the shell to reach the highest point of its path, assuming the explosion does not supervene, is

$$t = \frac{v_0 \sin\theta}{g} = \frac{(1500)(\frac{1}{2}3^{\frac{1}{2}})}{32} = 40.6 \text{ s.}$$

This is so close to 41 s that the explosion can be considered to take place at the peak of the trajectory. At this point, the vertical component of the momentum of the shell just before the explosion is zero, and therefore the total vertical component of the momentum of the fragments immediately after the explosion must be zero also. Since one fragment is at rest just after the catastrophe, the other must have been projected horizontally. Thus, the fragments strike the ground simultaneously and their center of mass must strike the ground as they do. But the fragments have the same mass and therefore the center of mass impacts at a point equidistant from the impact points of the fragments. As the center of mass moves as though no explosion took place, the situation is as presented in the sketch, with the fragments landing R/2 from the center of mass, R the range of the gun. The second fragment, then, lands from the gun a distance

$$\frac{3}{2}R = \frac{3}{2}(v_0^2 \sin 2\theta/g) = 91{,}340 \text{ ft} \quad \underline{\text{Ans.}}$$

9-35

Let

$\vec{V}$ = velocity of wedge relative to table;

$\vec{u}$ = velocity of block relative to wedge;

$\vec{v}$ = velocity of block relative to table.

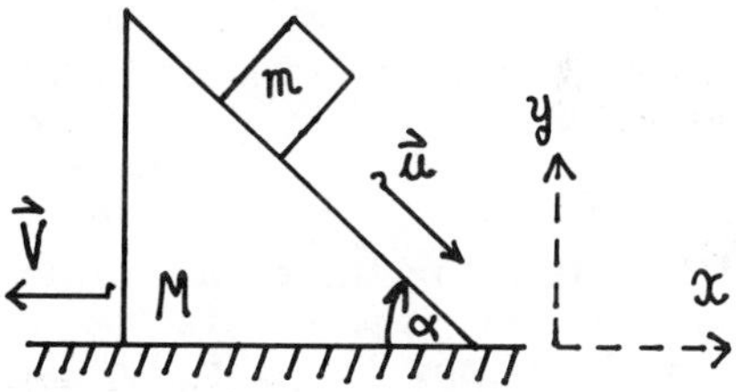

Then,

$$\vec{u} + \vec{V} = \vec{v}.$$

As the block remains in contact with the wedge, and referring to the coordinate system shown,

$$\vec{u} = (u\cos\alpha)\vec{i} - (u\sin\alpha)\vec{j};$$

since the block remains in contact with the table,

$$\vec{V} = -V\vec{i}.$$

Therefore,

$$\vec{v} = (u\cos\alpha - V)\vec{i} - (u\sin\alpha)\vec{j},$$

which implies that

$$v^2 = u^2 + V^2 - 2uV\cos\alpha.$$

Momentum is conserved in the horizontal direction:

$$MV = m(u\cos\alpha - V),$$

so that

$$u = \frac{m + M}{m\cos\alpha} V.$$

If u, v, V are now taken as applying at the instant the block makes contact with the table, conservation of energy gives

$$mgh = \tfrac{1}{2}MV^2 + \tfrac{1}{2}mv^2,$$

giving

$$v^2 = 2gh - \frac{M}{m} V^2.$$

The last equation and the second preceding equation give v and u in terms of V; substitute them into the equation for v^2 to obtain

$$V^2 = \frac{2ghm^2 \cos^2\alpha}{(M + m)(M + m\sin^2\alpha)} \quad \underline{\text{Ans}}.$$

9-40

First solve for the change in speed for only one man on the flatcar, who jumps off in the prescribed manner. Initially, the velocity of the car is v_0 to the right, and after the man has jumped off it is v (assuming that the man simply "runs off"). Then, since the man runs to the left, his velocity relative to the ground is $v - v_{rel}$ after jumping. The velocity of the center of mass of the system (car + man) is unaffected by this process and therefore

$$v_{cm} = v_0 = \frac{Wv + w(v - v_{rel})}{W + w}.$$

Therefore,

$$\delta v = v - v_0 = \frac{wv_{rel}}{W + w}.$$

(i) If there are n men and all go at once, use the result above with w replaced with nw:

$$\delta v_i = \frac{nwv_{rel}}{W + nw}.$$

(ii) If the n men go in succession, calculate the increment in speed imparted by each. For the first man,

$$\delta v_1 = \frac{wv_{rel}}{w + [W + (n - 1)w]},$$

since the original flatcar is now, essentially, replaced by the flatcar + the (n - 1) men who stay behind. The total change in speed is $\delta v_{ii} = \delta v_1 + \delta v_2 + \ldots + \delta v_n$, or

$$\delta v_{ii} = \frac{w\, v_{rel}}{W + nw}\left[1 + \frac{W + nw}{W + (n - 1)w} + \ldots + \frac{W + nw}{W + w}\right].$$

All of the terms inside the brackets are > 1 (except the first which equals one), and there are n of these terms. Hence the second method yields a greater change in speed than the first (which means that the answer to the question is YES).

9-43

(a) In ingesting the air the plane experiences a resisting force F_1 equal to

$$F_1 = v_{rel}\frac{dm}{dt} = (600\ \text{ft/s})(4.8\ \text{slug/s}) = 2880\ \text{lb}.$$

The mass ejected each second is the 4.8 slug of air plus 0.2 slug of fuel. The thrust imparted to the plane in ejecting this is

$$F_2 = v_{rel}\frac{dm}{dt} = (1600\ \text{ft/s})(5\ \text{slug/s}) = 8000\ \text{lb}.$$

The net thrust is $F_2 - F_1 = 5120$ lb Ans.

(b) The horsepower delivered is

$$P = (F_2 - F_1)v = \frac{(5120\ \text{lb})(600\ \text{ft/s})}{(550\ \text{ft·lb/s·hp})} = 5600\ \text{hp} \quad \underline{\text{Ans.}}$$

9-44

(a) The force F of the water on the truck equals the rate of change of momentum of the water measured relative to the earth:

$$F = \frac{dm}{dt}(v - V),$$

where dm/dt is the rate at which water strikes the truck, V is the speed of the truck and v the speed of the water: that is, dm/dt = 60 kg/s, v = 30 m/s. The acceleration of the truck, mass M = 2000 kg is

$$a = \frac{dV}{dt} = \frac{F}{M} = \frac{1}{M}\frac{dm}{dt}(v - V).$$

Putting in the numbers gives

$$\frac{dV}{dt} = 0.9 - 0.03\ V,$$

$$\int_0^V \frac{dV}{0.9 - 0.03V} = \int_0^t dt.$$

Carrying out the integrations, remembering to evaluate at the lower limits also, and taking anti-lns gives

$$V = 30(1 - e^{-0.03t}),$$

$$a = \frac{dV}{dt} = 0.9e^{-0.03t}.$$

Therefore, the acceleration at t = 0 is, from the equation above, $0.9\ \text{m/s}^2$ <u>Ans.</u>

(b) As the truck moves, the acceleration approaches zero exponentially.

(c) If the water remains inside the truck the mass M in (a) is variable:

$$M = M_0 + \frac{dm}{dt}\ t,$$

with M_0 now the mass of the empty truck. The equation of motion

$$M\frac{dV}{dt} = v_{rel}\frac{dM}{dt}$$

becomes

$$(M_0 + \frac{dm}{dt}t)\frac{dV}{dt} = (v - V)\frac{dm}{dt},$$

or, numerically,

$$(2000 + 60t)\frac{dV}{dt} = 60(30 - V).$$

Integrating as before,

$$\int_0^V \frac{dV}{30 - V} = \int_0^t \frac{60dt}{2000 + 60t},$$

$$\frac{30}{30 - V} = 1 + \frac{3}{100}t,$$

$$V = 0.9\frac{t}{1 + 0.03t}$$

so that

$$a = \frac{dV}{dt} = \frac{0.90}{(1 + 0.03t)^2}.$$

When $t = 0$, $a = 0.9$ m/s^2 as before and again approaches zero, but at a different rate from (a) and (b).

CHAPTER 10

10-7

Since the force increases uniformly,

$$\overline{F} = \tfrac{1}{2}(F_f - F_i) = \tfrac{1}{2}(50) = 25\ \text{N},$$

so that

$$\overline{F}(t_f - t_i) = m(v_f - v_i),$$

$$(25\ \text{N})(4\ \text{s}) = (10\ \text{kg})(v_f - 0),$$

$$v_f = 10\ \text{m/s}\quad \underline{\text{Ans}}.$$

10-12

Let v_{12}, v_{18} be the final speeds. Since the initial speed and hence initial momentum of each spacecraft is zero,

$$m_{12}v_{12} = m_{18}v_{18},$$

$$m_{12}v_{12} + m_{18}v_{18} = 600\ \text{N}\cdot\text{s}.$$

Putting $m_{12} = 1200$ kg and $m_{18} = 1800$ kg in the above equations, the final speeds are found to be

$$v_{12} = \frac{1}{4}\ \text{m/s},\quad v_{18} = \frac{1}{6}\ \text{m/s}.$$

Since the spacecraft go off in opposite directions, the relative speed is

$$v = v_{12} + v_{18} = \frac{1}{4} + \frac{1}{6} = \frac{5}{12}\ \text{m/s}\quad \underline{\text{Ans}}.$$

10-26

By conservation of momentum and mechanical energy,

$$mv_i = (m + M)V,$$

$$\tfrac{1}{2}mv_i^2 = \tfrac{1}{2}(m + M)V^2 + U_s = \tfrac{1}{2}(m + M)V^2 + f\cdot\tfrac{1}{2}mv_i^2.$$

Here V is the speed of the gun+ball system as the ball sticks, U_s is the potential energy of spring+ball when the ball is stuck and f is the desired fraction. Solve for V in the first equation and substitute into the second; a factor $\tfrac{1}{2}mv_i^2$ cancels and the rest yields for the fraction,

$$f = \frac{M}{m + M} \quad \underline{\text{Ans}}.$$

10-27

Let v be the speed of the marbles just before impact; since the collisions are completely inelastic, the marbles come to rest upon striking the box. Hence, the momentum transferred to the box by each marble is $p = m(v_f - v_i) = m(0 - v) = -mv$, or $p = mv$ in magnitude. As μ marbles strike each second, a time interval $T = \mu^{-1}$ elapses between impacts. Thus, the average force F imparted to the box by the colliding marbles is

$$F = \frac{p}{T} = \frac{mv}{1/\mu} = m\mu v = m\mu(2gh)^{\frac{1}{2}}.$$

But, after time t, μt marbles with total weight $(mg)\mu t$ already reside in the box. Hence, the scale reading R after time t is

$$R = m\mu(2gh)^{\frac{1}{2}} + mg\mu t = mg\mu\left[(2h/g)^{\frac{1}{2}} + t\right] \quad \underline{\text{Ans}}.$$

Numerically, in English units,

$$R = (0.01)(100)[(2\cdot 25/32)^{\frac{1}{2}} + 10] = 11.25 \text{ lb} \quad \underline{\text{Ans}}.$$

10-30

Let the x-axis be vertical and positive downward, with origin at the top of the shaft. If the ball is dropped at time $t = 0$,

$$x_b = \tfrac{1}{2}gt^2 \; ; \; x_e = h - vt,$$

where v is the speed of the elevator and $h = 60$ ft. If ball and elevator collide at time $t = t'$, then

$$x_b(t') = x_e(t'),$$

$$h - vt' = \tfrac{1}{2}gt'^2,$$

$$gt' = -v + (v^2 + 2gh)^{\frac{1}{2}},$$

picking, for the last equation, the positive root of the equation for t'. Relative to the elevator, the velocity of the ball is reversed by the collision (elastic collision): just before the collision the relative velocity is $+(v + gt')$ and just after it is $-(v + gt')$. Thus, the velocity of the ball relative to the shaft just after collision is $-(v + gt') - v = -(2v + gt')$. Therefore, the highest point $x = H$ subsequently reached by the ball is given from

$$0^2 = (2v + gt')^2 + 2g[H - x_e(t')],$$

where $x_e(t') = h - vt'$. Putting this in and rearranging gives

$$-4v^2 - 6v(gt') - (gt')^2 = 2g(H - h).$$

Finally, substitute for gt' from above and solve for H:

$$H = -2\left(\frac{v}{g}\right)(v^2 + 2gh)^{\frac{1}{2}}.$$

(a) Set $v = +6$ ft/s; $H = -23.3$ ft (i.e. above top of shaft) Ans.
(b) Put $v = -6$ ft/s; $H = +23.3$ ft (i.e. below top of shaft) Ans.

10-31

At the instant of maximum compression of the spring, the relative velocity of the two blocks is zero: that is, they move with a common speed V relative to the table. Hence, by conservation of momentum,

$$m_1v_1 + m_2v_2 = (m_1 + m_2)V,$$

$$(2)(10) + (5)(3) = (2 + 5)V,$$
$$V = 5 \text{ m/s}.$$

Then, by conservation of mechanical energy, with x the maximum compression of the spring,

$$\tfrac{1}{2}m_1 v_1^2 + \tfrac{1}{2}m_2 v_2^2 = \tfrac{1}{2}(m_1 + m_2)V^2 + \tfrac{1}{2}kx^2,$$
$$\tfrac{1}{2}(2)(10)^2 + \tfrac{1}{2}(5)(3)^2 = \tfrac{1}{2}(2 + 5)(5)^2 + \tfrac{1}{2}(1120)x^2,$$
$$x^2 = \frac{70}{1120} = \frac{1}{16},$$
$$x = \tfrac{1}{4} \text{ m} = 25 \text{ cm} \quad \underline{\text{Ans.}}$$

10-32

(a) In parenthesis are indicated the initial and final velocities for the specified mass in the collision under consideration. In the collision of the left (v_0, v') mass with the center (0, u) mass,

$$mv_0 + 0 = mv' + mu, \quad \tfrac{1}{2}mv_0^2 = \tfrac{1}{2}mv'^2 + \tfrac{1}{2}mu^2,$$

giving

$$v' = 0, \quad u = v_0.$$

Thus, the left mass comes to rest. In the subsequent collision between the center (u, u') mass with the right (0, V) mass,

$$mu + 0 = mu' + MV, \quad \tfrac{1}{2}mu^2 = \tfrac{1}{2}mu'^2 + \tfrac{1}{2}MV^2.$$

Solving gives

$$V = 2\frac{m}{m + M}v_0, \quad u' = \frac{m - M}{m + M}v_0.$$

If M is less than m, u' is greater than zero and the center mass moves to the right with a velocity u' smaller than V, so it never catches the right mass and, therefore, there are two collisions.
(b) If M is greater than m, the motion of the right mass is as above, but now u' is less than zero, so that the center mass moves to the left with a speed

$$-u' = \frac{M - m}{M + m} v_0.$$

This mass will collide subsequently with the stationary left mass and the two will exchange velocities.

<u>10-40</u>

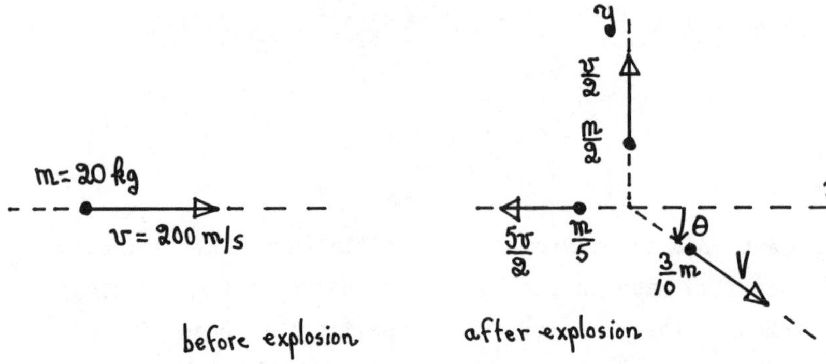

(a) Let m = 20 kg, v = 200 m/s be the mass and speed of the body before the explosion. From the numerical data given, the masses and speeds of the fragments after explosion can be written in terms of those before as indicated on the diagram. In order that the total momentum after explosion be entirely in the +x-direction, as it is before, the speed V of the third fragment must be directed into the fourth quadrant. Conservation of momentum in the x and y-directions gives

$$mv = -\left(\frac{m}{5}\right)\left(\frac{5v}{2}\right) + \left(\frac{3m}{10}\right)(V\cos\theta),$$

$$0 = \left(\frac{m}{2}\right)\left(\frac{v}{2}\right) - \left(\frac{3m}{10}\right)(V\sin\theta).$$

Simplifying

$$\frac{3}{2}\,mv = \frac{3}{10}\,mV\cos\theta,$$

$$\frac{1}{4}\,mv = \frac{3}{10}\,mV\sin\theta.$$

Dividing gives

$$\theta = \tan^{-1}\left(\frac{1}{6}\right) = 9.46° \quad \underline{\text{Ans}}.$$

Substituting this into the second equation, say, yields

$$V = \frac{5v}{6\sin\theta} = 1014 \text{ m/s} \quad \underline{\text{Ans}}.$$

(b) The energy E supplied by the explosion is the excess of total kinetic energy after explosion over that present before the explosion:

$$E = \tfrac{1}{2}\left(\frac{m}{2}\right)\left(\frac{v}{2}\right)^2 + \tfrac{1}{2}\left(\frac{m}{5}\right)\left(\frac{5v}{2}\right)^2 + \tfrac{1}{2}\left(\frac{3m}{10}\right)\left(\frac{5v}{6\sin\theta}\right)^2 - \tfrac{1}{2}mv^2,$$

$$E = \frac{97}{24}\,mv^2 = 3.23 \times 10^6 \text{ J} \quad \underline{\text{Ans}}.$$

10-41

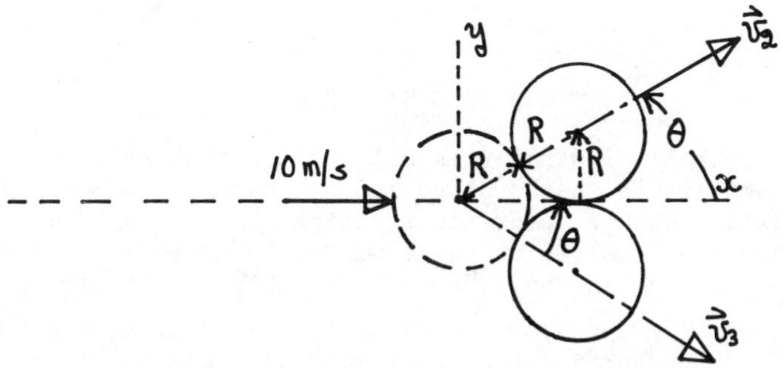

Since the balls are frictionless, the forces between them at the moment of collision are perpendicular to their surfaces, i.e., along a radius. The geometry at impact is shown in the diagram. Clearly,

$$\sin\theta = \frac{R}{2R} = \frac{1}{2}, \quad \cos\theta = \tfrac{1}{2}3^{\frac{1}{2}}.$$

Also, from the symmetry of the collision, $v_2 = v_3 = v$, say. This follows, also, from conservation of momentum in the y-direction. In addition, as the momentum in the y-direction before the impact is zero, it must be zero just after; it follows that the velocity of the first ball after collision must be in the x-direction; call this velocity V. By conservation of momentum,

$$m(10) = 2mv\cos\theta + mV,$$

$$10 = 3^{\frac{1}{2}}v + V.$$

By conservation of kinetic energy,

$$\tfrac{1}{2}m(10)^2 = 2\cdot\tfrac{1}{2}mv^2 + \tfrac{1}{2}mV^2.$$

This equation reduces to

$$100 = 2v^2 + V^2.$$

From the momentum equation,

$$3v^2 = (10 - V)^2 = 100 - 20V + V^2.$$

Substituting this into the equation preceding it yields

$$(5V + 10)(V - 10) = 0,$$

giving

$$V = -2 \text{ m/s} \quad \underline{\text{Ans}}.$$

That is, the ball originally in motion "bounces" and moves off after the collision in the direction from whence it came. The other solution of the quadratic equation simply reproduces the conditions before collision, as expected in cases of this type (i.e., elastic collisions). Hence, with this result for V,

$$v = 3^{-\frac{1}{2}}(10 + 2) = 6.93 \text{ m/s} \quad \underline{\text{Ans}}.$$

<u>10-42</u>

Let m be the mass of each object and v their common initial speed. Use the rules for addition of vectors by the geometric method to arrange the initial and final velocities so that the vector sum of the initial momenta equals the momentum of the pair after sticking together in the collision. By conservation of momentum in the x and y-directions respectively,

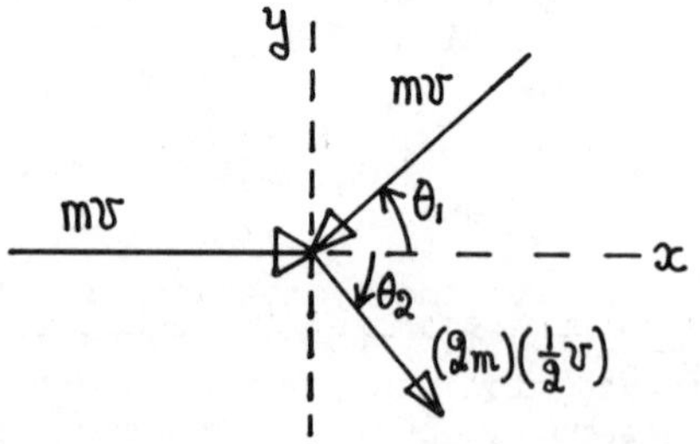

$$mv - mv\cos\theta_1 = (2m)(\tfrac{1}{2}v)\cos\theta_2,$$

$$mv\sin\theta_1 = (2m)(\tfrac{1}{2}v)\sin\theta_2.$$

Dividing out the mv gives

$$1 - \cos\theta_1 = \cos\theta_2,$$
$$\sin\theta_1 = \sin\theta_2.$$

The second equation above implies that $\theta_1 = \theta_2$. Using this in the previous equation gives

$$1 - \cos\theta_1 = \cos\theta_1,$$
$$\theta_1 = \cos^{-1}(\tfrac{1}{2}) = \pi/3,$$

so that the angle between the initial velocities ("tail to tail") is $2\pi/3$.

10-43

Conservation of momentum and kinetic energy give

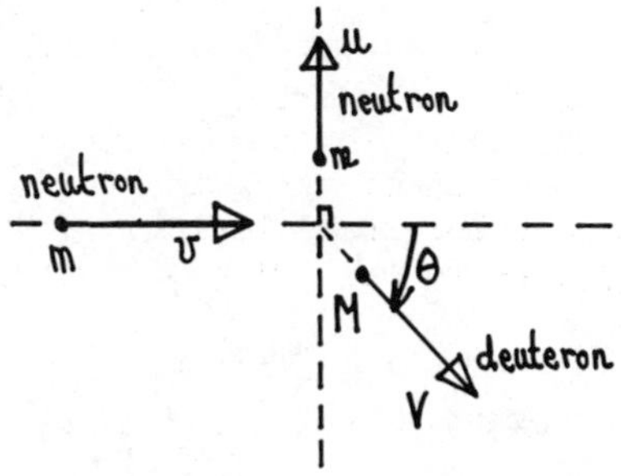

$$mv = MV\cos\theta,$$
$$mu = MV\sin\theta,$$
$$\tfrac{1}{2}mv^2 = \tfrac{1}{2}mu^2 + \tfrac{1}{2}MV^2,$$

m the mass of the neutron, M the mass of the deuteron. From the first two equations,

$$M^2V^2 = m^2v^2 + m^2u^2,$$

and from the third equation,

$$m^2u^2 = m^2v^2 - MmV^2.$$

Now combine the last two equations to get the kinetic energy of the deuteron:

$$\tfrac{1}{2}MV^2 = \frac{m^2}{m + M}v^2.$$

But M = 2m and therefore

$$\tfrac{1}{2}MV^2 = \frac{m}{m + M}(mv^2) = \frac{1}{3}mv^2 = \frac{2}{3}(\tfrac{1}{2}mv^2),$$

establishing the result quoted in the problem.

10-44

By conservation of the two components of momentum and of kinetic energy,

$$m_1v_1 + 0 = m_1u_1\cos\theta_1 + m_2u_2\cos\theta_2,$$
$$0 = m_1u_1\sin\theta_1 + m_2u_2\sin\theta_2,$$
$$\tfrac{1}{2}m_1v_1^2 + 0 = \tfrac{1}{2}m_1u_1^2 + \tfrac{1}{2}m_2u_2^2.$$

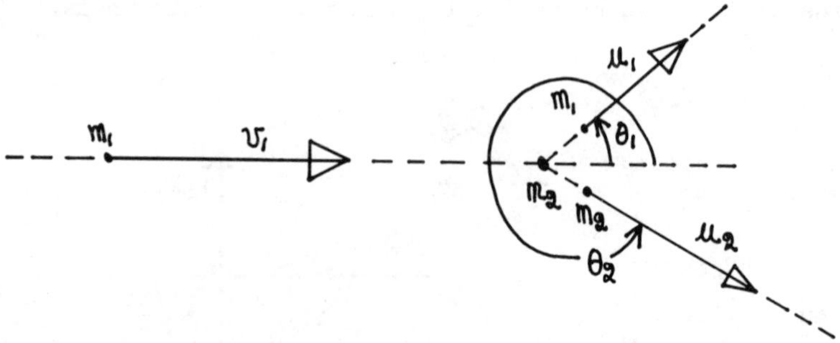

Eliminating u_1 and u_2, the velocities after collision (say, by solving for u_1 from the middle equation, substituting into the first and third equations, then eliminating u_2 between the two resulting equations), yields

$$\tan\theta_1 = \sin 2\theta_2/(\cos 2\theta_2 - m_1/m_2).$$

To find the maximum of $\tan\theta_1$ set

$$d\tan\theta_1/d\theta_2 = 0$$

to obtain

$$\cos 2\theta_2 = m_2/m_1.$$

Substitute this into the equation for $\tan\theta_1$; the resulting $\theta_1 = \theta_m$ is given by

$$\tan^2\theta_m = m_2^2/(m_1^2 - m_2^2).$$

Since

$$\cos^2\theta_m = 1/(1 + \tan^2\theta_m),$$

it follows that

$$\cos^2\theta_m = 1 - m_2^2/m_1^2.$$

(a) For any $m_1 > m_2$, the equation above yields a value for θ_m that, depending on m_2/m_1, will be somewhere between 0 and $\pi/2$.

(b) If $m_1 = m_2$, the equation above cannot be used, since zero denominators would then have occured in its derivation. Return to the original equation for $\tan\theta_1$ and set $m_1/m_2 = 1$:

$$\tan\theta_1 = \sin2\theta_2/(\cos2\theta_2 - 1) = -\cot\theta_2$$

since $\cos2\theta = 1 - 2\sin^2\theta$ and $\sin2\theta = 2\sin\theta\cos\theta$. If θ_2 was taken clockwise instead of counterclockwise, θ_2 above would be replaced with $2\pi - \theta_2$; but $\cot\theta = -\cot(2\pi - \theta)$ so that, for this new θ_2,

$$\tan\theta_1 = \cot\theta_2,$$
$$\theta_1 + \theta_2 = \pi/2.$$

(c) If $m_2 > m_1$, $\cos2\theta_2$ can take on the value m_1/m_2 since this is now less than one. But then, from the original equation for $\tan\theta_1$, $\tan\theta_1$ approaches either $+\infty$ or $-\infty$, depending on the sign of $\sin2\theta_2$. This implies that θ_1 can take on any value.

10-45

(a) Directly from the given information,

$$\delta(mc^2) = c^2(236 - 132 - 98)u = (6\ u)(1.66 \times 10^{-27}\ kg/u)c^2;$$

Putting $c = 3 \times 10^8$ m/s gives

$$\delta(mc^2) = 5603\ MeV,$$

since 1 eV = 1.6 X 10^{-19} J. The energy lost is 5603 - 192 = 5411 MeV.

(b) By conservation of momentum and the definition of Q,

$$m_A v_A = m_B v_B,$$

$$\tfrac{1}{2}m_A v_A^2 + \tfrac{1}{2}m_B v_B^2 - 0 = Q = 192 \text{ MeV} = 307 \times 10^{-13} \text{ J},$$

since the U^{236} nucleus was at rest. With m_A = 132 u, m_B = 98 u,

$$v_A = 1.09 \times 10^7 \text{ m/s}, \quad v_B = (m_A/m_B)v_A = 1.47 \times 10^7 \text{ m/s} \quad \underline{\text{Ans.}}$$

(c) By straightforward calculation,

$$Q_A = \tfrac{1}{2}m_A v_A^2 = 81.7 \text{ MeV}, \quad Q_B = \tfrac{1}{2}m_B v_B^2 = 110 \text{ Mev} \quad \underline{\text{Ans.}}$$

<u>10-50</u>

The cross section is given from

$$R_x = R_0 N\sigma/A.$$

Here A = 1 m^2. Now 65 g of copper contains N_0 = 6.02 X 10^{23} atoms (Avogadro's number), so that 0.005 g must contain 4.63 X 10^{19} atoms, by direct proportion. Hence,

$$(4.6 \times 10^{11} \text{ s}^{-1}) = (1.1 \times 10^{18} \text{ /m}^2\cdot\text{s})(4.63 \times 10^{19})\sigma \times 10^{-28} \text{ m}^2,$$

$$\sigma = 90 \text{ b} \quad \underline{\text{Ans.}}$$

CHAPTER 11

11-3

(a) A wheel with 500 teeth has 500 slots also, so that the angular separation between adjacent slots is $2\pi/500$ rad. Thus the wheel rotates through an angle $\pi/250$ rad in the time t taken for light to travel a distance 2l; this time is 2l/c and therefore,

$$\omega t = \frac{\pi}{250} \text{ rad},$$

$$t = \frac{2l}{c};$$

eliminating t,

$$\omega = \frac{c\pi}{500l} = 3770 \text{ rad/s} \quad \underline{\text{Ans}}.$$

(b) The linear speed at the rim is

$$v = r\omega = (0.05 \text{ m})(3770 \text{ rad/s}) = 188.5 \text{ m/s} \quad \underline{\text{Ans}}.$$

11-9

(a) The initial angular speed is

$$\omega_0 = \frac{(78 \text{ rev})(2\pi \text{ rad/rev})}{60 \text{ s}} = 2.6\pi \text{ rad/s},$$

and the final angular speed is zero. Thus,

$$\omega = \omega_0 + \alpha t,$$

$$0 = 2.6\pi + \alpha(30),$$

$$\alpha = -\frac{0.26}{3}\pi = -0.272 \text{ rad/s}^2 \quad \underline{\text{Ans}}.$$

(b) The angle θ turned through in this time is

$$\theta = \tfrac{1}{2}\alpha t^2 + \omega_0 t,$$

$$\theta = \tfrac{1}{2}(-\frac{0.26}{3}\pi)(30)^2 + (2.6\pi)(30) = 39\pi \text{ rad.}$$

The number of revolutions is $\theta/2\pi = 19.5$ rev Ans.

11-11

(a) The initial and final angular speeds are

$$\omega_i = \frac{(300 \text{ rev})(2\pi \text{ rad/rev})}{60 \text{ s}} = 10\pi \text{ rad/s},$$

$$\omega_f = \frac{(225 \text{ rev})(2\pi \text{ rad/rev})}{60 \text{ s}} = 7.5\pi \text{ rad/s}$$

so that

$$\omega_f = \omega_i + \alpha t,$$

$$7.5\pi = 10\pi + \alpha(60),$$

$$\alpha = -\frac{\pi}{24} = -0.131 \text{ rad/s}^2 \quad \text{Ans.}$$

(b) Measuring the time from the moment of the second observation gives

$$\omega = \omega_0 + \alpha t,$$

$$0 = 7.5\pi + (-\frac{\pi}{24})t,$$

$$t = 180 \text{ s} = 3 \text{ min} \quad \text{Ans},$$

after the second observation, or 4 min after the first.

(c) The angle θ turned through, after the second observation, in coming to rest is

$$\omega^2 = \omega_0^2 + 2\alpha\theta,$$

$$0 = (7.5\pi)^2 + 2(-\frac{\pi}{24})\,\theta,$$

$$\theta = 675\pi \text{ rad.}$$

The number of revolutions is $\theta/2\pi = 337.5$ rev Ans.

11-12

Let ω_1 be the angular velocity at the start of the 4 s interval and θ the angle turned through during the interval. Then,

$$\theta = \tfrac{1}{2}\alpha t^2 + \omega_1 t + \theta_0,$$

$$120 = \tfrac{1}{2}(3)(4)^2 + \omega_1(4),$$

$$\omega_1 = 24 \text{ rad/s.}$$

If T = time the wheel had been rotating before the start of the 4 s interval, and the wheel started from rest,

$$\omega_1 = \alpha T,$$

$$24 = (3)T,$$

$$T = 8.00 \text{ s} \quad \text{Ans.}$$

11-16

If the belt does not slip, the linear speeds at the rims of the two wheels are the same; that is,

$$r_A\omega_A = r_C\omega_C.$$

Differentiating with respect to time gives

$$\alpha_C = (r_A/r_C)\alpha_A.$$

Now $\omega_C = \alpha_C t + \omega_{0C}$ and $\omega_{0C} = 0$. Hence,

$$\omega_C = (r_A/r_C)\alpha_A t.$$

Since $\omega_C = 100$ rpm corresponds to $(100)(2\pi)/(60)$ rad/s,

$$\frac{10\pi}{3} = \left(\frac{10}{25}\right)\left(\frac{\pi}{2}\right)t,$$

$$t = \frac{50}{3} = 16.7 \text{ s} \quad \underline{\text{Ans.}}$$

11-18

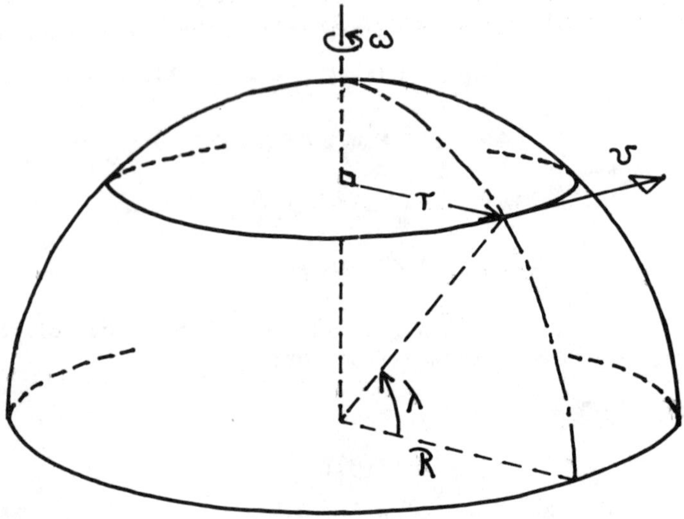

The angular speed ω is 2π/period; the period of rotation is 24 h = 86,400 s, giving

$$\omega = \frac{2\pi}{86400} = 7.27 \times 10^{-5} \text{ rad/s}$$

everywhere. If r is the radius of the daily circular path, v = 2πr/T. But r = Rcosλ, R the radius of the earth and λ the latitude. Hence,

$$v = \frac{2\pi R}{T} \cos\lambda.$$

At the equator $v = 2\pi R/T = \omega R = (7.27 \times 10^{-5})(6.37 \times 10^{6}) = 463$ m/s. Since cos45° = 0.707, the linear speed at latitude 45° = 327 m/s.

11-24

(a) The initial speed v = 80 km/h = (80)(1000)/(3600) = 200/9 = 22.22 m/s. If r is the radius of each wheel, then assuming no slipping,

$$\omega = \frac{v}{r} = \frac{(200/9\ \text{m/s})}{(0.75/2\ \text{m})} = 59.26\ \text{rad/s}\quad \underline{\text{Ans.}}$$

(b) The angle θ turned through in being brought to rest is $(2\pi)(30) = 60\pi$ rad; the final angular speed is zero; hence,

$$\omega_f^2 = \omega_0^2 + 2\alpha\theta,$$

$$0 = (59.26)^2 + 2(\alpha)(60\pi),$$

$$\alpha = -9.315\ \text{rad/s}^2\quad \underline{\text{Ans.}}$$

(c) The linear equation parallel to (b) is

$$v_f^2 = v_0^2 + 2ax.$$

The final linear speed is zero; with no slipping $a = \alpha r = (-9.315)\cdot(0.375) = -3.493\ \text{m/s}^2$. Thus,

$$0 = (22.22)^2 + 2(-3.493)x,$$

$$x = 70.7\ \text{m}\quad \underline{\text{Ans.}}$$

11-26

The translational speed V of the car is 80 km/h = 22.22 m/s. The radius r of the wheel is 0.33 m. If there is no slipping, the linear speed v due to rotation = V, the linear speed due to translation. An observer seated in the car sees only the velocity and acceleration due to the car's rotation, since the car is at rest relative to her as far as translation is concerned. The acceleration due to uniform circular motion is

$$a = \frac{v^2}{r} = \frac{(22.22)^2}{0.33} = 1496\ \text{m/s}^2,$$

toward the center of the wheel. Let the car be moving to the right.

(a) At the top, $\vec{v} = 22.22$ m/s to the right, $\vec{a} = 1496\ \text{m/s}^2$ down.
(b) At the bottom, the vectors $\vec{v}$ and $\vec{a}$ are reversed relative to the top.
(c) The center of the wheel appears to be at rest: v = 0, a = 0.

To obtain the velocities relative to an observer standing on the road, the velocity $\vec{V}$ of the car must be added to the velocities relative to the car. Note that V = v in magnitude. Since $\vec{V}$ is assumed constant, the acceleration of the car relative to the ground is zero. Hence, the accelerations of all three points are the same for observers in the car as for observers on the road. For the velocities:

(a) at the top, 2v = 44.44 m/s to the right;
(b) at the bottom, v - v = 0.
(c) in the center, v = 22.22 m/s to the right.

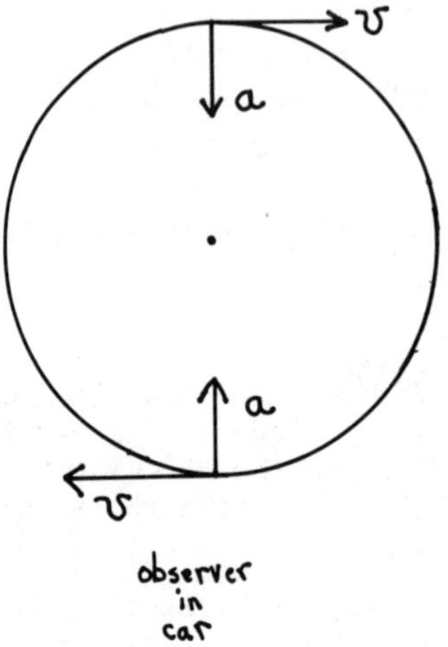

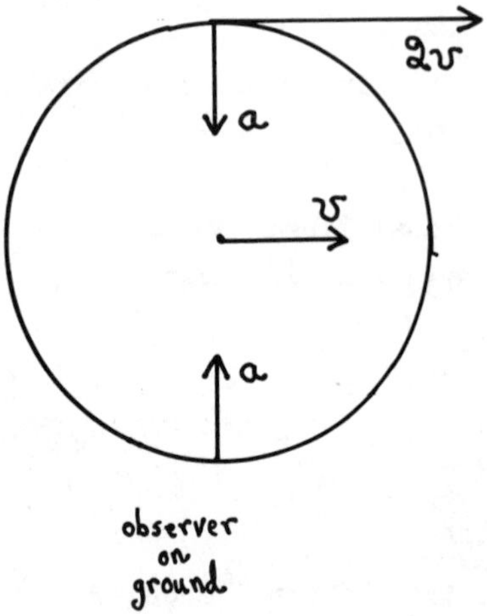

CHAPTER 12

12-10

By straightforward calculation,

(a)

$$I = \sum m_i r_i^2 = ml^2 + m(2l)^2 + m(3l)^2 = 14ml^2 \quad \underline{\text{Ans}};$$

(b)

$$L_m = I_m \omega = (4ml^2)\omega = 4ml^2\omega \quad \underline{\text{Ans}};$$

(c)

$$L = I\omega = 14ml^2\omega \quad \underline{\text{Ans}}.$$

12-19

If friction at the end of the stick in contact with the floor can be neglected, then conservation of energy applies. In the vertical position the height h of the center of mass of the stick above the floor is $\frac{1}{2}L$, L the length of the stick (1 m). As it strikes the floor, h = 0. Initially the kinetic energy K = 0; upon striking the floor it is

$$K = \tfrac{1}{2}I\omega^2 = \tfrac{1}{2}(\tfrac{1}{3}\,ml^2)(\tfrac{v}{l})^2 = \tfrac{1}{6}\,mv^2,$$

v the speed at the end of the stick. Hence,

$$mg(\tfrac{1}{2}L) = \tfrac{1}{6}\,mv^2,$$

$$v = (3gL)^{\frac{1}{2}} = 5.42 \text{ m/s} \quad \underline{\text{Ans}}.$$

12-20

(a) The radial acceleration is $a_r = h\omega^2$. By energy conservation,

$$mg(\tfrac{1}{2}h) = \tfrac{1}{2}I\omega^2 + mg(\tfrac{1}{2}h\cos\theta),$$

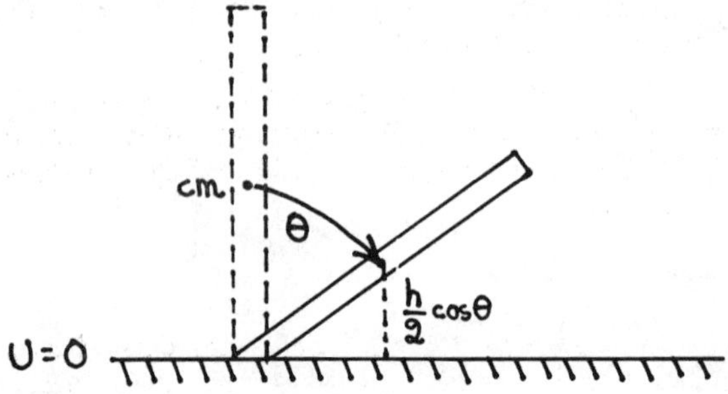

where $I = mh^2/3$ since the chimney, in effect, is rotating about an axis through its base. Substituting the expression for I yields

$$mgh(1 - \cos\theta) = \frac{1}{3} mh^2\omega^2,$$

and therefore

$$a_r = h\omega^2 = 3g(1 - \cos\theta) \quad \underline{\text{Ans}}.$$

(b) The tangential acceleration a_t is

$$a_t = \frac{dv}{dt} = \frac{d}{dt}(h\omega) = h\,\frac{d\omega}{dt}.$$

Using the result from (a),

$$\frac{d\omega}{dt} = \tfrac{1}{2}\omega^{-1}\left(\frac{3g}{h}\sin\theta\right)\frac{d\theta}{dt} = \frac{3g}{2h}\sin\theta$$

since $\omega = \frac{d\theta}{dt}$. Thus the tangential acceleration is

$$a_t = \frac{3g}{2}\sin\theta \quad \underline{\text{Ans}}.$$

(c) The total linear acceleration a is

$$a = (a_r^2 + a_t^2)^{\frac{1}{2}} = \frac{3g}{2}[(1 - \cos\theta)(5 - 3\cos\theta)]^{\frac{1}{2}}.$$

This is zero at $\theta = 0^\circ$ and increases as θ increases, passing g near $\theta = 34^\circ$. It is easily seen that $a > g$ near $\theta = 90^\circ$, for then $\cos\theta = 0$ approximately. Hence, the answer is yes.

(d) As the chimney tips over, the weight of the upper part acts transverse to (across) the column. Chimneys standing upright are not subject to this force and are not specifically designed to withstand it. See Fig. 11-6.

12-21

Let I be the rotational inertia of the body about the axis (leg) of rotation. Clearly, the center of mass of the body lies at the midpoint of the cross-piece and falls through a vertical distance $\frac{1}{2}l$. By conservation of energy,

$$(3m)g(\tfrac{1}{2}l) = \tfrac{1}{2}I\omega^2,$$

m the mass of each rod. It remains to calculate I. The rotational inertia of the leg about which the rotation takes place is zero, for the moment arm is zero. The cross-piece is rotating about an axis through its end, so that its rotational inertia is $ml^2/3$. The other leg has a rotational inertia of ml^2 about the axis for each part of the leg is at a distance l from the axis. Therefore,

$$I = \frac{1}{3}\,ml^2 + ml^2 = \frac{4}{3}\,ml^2.$$

Hence,

$$(3m)g(\tfrac{1}{2}\,l) = \tfrac{1}{2}(\tfrac{4}{3}\,ml^2)\omega^2,$$

$$\omega = \frac{3}{2}\,(g/l)^{\frac{1}{2}} \quad \underline{\text{Ans}}.$$

12-22

Let the original level of the small object m be the zero level of gravitational potential energy U, so that the initial potential energy $U_i = 0$. After the object has fallen a distance h, its potential energy is $U_f = -mgh$. As the system is released from rest its initial kinetic energy $K_i = 0$. After falling a distance h the object has kinetic energy $\frac{1}{2}mv^2$. If the string does not slip, the tangential linear speeds of the rim of the pulley and the equator of the sphere each are v also; the respective angular speeds are $\frac{v}{r}$ and $\frac{v}{R}$. The rotational inertias are I for the pulley and $\frac{2}{3}MR^2$ for the shell. By conservation of energy,

$$K_i + U_i = K_f + U_f,$$

$$0 + 0 = \tfrac{1}{2}mv^2 + \tfrac{1}{2}I(\tfrac{v}{r})^2 + \tfrac{1}{2}(\tfrac{2}{3}MR^2)(\tfrac{v}{R})^2 - mgh,$$

$$v = \left(\frac{2gh}{1 + 2M/3m + I/mr^2}\right)^{\frac{1}{2}} \quad \underline{\text{Ans}}.$$

<u>12-27</u>

(a) An angular speed of 39 rev/s = 39(2π) = 245 rad/s. For a constant angular acceleration

$$\omega = \alpha t + \omega_0,$$

$$0 = \alpha(32) + 245,$$

$$\alpha = -7.66 \text{ rad/s}^2 \quad \underline{\text{Ans}}.$$

(b) The torque $\tau = I\alpha$. The rotational inertia I is given by

$$I = I_{rod} + 2I_{ball},$$

$$I = \frac{1}{12}ML^2 + 2\cdot m(\tfrac{L}{2})^2,$$

$$I = \tfrac{1}{2}L^2(\tfrac{M}{6} + m) = 1.5312 \text{ kg}\cdot\text{m}^2.$$

Therefore, $\tau = (1.5312)(-7.6576) = -11.7$ J <u>Ans</u>.

(c) The work W done equals the change in kinetic energy. Since the final kinetic energy is zero,

$$W = -\tfrac{1}{2}I\omega_0^2 = -\tfrac{1}{2}(1.5312)(245)^2 = -45955 \text{ J} \quad \underline{\text{Ans}}.$$

(d) The number N of revolutions is $\theta/2\pi$, where θ is the angle turned through in being brought to rest:

$$\omega^2 = \omega_0^2 + 2\alpha\theta,$$

$$0 = (245)^2 + 2(-7.6576)(\theta),$$

$$\theta = 3919.3 \text{ rad},$$

$$N = 624 \quad \underline{\text{Ans}}.$$

(e) The calculation in (a) assumes constant angular acceleration and therefore constant torque. If the torque varies, a single value such as in (b) cannot be assigned to it. The equation in (d) for θ likewise requires constant angular acceleration. Only (c), relying on the work-energy theorem, survives dropping the assumption of a constant torque.

<u>12-28</u>

(a) By definition of average torque $\bar{\tau}$,

$$\bar{\tau} = \frac{\delta L}{\delta t} = \frac{-1}{1.5} = -0.667 \text{ J} \quad \underline{\text{Ans}}.$$

(b) The angular acceleration is

$$\alpha = \frac{\bar{\tau}}{I} = \frac{-0.667}{0.125} = -5.336 \text{ rad/s}^2.$$

The initial and final angular velocities are

$$\omega_i = L_i/I = 3/0.125 = 24 \text{ rad/s},$$

$$\omega_f = L_f/I = 2/0.125 = 16 \text{ rad/s}.$$

Hence,

$$\omega_f^2 = \omega_i^2 + 2\alpha\theta,$$

$$16^2 = 24^2 + 2(-5.336)\theta,$$

$$\theta = 29.985 \text{ rad},$$

and the number N of revolutions is

$$N = \theta/2\pi = 4.77 \quad \underline{\text{Ans}}.$$

(c) By the work-energy theorem,

$$W = K_f - K_i = \tfrac{1}{2}I(\omega_f^2 - \omega_i^2) = \tfrac{1}{2}(0.125)(16^2 - 24^2),$$

$$W = -20\ \text{J}\quad \underline{\text{Ans.}}$$

(d) The power P supplied by the wheel is $P = (-W)/t = 20/1.5 = 13.3$ W Ans.

12-30

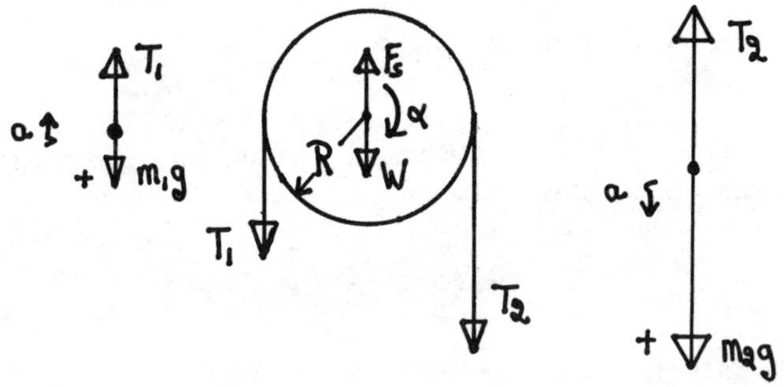

Let m_2 be the heavier block. The equations of motion for the blocks are

$$m_1g - T_1 = -m_1a,$$

$$m_2g - T_2 = m_2a.$$

Subtracting gives

$$(m_1 + m_2)a = (m_2 - m_1)g - (T_2 - T_1).$$

For the pulley, neither the force F_s exerted by the support, nor the weight W of the pulley, exert a torque about the axis of rotation. The angular acceleration α of the pulley, in terms of the linear acceleration a, is given by

$$\alpha = \frac{a}{R},$$

R the pulley radius. Hence,

$$(T_2 - T_1)R = I\left(\frac{a}{R}\right),$$

$$T_2 - T_1 = Ia/R^2.$$

Substituting this into the third equation above and solving the result for I yields

$$I = [(m_2 - m_1)\frac{g}{a} - (m_2 + m_1)]R^2.$$

As m_2 is observed to fall a distance s = 0.75 m in 5 s,

$$a = 2s/t^2 = 0.06\ \text{m/s}^2.$$

With R = 0.05 m, m_1 = 0.46 kg and m_2 = 0.50 kg, the rotational inertia I is found to be

$$I = 0.013933\ \text{kg}\cdot\text{m}^2$$ Ans.

12-31

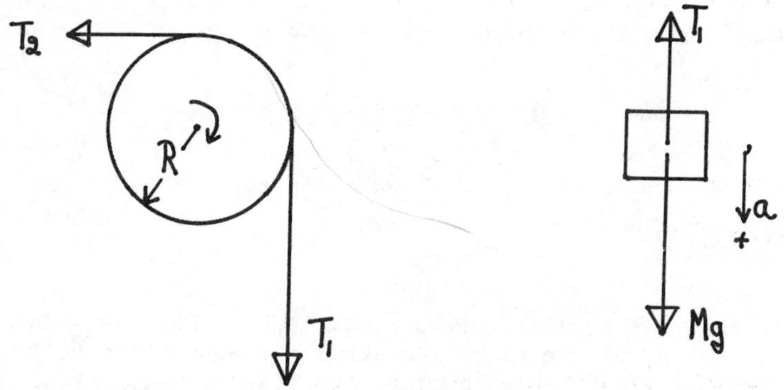

(a) Since $\theta = \frac{1}{2}\alpha t^2 + \omega_0 t$, and the system, presumably, is released from rest,

$$\theta = \tfrac{1}{2}\alpha t^2,$$

$$\alpha = 2\theta/t^2 \quad \underline{\text{Ans.}}$$

(b) If the string does not slip, the tangential linear speed v of the rim of the pulley equals the speed of the string and this, in turn, equals the speed of the blocks. But $v = R\omega$; differentiating with respect to time gives

$$a = R\alpha = 2R\theta/t^2 \quad \underline{\text{Ans.}}$$

(c) Since it is not known if friction is present between the upper block and the plane, examine the other block. Its equation of motion is

$$Mg - T_1 = Ma.$$

Using the result from (b),

$$T_1 = M(g - a) = M(g - 2R\theta/t^2) \quad \underline{\text{Ans.}}$$

Turn now to the pulley. Its weight and the force of support exert no torque, and therefore

$$(T_1 - T_2)R = I\alpha \ .$$

Substitute T_1 from above and α from (a) and solve for T_2 to get

$$T_2 = T_1 - I\frac{\alpha}{R} = Mg - 2R\theta M/t^2 - 2I\theta/Rt^2,$$

$$T_2 = Mg - (2\theta/t^2)(MR + \frac{I}{R}) \quad \underline{\text{Ans.}}$$

<u>12-34</u>

The work W required to stop the hoop equals, in absolute value, the kinetic energy of the hoop. The angular speed of the hoop is v/R, and the rotational inertia about the point of contact with the floor (about which the motion can be considered as pure rotation) is, by virtue of the parallel-axis theorem),

$$I = I_{cm} + MR^2 = MR^2 + MR^2 = 2MR^2.$$

The work required is

$$W = \tfrac{1}{2}(2MR^2)(\tfrac{v}{R})^2 = Mv^2 = 3.375\ \text{J} \quad \underline{\text{Ans}}.$$

Strictly speaking, the work to be done against the hoop is -3.375 J since, to stop the hoop, a force must be exerted opposite to its direction of motion (cos180° = -1).

12-39

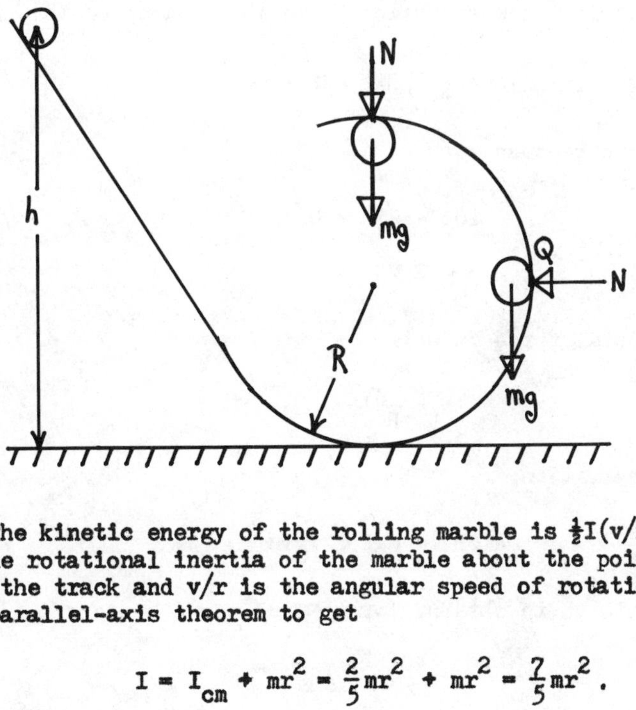

(a) The kinetic energy of the rolling marble is $\tfrac{1}{2}I(v/r)^2$, where I is the rotational inertia of the marble about the point of contact with the track and v/r is the angular speed of rotation. For I, use the parallel-axis theorem to get

$$I = I_{cm} + mr^2 = \tfrac{2}{5}mr^2 + mr^2 = \tfrac{7}{5}mr^2.$$

The kinetic energy of rolling, then, is

$$K = \tfrac{1}{2}(\tfrac{7}{5}mr^2)(\tfrac{v}{r})^2 = \tfrac{7}{10}mr^2.$$

Applying energy conservation,

$$mgh = mg(2R - r) + \frac{7}{10}mv^2.$$

Newton's second law implies that

$$N + mg = m\frac{v^2}{R - r}.$$

Eliminating v^2 between these equations yields

$$N = \frac{mg}{7}\left(\frac{10h - 27R + 17r}{R - r}\right).$$

If the marble barely makes contact, N is about zero, so that

$$10h - 27R + 17r = 0.$$

For $r \ll R$, this becomes

$$10h - 27R = 0,$$

$$h = 2.7R \quad \underline{\text{Ans}}.$$

(b) The horizontal force at Q is

$$N = \frac{mv^2}{R - r}.$$

By energy conservation,

$$mg(6R) = mgR + \frac{7}{10}mv^2.$$

Again, eliminate v^2 to obtain, for $r \ll R$,

$$N = \frac{50}{7}mg \quad \underline{\text{Ans}}.$$

12-46

The initial angular momentum of the system wheel+train is zero; by the conservation of angular momentum, the angular momentum with the power on must be zero also. Thus, the wheel and train rotate in opposite directions relative to the earth. If the sense of rotation of the train is positive, then for the magnitudes of the momenta,

$$L_{train} - L_{wheel} = 0.$$

Since the rotational inertia of the wheel is $I = MR^2$,

$$L_{wheel} = MR^2\omega.$$

The speed V of the train relative to the earth is

$$V = v - R\omega$$

since $R\omega$ is the speed of the track. Thus,

$$MR^2\omega = mVR = m(v - R\omega)R,$$

$$\omega = \frac{m}{M + m}\left(\frac{v}{R}\right) \quad \underline{\text{Ans}}.$$

12-51

(a) The total rotational inertia, initially, is

$$I_0 = I_{rod} + 2I_{girl},$$

$$I_0 = \frac{1}{12}ML^2 + 2\cdot M(\tfrac{1}{2}L)^2 = \frac{7}{12}ML^2 \quad \underline{\text{Ans}}.$$

(b) The initial angular momentum is $L_0 = I_0\omega_0 = \frac{7}{12}ML^2\omega_0$. By the right-hand rule, the vector points vertically down.

(c) Each girl is now at a distance $\frac{1}{4}L$ from the axis of rotation; hence, the new rotational inertia I is

$$I = \frac{1}{12}ML^2 + 2\cdot M(\tfrac{1}{4}L)^2 = \frac{5}{24}ML^2 = \frac{5}{14}I_0 .$$

The new angular speed ω can be found from the conservation of angular momentum:

$$I_0\omega_0 = I\omega,$$

$$\frac{7}{12}ML^2\omega_0 = \frac{5}{24}ML^2\omega,$$

$$\omega = \frac{14}{5}\omega_0 \quad \underline{\text{Ans.}}$$

(d) The change in kinetic energy is

$$\delta K = K_f - K_i = \tfrac{1}{2}I\omega^2 - \tfrac{1}{2}I_0\omega_0^2 = \tfrac{1}{2}(\tfrac{5}{14}I_0)(\tfrac{14}{5}\omega_0)^2 - \tfrac{1}{2}I_0\omega_0^2,$$

$$\delta K = \tfrac{1}{2}I_0\omega_0^2(\tfrac{14}{5} - 1) = \tfrac{9}{10}(\tfrac{7}{12}ML^2)\omega_0^2,$$

$$\delta K = \frac{63}{120}ML^2\omega_0^2 \quad \underline{\text{Ans.}}$$

This represents the work that the girls must do to pull themselves towards to rotation axis.

12-52

(a) By Newton's third law, the friction forces exerted by each wheel on the other are equal in magnitude, f say, and oppositely directed. Let ω_1 be the angular velocity of the larger wheel when

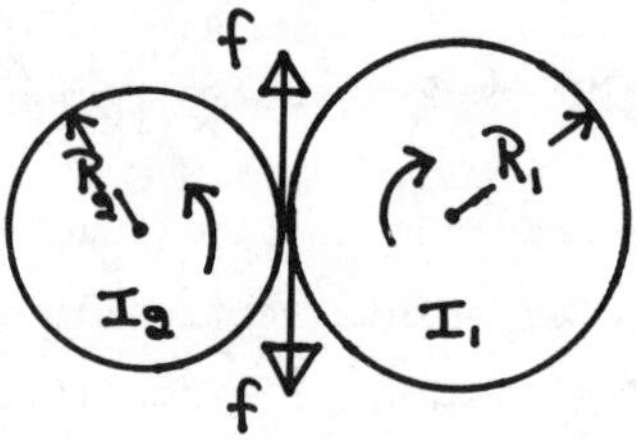

slipping ceases. The frictional torques are exerted over a finite time, during which f may change. Now,

$$\tau = \frac{dL}{dt}$$

so that

$$\int \tau\, dt = L_f - L_i,$$

the integral extending over the time the frictional forces act. Applying this to each wheel gives

$$\int fR_2 dt = I_2\omega_2 - I_2(0),$$
$$-\int fR_1 dt = I_1\omega_1 - I_1\omega_0,$$

the torques being in the opposite sense. The radii can be factored out of the left-hand sides above and the resulting integral $\int fdt$ eliminated between the two equations, say, by dividing them:

$$-\frac{R_2}{R_1} = \frac{I_2\omega_2}{I_1(\omega_1 - \omega_0)}.$$

When slipping ceases, the linear speeds of the rims of the wheels must be equal:

$$R_1\omega_1 = R_2\omega_2.$$

Now eliminate ω_1 from the last two equations to obtain

$$\omega_2 = \omega_0\left(\frac{R_1 I_2}{R_2 I_1} + \frac{R_2}{R_1}\right)^{-1} \quad \underline{\text{Ans.}}$$

(b) It is easy to verify that

$$I_2\omega_2 + I_1(\omega_1 - \omega_0) = 0$$

is not satisfied (unless $R_1 = R_2$): externally applied torques must be provided to hold the wheels in place.

12-53

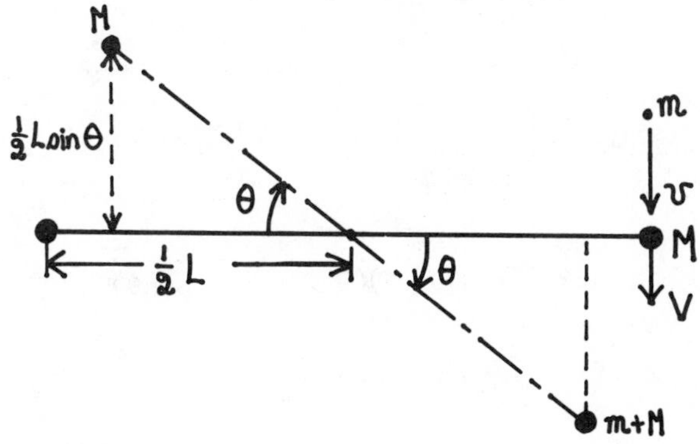

(a) Apply conservation of angular momentum about the axis of the rod of length ℓ; treating all masses as point masses,

$$mv\left(\frac{\ell}{2}\right) = \left[M\left(\frac{\ell}{2}\right)^2 + (M + m)\left(\frac{\ell}{2}\right)^2\right]\omega,$$

$$\omega = \frac{m\,v}{(2M + m)(\ell/2)} = \frac{(0.05)\,(3)}{(4.05)(0.05/2)} = 1.4815 \text{ rad/s} \quad \underline{\text{Ans}}.$$

(b) The desired ratio is

$$\frac{K_{aft}}{K_{bef}} = \frac{\frac{1}{2}I\omega^2}{\frac{1}{2}mv^2} = \frac{\frac{1}{2}(2M + m)(\ell/2)^2\omega^2}{\frac{1}{2}mv^2} = \left(\frac{2M}{m} + 1\right)\left(\frac{\ell\omega}{2v}\right)^2,$$

$$\frac{K_{aft}}{K_{bef}} = \left[2\left(\frac{2}{0.05}\right) + 1\right]\left[\frac{(0.05)(1.4815)}{(2)(3)}\right]^2 = 0.0123 \quad \underline{\text{Ans}}.$$

(c) Conservation of energy can be applied to the motion of the system subsequent to the completely inelastic collision between the putty and stick. It is assumed that the rod has not had a chance to move during the collision so that it is still horizontal immediately after the impact. In this position, set the gravitational potential energy $U = 0$. At the extreme position of its swing, the kinetic energy is momentarily zero. Since part of the rod is above the horizontal and part below in this position, energy conservation gives

$$K_i = U_f,$$

$$(0.0123)\left(\tfrac{1}{2}mv^2\right) = [M - (M + m)]g\left[\frac{\ell}{2}\sin\theta\right].$$

But

$$\tfrac{1}{2}mv^2 = \tfrac{1}{2}(0.05)(3)^2 = 0.225 \text{ J}.$$

Hence,

$$(0.0123)(0.225) = -(0.05)(9.8)\left(\frac{0.05}{2}\sin\theta\right),$$

$$\sin\theta = -0.22592,$$

$$\theta = 193^\circ \quad \underline{\text{Ans}},$$

that is, a little more than one-half a revolution.

CHAPTER 13

13-6

(a) Resolve the forces shown into horizontal and vertical components:

$$N - T\cos\theta = 0,$$

$$w - T\sin\theta = 0.$$

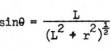

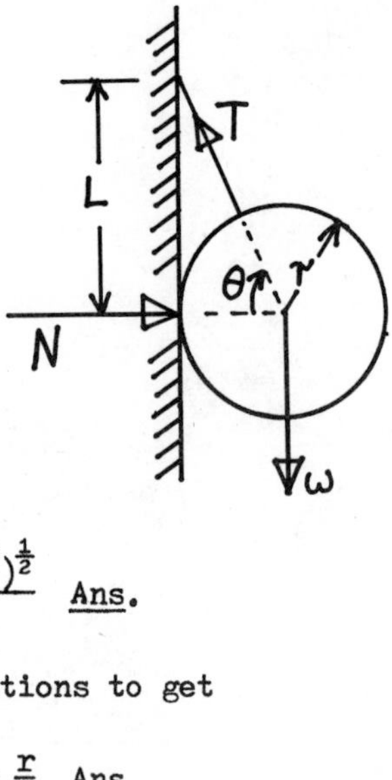

The second equation gives for the tension

$$T = \frac{w}{\sin\theta}.$$

From the geometry of the situation,

$$\sin\theta = \frac{L}{(L^2 + r^2)^{\frac{1}{2}}}$$

and therefore

$$T = w\,\frac{(L^2 + r^2)^{\frac{1}{2}}}{L} \quad \underline{\text{Ans}}.$$

(b) Divide the original pair of equations to get

$$N = w\cot\theta = w\,\frac{r}{L} \quad \underline{\text{Ans}}.$$

13-9

The frictional force f exerted by the air is, by assumption,

$$f = kv$$

upward, k a positive constant. Upon opening her parachute,

$$kv - mg = ma,$$

taking up as the positive direction. Substituting the given numbers yields for k, in English units,

$$k = \frac{m(32 + 30)}{70} = \frac{31}{35} m.$$

At the terminal speed u, her acceleration is zero:

$$ku - mg = 0,$$

$$k = \frac{32}{u} m.$$

Equating the two expressions for k,

$$\frac{32}{u} m = \frac{31}{35} m,$$

$$u = 36.13 \text{ ft/s} \quad \underline{\text{Ans.}}$$

13-10

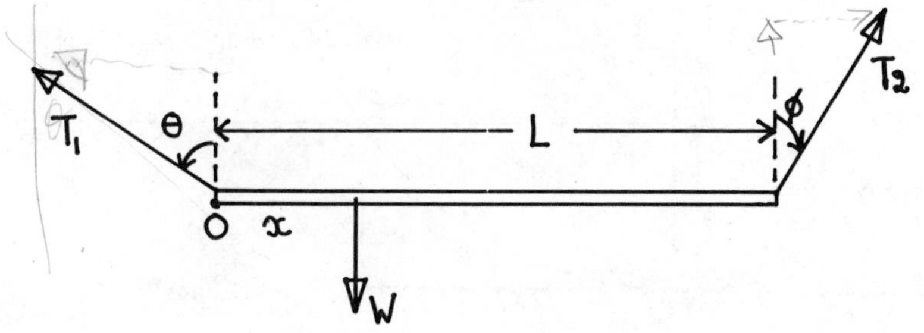

For translational equilibrium, and taking moments about 'O',

$$T_1 \cos\theta + T_2 \cos\phi = W,$$

$$T_1 \sin\theta = T_2 \sin\phi,$$

$$Wx = (T_2 \cos\phi) L.$$

Multiply the first equation by sinθ, the second by cosθ and subtract, to obtain for T_2

$$T_2 = \frac{W \sin\theta}{\sin(\theta + \phi)}.$$

But $\theta + \phi = 90°$ so that

$$T_2 = W \sin\theta.$$

Substitute this into the third equation to find

$$x = L\sin\theta\cos\phi = L\sin^2\theta,$$

$$x = 2.20 \text{ m} \quad \underline{\text{Ans.}}$$

13-17

The normal force vanishes as the wheel lifts. Taking moments about '0' as this happens,

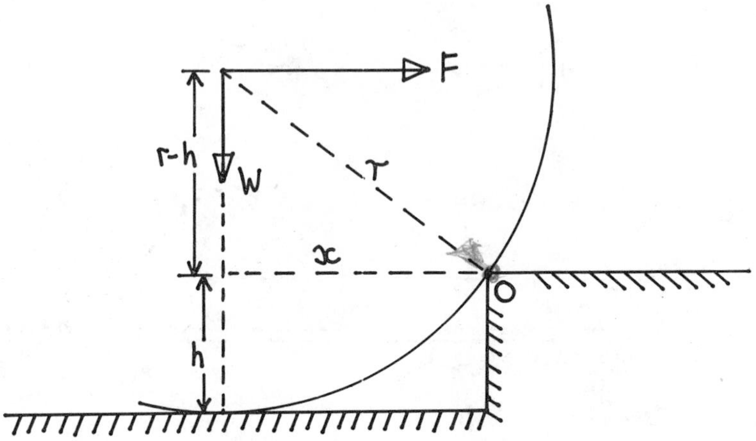

$$Wx = F(r - h).$$

But,

$$x^2 = r^2 - (r - h)^2,$$

$$x = (2rh - h^2)^{\frac{1}{2}},$$

so that

$$F = W\,\frac{(2rh - h^2)^{\frac{1}{2}}}{r - h} \quad \underline{\text{Ans.}}$$

13-19

Take moments about the pivot:

$$Mg(50 - 45.5) = 2mg(45.5 - 12),$$

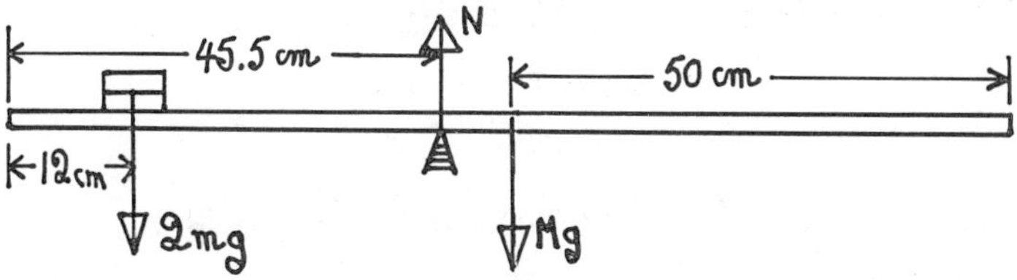

$$M = \frac{(2)(5)(33.5)}{45},$$

$$M = 74.44 \text{ g} \quad \underline{\text{Ans.}}$$

13-26

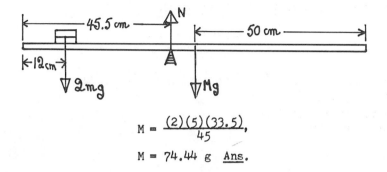

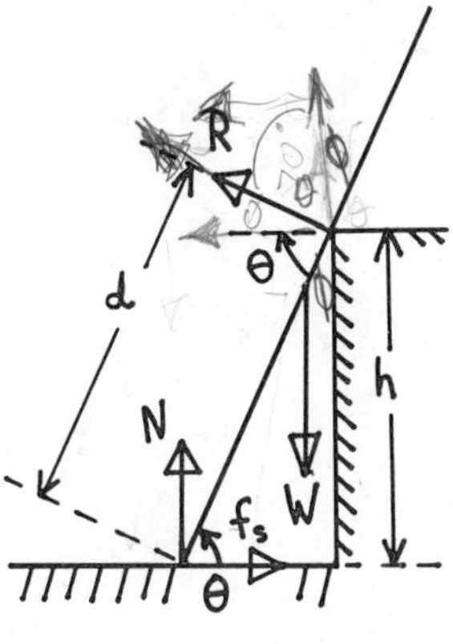

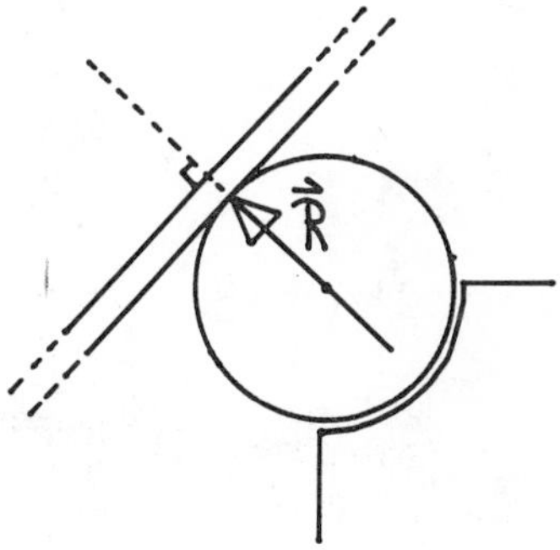

The force $\vec{f}_s$ of static friction opposes the tendency of the plank to slip backwards to the left and therefore points to the right. The roller, being frictionless, exerts only a normal force (called R to distinguish it from the normal force N exerted by the floor) on the plank. Examine the situation at $\theta = 70°$, where $f_s = \mu_s N$, its limiting value. The translational and rotational equations, the last taken about the bottom end of the plank, are

$$R\cos\theta + N - W = 0,$$

$$R\sin\theta - \mu_s N = 0,$$

$$Rd - W\frac{\ell}{2}\cos\theta = 0.$$

Multiply the first equation by μ_s and add to the second to obtain

$$R(\sin\theta + \mu_s\cos\theta) = \mu_s W.$$

The third equation gives $R = (W\ell/2d)\cos\theta$; substitute this into the equation above to find that

$$\sin\theta\cos\theta + \mu_s\cos^2\theta = \frac{2d}{\ell}\mu_s.$$

But

$$d = \frac{h}{\sin\theta};$$

Therefore,

$$\sin^2\theta\cos\theta + \mu_s\sin\theta\cos^2\theta = \frac{2h}{\ell}\mu_s,$$

$$\mu_s = \frac{\sin^2\theta\cos\theta}{2h - \ell\sin\theta\cos^2\theta}\ell.$$

Numerically, $2h = \ell$ and $\theta = 70^\circ$ giving

$$\mu_s = \frac{\sin^2\theta\cos\theta}{1 - \sin\theta\cos^2\theta} = 0.339 \quad \underline{\text{Ans.}}$$

<u>13-31</u>

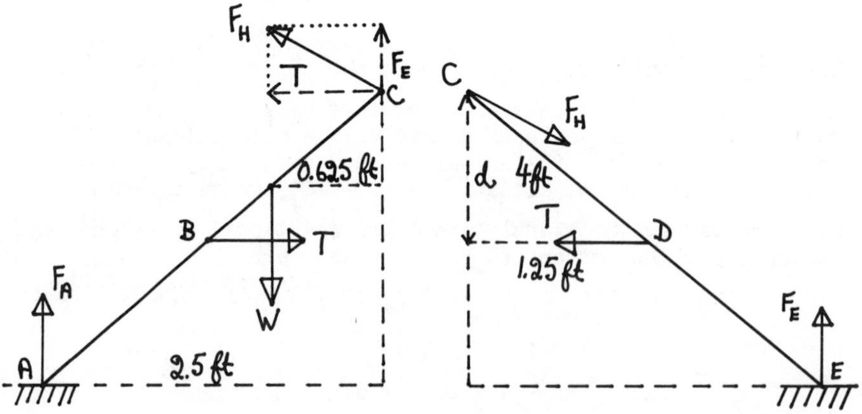

The distance d is

$$d = (4^2 - 1.25^2)^{\frac{1}{2}} = 3.80 \text{ ft.}$$

Note that the vertical component of the force F_H exerted by the hinge on the left leg equals the force F_E exerted by the ground on the right leg, for the equations of translational equilibrium for the left leg and for the ladder as a whole are identical:

$$F_A + F_E - W = 0.$$

Taking moments about C gives, first for the right leg,

$$2.5F_E - 3.8T = 0,$$

and for the left leg

$$0.625W + 3.8T - 2.5F_A = 0.$$

Set W = 192 lb; solution of the three equations gives

(a) $T = 47$ lb <u>Ans.</u>

(b) $F_A = 120$ lb, $F_E = 72$ lb <u>Ans.</u>

<u>13-32</u>

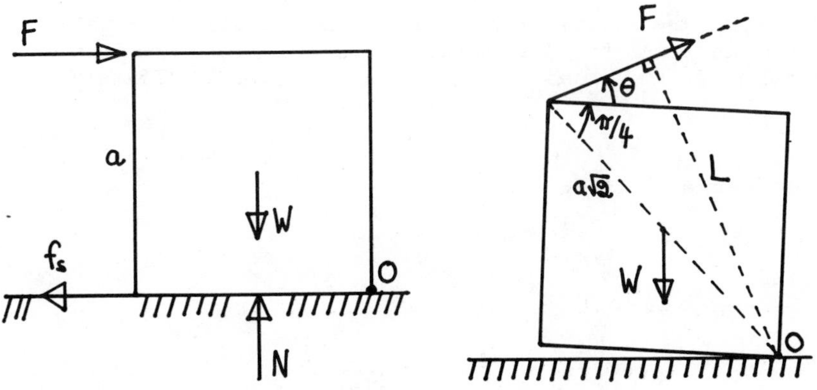

(a) The normal force vanishes as the box begins to roll. Taking moments about 0,

$$W\frac{a}{2} = Fa,$$

$$F = \frac{W}{2} = 445\ \text{N} \quad \underline{\text{Ans.}}$$

(b) For the minimum force, assume the box was on the verge of slipping, so that $f_s = \mu_s N$. Translational equilibrium requires that $f_s = F$; thus

$$F = \mu_s N = \mu_s W,$$

$$\mu_s = \frac{F}{W} = 0.5 \quad \underline{\text{Ans.}}$$

(c) If F is applied at an angle θ above the horizontal at the top of the box (for the greatest moment arm) then, again taking moments about 0,

$$F(2^{\frac{1}{2}}a)\sin(\theta + \tfrac{\pi}{4}) = (890)\tfrac{a}{2},$$

$$F\sin(\theta + \tfrac{\pi}{4}) = 315.$$

For F to be a minimum, set $\sin(\theta + \frac{\pi}{4}) = 1$, its maximum possible value. Hence, $\theta = \pi/4$ and

$$F = 315\ \text{N} \quad \underline{\text{Ans.}}$$

<u>13-33</u>

By symmetry, the forces in the sides AD and BC will be equal, as will the forces in the diagonals. Assume all forces are tensions. Inspection of the forces acting on A and D gives immediately

$$T = T''.$$

But, looking at D, say,

$$F\cos 45^\circ + T'' = 0,$$

$$F\sin 45^\circ + T' = 0.$$

These give

$$F = -T''/\cos 45^\circ = -2^{\frac{1}{2}}T'' = -2^{\frac{1}{2}}T,$$

and

$$T' = T'' = T.$$

The negative sign for F indicates compression, rather than the tension that was assumed.

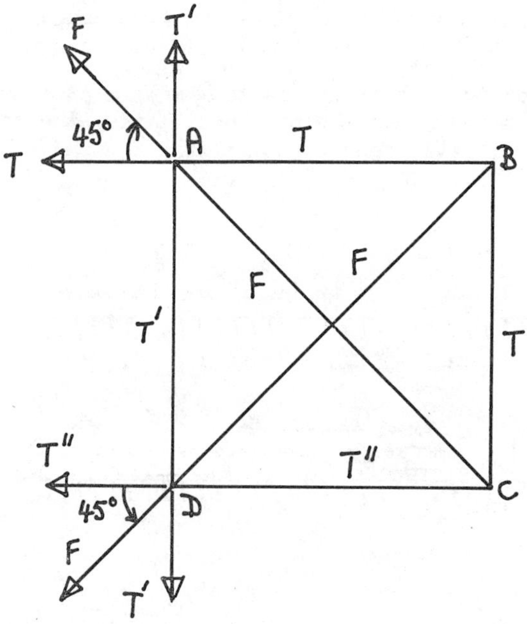

CHAPTER 14

14-2

Attach the cut pieces to each other and to a mass m on a smooth horizontal surface as shown. As the mass oscillates, the piece of force constant k_1 is stretched x_1; similarly for the other

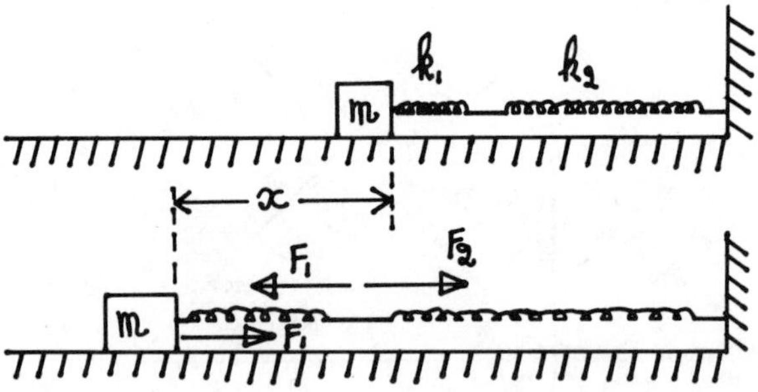

piece. The forces exerted by the pieces on each other at the point where they join are

$$F_1 = k_1x_1, \quad F_2 = k_2x_2.$$

By Newton's third law $F_1 = F_2$, so that

$$k_1x_1 = k_2x_2.$$

Now,

$$x = x_1 + x_2;$$

hence

$$x_1 = x - x_2 = x - \frac{k_1}{k_2}x_1,$$

$$x_1 = \frac{k_2}{k_1 + k_2} x.$$

Now, the force on the mass m is $F_1 = k_1 x_1$; in terms of its displacement x this is

$$F_1 = \frac{k_1 k_2}{k_1 + k_2} x.$$

Thus, the pieces joined in this way act like a spring with a force constant $k_1 k_2/(k_1 + k_2)$. But, joined as they are, they are indistinguishable from the spring from which they were made; hence

$$k = \frac{k_1 k_2}{k_1 + k_2}.$$

This can be written

$$\frac{1}{k} = \frac{1}{k_1} + \frac{1}{k_2}.$$

The lengths are related by

$$l = l_1 + l_2.$$

Examination of the last two equations makes it reasonable to suppose that

$$k = \frac{C}{l},$$

C a constant; i.e., the force constant of a spring is inversely proportional to its length. Thus,

$$k_1 = \frac{C}{l_1}, \; k_2 = \frac{C}{l_2}$$

with $C = kl$. If $l_1 = nl_2$, then

$$l_1 + l_2 = nl_2 + l_2 = l,$$

$$l_2 = \frac{1}{n + 1}, \; l_1 = \frac{n}{n + 1} l.$$

Therefore,

$$k_1 = \frac{n + 1}{n} k, \; k_2 = (n + 1)k \quad \underline{\text{Ans.}}$$

<u>14-7</u>

The frequency ν of motion of the piston is

$$\nu = \frac{180 \text{ osc}}{60 \text{ s}} = 3 \text{ Hz.}$$

The maximum speed v_m is

$$v_m = 2\pi\nu A = 2\pi(3)\left(\frac{0.76}{2}\right) = 7.16 \text{ m/s} \quad \underline{\text{Ans.}}$$

<u>14-14</u>

(a) The amplitude A of oscillation is $\frac{1}{2}$(10 cm) = 5 cm. As the object passes through the point 5 cm below its initial position, the resultant force on it must be zero, for the acceleration midway between the turning points vanishes. The two forces acting are the weight mg of the object and the spring force kA directed upward. Therefore,

$$mg = kA.$$

The frequency ν of oscillation is

$$\nu = \frac{1}{2\pi}(k/m)^{\frac{1}{2}} = \frac{1}{2\pi}(g/A)^{\frac{1}{2}} = \frac{1}{2\pi}(9.8/0.05)^{\frac{1}{2}},$$

$$\nu = \frac{7}{\pi} = 2.228 \text{ Hz} \quad \underline{\text{Ans.}}$$

(b) Let the position x of the object be

$$x = A\sin 2\pi\nu t$$

so that the object passes through its equilibrium position x = 0 at t = 0. From (a), $2\pi\nu$ = 14 Hz, A = 5 cm so that

$$x = 5\sin 14t.$$

When the object is 8 cm below its original position it is at x = 3 cm, 3 cm below its equilibrium position. This occurs at a time t' given by

$$3 = 5\sin 14t',$$

$$\sin 14t' = 0.6.$$

The velocity v at any time is

$$v = \frac{dx}{dt} = 70\cos 14t,$$

so that at t = t',

$$v = 70\cos 14t' = 70(0.8) = 56 \text{ cm/s} \quad \underline{\text{Ans}}.$$

(c) Let the added mass be M. The new frequency is

$$\nu = \frac{1}{2\pi}\left(\frac{k}{M + m}\right)^{\frac{1}{2}}.$$

This is half the original frequency; hence,

$$\frac{1}{2\pi}\left(\frac{k}{M + m}\right)^{\frac{1}{2}} = \frac{1}{2}\cdot\frac{1}{2\pi}\left(\frac{k}{m}\right)^{\frac{1}{2}},$$

$$4m = M + m,$$

$$m = \frac{M}{3} = 100 \text{ g} \quad \underline{\text{Ans}}.$$

(d) At the new equilibrium position the resultant force is zero:

$$(m + M)g = kA'$$

with A' the new amplitude. From (a) this becomes

$$(m + M)g = \frac{mg}{A}A',$$

$$A' = \frac{m + M}{m}A = \frac{100 + 300}{100}(5) = 20 \text{ cm} \quad \underline{\text{Ans}},$$

or 20 cm below the original position of the first mass in (a).

14-18

The mass m of a single silver atom is

$$m = \frac{108}{6.02 \times 10^{23}} = 1.794 \times 10^{-22}\ \text{g},$$

$$m = 1.794 \times 10^{-25}\ \text{kg}.$$

Along any axis of the cubic lattice passing through atoms, each atom has two nearest neighbors. Hence,

$$\nu = \frac{1}{2\pi}\left(\frac{2k}{m}\right)^{\frac{1}{2}},$$

k the force constant of a single spring. This gives

$$k = 2\pi^2\nu^2 m = 354\ \text{N/m}\quad \underline{\text{Ans}}.$$

14-21

See Problem 14-2.

14-22

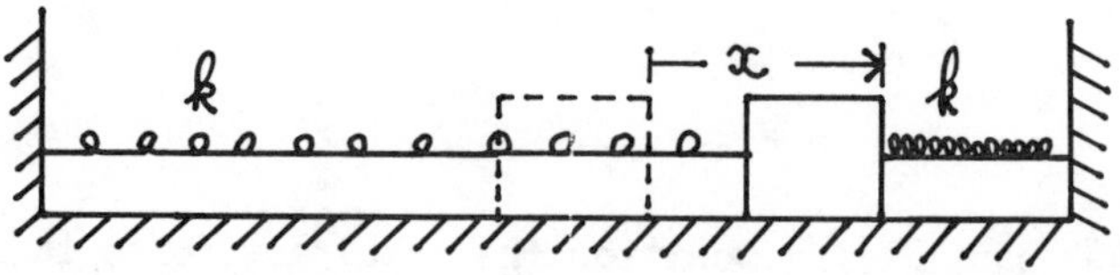

If the block is displaced x to the right, say, from the position of equilibrium, then the right-hand spring is compressed by x and exerts a force kx on the block directed to the left (push); the left-hand spring is stretched the same distance x and pulls on the block with a force kx also to the left. Thus, the resultant force F on the block is $F = 2kx = (2k)x = Kx$, K the effective force constant, and $K = 2k$. The frequency, then, is

$$\nu = \frac{1}{2\pi}\left(\frac{K}{m}\right)^{\frac{1}{2}} = \frac{1}{2\pi}\left(\frac{2k}{m}\right)^{\frac{1}{2}}.$$

14-24

(a) The amplitude of each is $\frac{1}{2}A$ and their periods are T = 1.5 s. If their displacements be x_1 and x_2, then,

$$x_1 = \tfrac{1}{2}A\cos(\frac{2\pi}{T}t),$$
$$x_2 = \tfrac{1}{2}A\cos(\frac{2\pi}{T}t - \frac{\pi}{6}),$$

so that particle 1 is at one end of the line at time t = 0 and particle 2 is at this same point at time

$$t = \frac{\pi/6}{2\pi/T} = \frac{1}{8}\ s,$$

since T = 1.5 s. Hence, particle 1 leads particle 2. It is required to find their distance apart at time t = 0.125 + 0.5 = 0.625 s. Using the first two equations, it is found that at this time,

$$x_1 = \tfrac{1}{2}A\cos(5\pi/6) = -0.4330A;$$
$$x_2 = \tfrac{1}{2}A\cos(2\pi/3) = -0.2500A.$$

Hence, their distance apart, a positive number, is

$$s = x_2 - x_1 = 0.1830A \quad \underline{\text{Ans}}.$$

(b) To establish the directions of motion, examine the velocities at t = 0.625 s. Since v = dx/dt, these involve the sines of the angles encountered in (a). But, sin(5π/6) and sin(2π/3) both are positive, indicating that the particles are moving in the same direction at this time.

14-25

If it does not slip on the plane, the block too executes SHM. Under these conditions, the maximum force on the block in the horizontal direction is

$$F_{max} = ma_{max} = m(4\pi^2\nu^2A).$$

This force is supplied by static friction. If the amplitude A increases, F_{max} increases also. But the force of static friction cannot increase beyond $\mu_s N$, where $N = mg$ in this case. Hence, the maximum amplitude can be found from

$$\mu_s mg = 4\pi^2 m\nu^2 A_{max},$$

$$A_{max} = \mu_s g/(4\pi^2\nu^2) = 0.031 \text{ m} = 3.1 \text{ cm} \quad \underline{\text{Ans}},$$

using $\mu_s = 0.5$ and $\nu = 2$ Hz.

14-34

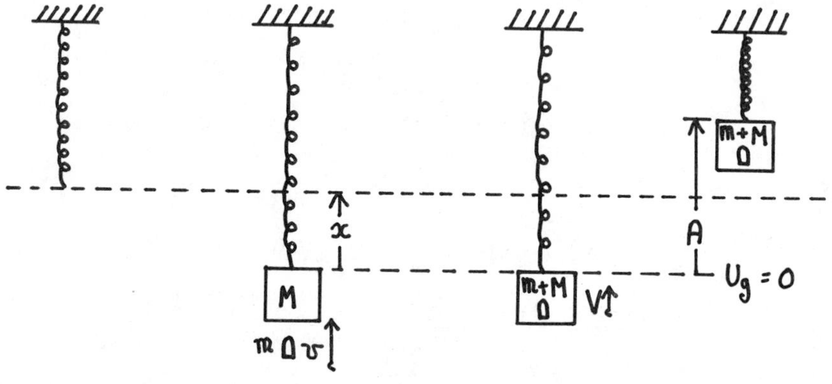

(a) Momentum conservation gives

$$mv = (m + M)V,$$

$$V = \frac{m}{m + M} v.$$

After the bullet comes to rest, energy conservation implies that

$$\tfrac{1}{2}(m + M)V^2 + \tfrac{1}{2}kx^2 = \tfrac{1}{2}k(A - x)^2 + (m + M)gA.$$

The equilibrium position of the block before the bullet is fired is taken as the zero-level of gravitational potential energy; x is the distance the block stretches the spring before the bullet is fired. The difference in equilibrium positions for the block (mass M) before the bullet is fired and the block+bullet (mass m + M) after the bullet impacts is neglected. This last also indicates that

$$(m + M)g = kx.$$

Substituting this into the previous equation gives

$$\tfrac{1}{2}(m + M)V^2 = \tfrac{1}{2}kA^2,$$

that is, the gravitational field appears to "drop out." Putting in the expression found for V and solving for A results in

$$A = \frac{mv}{[k(m + M)]^{\frac{1}{2}}} = \frac{7.5}{45}\ \text{m},$$

$$A = 16.7\ \text{cm}\quad \underline{\text{Ans.}}$$

(b) The desired fraction is

$$\frac{\frac{1}{2}kA^2}{\frac{1}{2}mv^2} = \frac{\frac{1}{2}(m + M)V^2}{\frac{1}{2}mv^2} = \frac{m}{m + M} = \frac{1}{81}$$

or 1.23% <u>Ans</u>.

<u>14-36</u>

The translational kinetic energy is

$$K_t = \tfrac{1}{2}Mv^2.$$

Since the rotational inertia of a cylinder about its axis is $\frac{1}{2}MR^2$ and the cylinder is rolling, the rotational kinetic energy is

$$K_r = \tfrac{1}{2}(\tfrac{1}{2}MR^2)(\tfrac{v}{R})^2 = \tfrac{1}{4}Mv^2;$$

finally, the potential energy is $\frac{1}{2}kx^2$. Hence, the total energy E is

$$E = \frac{3}{4}Mv^2 + \frac{1}{2}kx^2.$$

Evaluate E at the moment of release; as the cylinder was released from rest,

$$E = \tfrac{1}{2}kx_1^2 = \tfrac{1}{2}(3)(0.25)^2 = \frac{3}{32}\ \text{J}.$$

At equilibrium $x = 0$ and $v = v_{max}$; therefore energy conservation gives

$$E = \frac{3}{4}Mv_{max}^2 = \frac{3}{32},$$

$$Mv_{max}^2 = \frac{1}{8}.$$

Hence, (a) $K_t = \frac{1}{16}$ J and (b) $K_r = \frac{1}{32}$ J.

(c) As before,

$$E = \frac{3}{4}Mv^2 + \frac{1}{2}kx^2.$$

Differentiate with respect to time; note that $dE/dt = 0$ and then factor out $2dx/dt = 2v$ to get

$$\frac{d^2x}{dt^2} + \frac{2k}{3M}x = 0.$$

This is the differential equation for SHM: the coefficient of x is the square of the angular speed; hence,

$$T = 2\pi\left(\frac{3M}{2k}\right)^{\frac{1}{2}}.$$

14-39

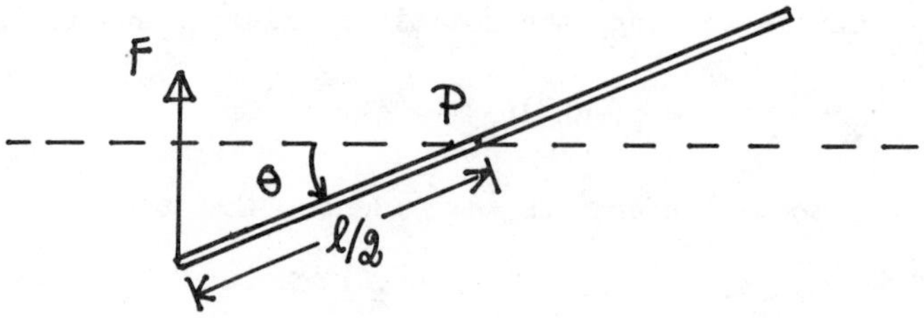

The torque exerted by the spring force F about the pivot P is

$$\frac{\ell}{2}\,F = \frac{\ell}{2}(-kx) = -k(x)\frac{\ell}{2} = -k\left(\frac{\ell}{2}\,\theta\right)\frac{\ell}{2},$$

$$-\tfrac{1}{4}k\ell^2\theta = I\,\frac{d^2\theta}{dt^2}$$

if θ is small. Since $I = \frac{1}{12}\,m\ell^2$,

$$-\tfrac{1}{4}k\ell^2\theta = \frac{1}{12}\,m\ell^2\,\frac{d^2\theta}{dt^2},$$

$$\frac{d^2\theta}{dt^2} + \frac{3k}{m}\,\theta = 0.$$

This is the standard equation for SHM and therefore

$$T = 2\pi\left(\frac{m}{3k}\right)^{\frac{1}{2}}\ \underline{\text{Ans.}}$$

<u>14-40</u>

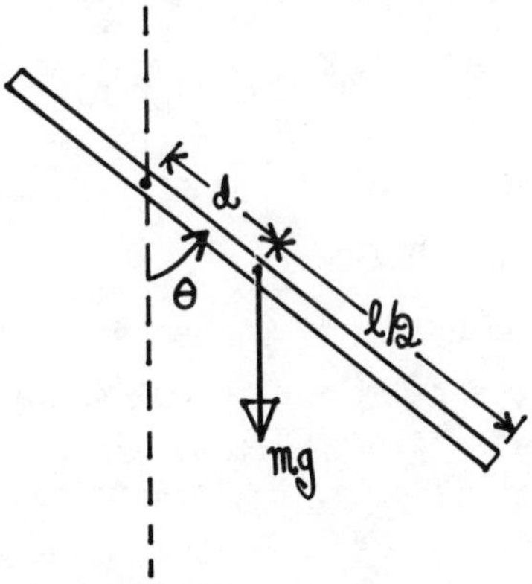

The restoring torque is $-mgd\sin\theta = -(mgd)\theta$ for small θ. Hence,

$$-(mgd)\theta = I\,\frac{d^2\theta}{dt^2}.$$

The rotational inertia I about the pivot is found from the parallel-axis theorem:

$$I = I_{cm} + md^2 = \frac{1}{12} m\ell^2 + md^2.$$

Hence,

$$-(mgd)\theta = m(\frac{1}{12}\ell^2 + d^2)\frac{d^2\theta}{dt^2},$$

$$\frac{d^2\theta}{dt^2} + \frac{12gd}{\ell^2 + 12d^2}\theta = 0.$$

Comparing with the standard SHM equation gives

$$T = 2\pi(\frac{\ell^2 + 12d^2}{12gd})^{\frac{1}{2}} \quad \underline{\text{Ans}}.$$

14-42

(a) Since the period T = 0.5 s, the frequency $\nu = 1/T = 2$ Hz; combining this with the given amplitude of π rad gives

$$\theta = \pi\cos(4\pi t + \phi),$$

for the angular displacement as a function of time. The angular velocity is

$$\frac{d\theta}{dt} = -4\pi^2\sin(4\pi t + \phi).$$

Since the sine varies between -1 and +1, the maximum value of $\frac{d\theta}{dt}$ is $4\pi^2$ = 39.5 rad/s Ans.

(b) When the displacement $\theta = \pi/2$, then $\cos(4\pi t + \phi) = \frac{1}{2}$, by the first equation. Hence $\sin(4\pi t + \phi) = \pm 3^{\frac{1}{2}}/2$, so that $\frac{d\theta}{dt} = \pm 2\pi^2 3^{\frac{1}{2}} = \pm 34.2$ rad/s at this instant. In either case, the angular speed, as opposed to angular velocity, is 34.2 rad/s Ans.

(c) When $\theta = \pi/4$, $\cos(4\pi t + \phi) = \frac{1}{4}$. But the angular acceleration is

$$\frac{d^2\theta}{dt^2} = -16\pi^3\cos(4\pi t + \phi),$$

giving, for the instant $\theta = \pi/4$, an angular acceleration of $-4\pi^3$

$= -124\ \text{rad/s}^2$ Ans.

14-43

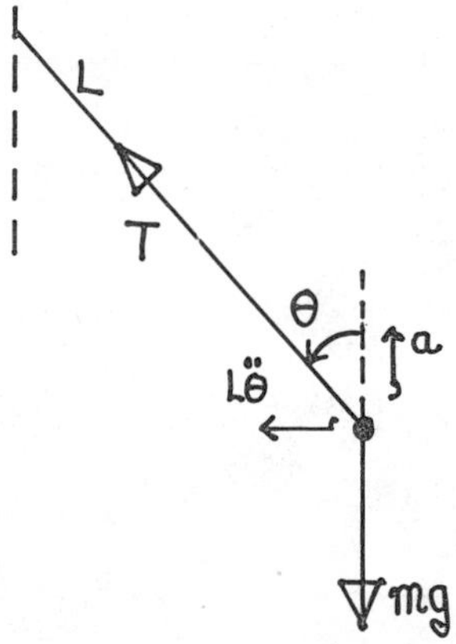

Writing Newton's second law in the vertical and horizontal directions gives

$$T\cos\theta - mg = ma,$$
$$T\sin\theta = ma_h.$$

Call the length of the string L. For small θ the equations become

$$T - mg = ma,$$
$$T\theta = -\ mL\,\frac{d^2\theta}{dt^2}.$$

Eliminating the tension T between these gives

$$m(g + a) = -mL\,\frac{d^2\theta}{dt^2},$$

$$\frac{d^2\theta}{dt^2} + \left(\frac{g + a}{L}\right)\theta = 0.$$

(a) Here a = 0 and therefore

$$\nu = \frac{1}{2\pi}\left(\frac{g}{L}\right)^{\frac{1}{2}} = 0.352\ \text{Hz} \quad \text{Ans.}$$

(b) In this case $a = +2\ m/s^2$ giving

$$\nu = \frac{1}{2\pi}\left(\frac{g + a}{L}\right)^{\frac{1}{2}} = 0.387\ \text{Hz} \quad \underline{\text{Ans}}.$$

(c) For free-fall, put $a = -g$ giving $\nu = 0$ (no tendency to oscillate).

14-49

To obtain the path, i.e. y vs. x, it is necessary to eliminate the time t. To do this, write

$$y = A\cos(\omega t + \alpha) = A(\cos\omega t\cdot\cos\alpha - \sin\omega t\cdot\sin\alpha),$$

$$y = A[\tfrac{x}{A}\cdot\cos\alpha - (1 - x^2/A^2)^{\frac{1}{2}}\sin\alpha].$$

Rearrange and square:

$$y - x\cos\alpha = -A(1 - x^2/A^2)^{\frac{1}{2}}\sin\alpha,$$

$$y^2 - 2xy\cos\alpha + x^2\cos^2\alpha = A^2(1 - x^2/A^2)\sin^2\alpha,$$

$$y^2 - 2xy\cos\alpha + x^2 = A^2\sin^2\alpha,$$

the last equation being for the path. It is the equation of an ellipse and, since it remains unchanged if x and y are exchanged, the axes of the ellipse must bisect the x and y-coordinate axes.

(a) If $\alpha = 0$, the equation reduces to $x = y$, and the ellipse to a straight line.

(b) With $\alpha = 30°$, the full ellipse appears; the angle α determines the elongation of the ellipse.

(c) For $\alpha = 90°$, the ellipse becomes the circle $x^2 + y^2 = A^2$.

14-53

The displacement x of the block is given by

$$x = Ae^{-bt/2m}\cos(\omega' t + \delta).$$

Now let t be the time required for the amplitude to fall to one-third of its initial value, which is A; then,

$$\frac{1}{3} A = Ae^{-bt/2m},$$

$$t = \frac{2m}{b} \ln 3.$$

The number N of oscillations completed in this time is

$$N = \nu' t = \frac{\omega'}{2\pi} \frac{2m}{b} \ln 3 = \frac{\omega' m}{\pi b} \ln 3.$$

Evaluating ω':

$$\omega' = (k/m - b^2/4m^2)^{\frac{1}{2}} = 2.31 \text{ rad/s}.$$

Using this, the preceding equation gives $N = 5.26$ Ans.

CHAPTER 15

15-5

The gravitational force on m of the hollowed sphere plus the force on m that was exerted by the portion that was subsequently removed must equal the force on m exerted by the uniform sphere, mass M, before hollowing. Hence, subtraction of these last two forces gives the desired attraction of the hollowed sphere. The force F* on m by M before hollowing is

$$F^* = G\,\frac{mM}{d^2}.$$

The mass M_h of the portion hollowed out is

$$M_h = \frac{M}{\frac{4}{3}\pi R^3}\,\frac{4}{3}\pi\left(\frac{R}{2}\right)^3 = \frac{1}{8}\,M$$

since the radius of the hollow is R/2. The force produced by this hollowed portion is

$$F_h = G\,\frac{m\,M_h}{(d - R/2)^2} = \frac{GmM_h/d^2}{(1 - R/2d)^2} = \frac{1}{8}\,\frac{G\,mM/d^2}{(1 - R/2d)^2}\,.$$

Thus, the force exerted by the remaining portion of the original sphere is

$$F = F^* - F_h = \frac{GmM}{d^2}\left[1 - \frac{1}{8(1 - R/2d)^2}\right].$$

15-7

(a) The component of the gravitational force F on the object along the chord is $F\cos\theta = F(x/r)$, towards x = 0, r the distance of the object from the center of the earth. If M* is the mass of that part of the earth interior to r and δ is the density of the earth (assumed uniform), then

$$F = GM^*m/r^2 = G\left(\frac{4}{3}\,\pi r^3\delta\right)m/r^2 = \frac{4}{3}\,\pi G\delta mr.$$

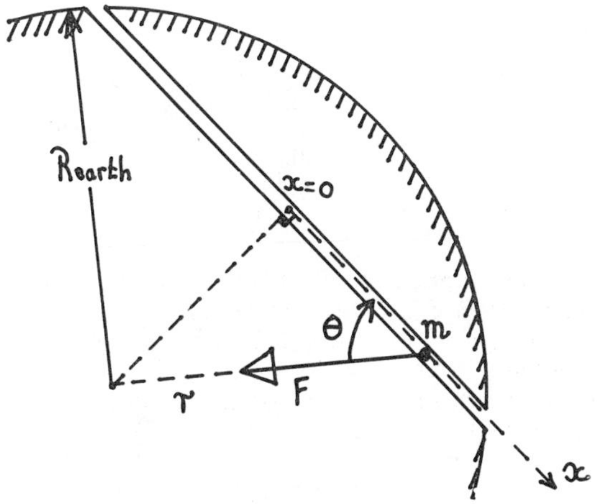

Therefore,

$$F\cos\theta = \left(\frac{4}{3}\pi G\delta m\right)x = -m\frac{d^2x}{dt^2},$$

the minus sign since the force is always directed toward x = 0. This is an equation describing simple harmonic motion with a frequency ν given by

$$\nu^2 = \frac{G\delta}{3\pi}.$$

(b) By comparison with the case of a chute dug along a diameter, the frequencies are seen to be equal; hence, the periods $T = 1/\nu = 84.2$ min are the same also.

(c) In simple harmonic motion, $v_{max} = 2\pi A/T$. In the present case the amplitude $A = \frac{1}{2}$(chord length) $< R_{earth}$, the amplitude for a chute along a diameter. Since the periods T are the same along a chord and diameter, v_{max} will be smaller for the former situation.

<u>15-8</u>

(a) Let the mass of the particle at P be m, and dw the solid angle of the narrow double cone constructed with apex at P. The cone intercepts areas dA_1 and dA_2 on the surface of the shell (which is of uniform density σ). These areas are not perpendicular to the

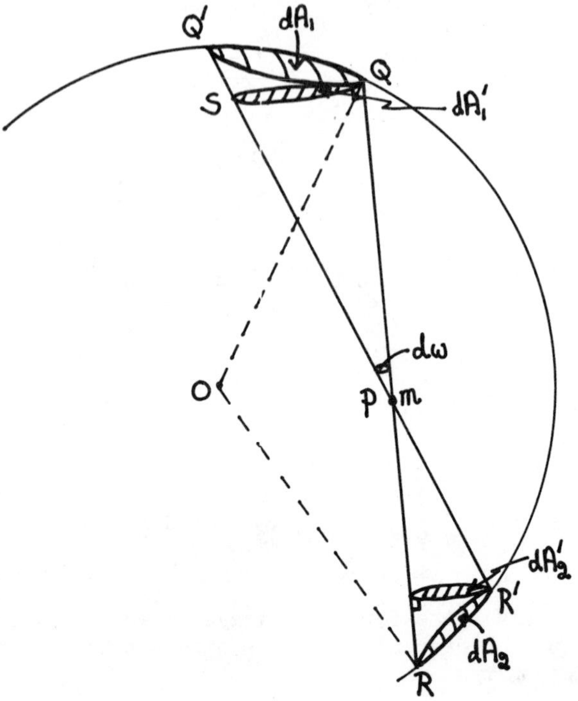

generators of the cone, so construct areas dA_1' and dA_2' that are perpendicular. The areas dA_1 and dA_2 are so small that Newton's law of universal gravitation can be applied to them. The force on m due to dA_1 is

$$dF_1 = Gm\sigma(dA_1)/(QP)^2, \text{ along PQ.}$$

The force on m due to dA_2 is

$$dF_2 = Gm\sigma(dA_2)/(RP)^2, \text{ along PR.}$$

But

$$dw = dA_1'/(QP)^2 = dA_1 \cos\measuredangle Q'QS/(QP)^2.$$

Since $\measuredangle OQQ' = \measuredangle SQP = \pi/2$, then $\measuredangle Q'QS = \measuredangle OQP$ and therefore

$$dw = dA_1 \cos\measuredangle OQP/(QP)^2,$$

$$dA_1 = (QP)^2 \sec\measuredangle OQP\ dw.$$

Similarly $dA_2 = (RP)^2 \sec\measuredangle ORP\ dw$. Thus,

$$dF_1 = Gm\sigma(QP)^2 \sec\measuredangle OQP\ dw/(QP)^2 = Gm\sigma \sec\measuredangle OQP\ dw,$$

$$dF_2 = Gm\sigma(RP)^2 \sec\measuredangle ORP\ dw/(RP)^2 = Gm\sigma \sec\measuredangle ORP\ dw.$$

But $OR = OQ$ and thus $\measuredangle OQP = \measuredangle ORP$ and $dF_1 = dF_2$, or $d\vec{F}_1 + d\vec{F}_2 = 0$.

(b) The entire shell may be divided into similar pairs of areas by taking cones in all directions about P. Thus, the net force on m is zero.

<u>15-16</u>

(a) Let the suspended body have mass m, the spring a force constant k. The forces on m in suspension are its true weight mg directed down and the spring force kx directed up. This last is equal in magnitude to the force exerted by m on the spring and equals the scale reading. The mass m is moving on the equator (a circle) with linear speed V relative to space; hence,

$$mg - kx = m\,\frac{V^2}{R}.$$

R is the radius of the earth. The speed V is made up of the speed $R\omega$ due to the earth's rotation and the speed v of the ship relative to the earth; that is,

$$V = R\omega \pm v,$$

The + sign if the ship is sailing in the direction of the earth's rotation, i.e. to the east, and the - sign if in the opposite direction against the rotation of the earth. The scale reading, then, is

$$kx = mg - m\,\frac{(R\omega \pm v)^2}{R},$$

$$kx = mg - mR\omega^2 \pm 2m\omega v - m\,\frac{v^2}{R},$$

the + sign now for a ship sailing west. Now, $v \ll R\omega$, that is, the speed of the ship relative to the earth is, very likely, much smaller than the speed due to the earth's rotation (about 1000 mph at the equator). This means that the last term in the equation above is numerically much smaller than the others and can be dropped with little error. When the ship is at rest, the scale reading is W_0 given by

$$W_0 = mg - mR\omega^2$$

so that

$$W = W_0 \pm 2m\omega v = W_0\left(1 \pm \frac{2m\omega v}{W_0}\right).$$

Finally, in the last term on the right above,

$$\frac{2m\omega v}{W_0} = \frac{2m\omega v}{mg - mR\omega^2} = \frac{2\omega v}{g}\left(1 + \frac{R\omega^2}{g} + \ldots\right).$$

$R\omega^2$ is the acceleration imparted by the rotation of the earth and is much less than g, the acceleration due to gravity. Thus, to a good approximation,

$$W = W_0(1 \pm 2\omega v/g).$$

(b) Since the term mV^2/R was transposed during the analysis, in the final equation above the - sign applies when sailing east, the + sign when sailing west along the equator.

15-29

Start with the second law and substitute in Kepler's third law $T^2 = kR^3$:

$$F = m\frac{v^2}{R} = m\frac{(2\pi R/T)^2}{R} = 4\pi^2 m\frac{R}{T^2} = 4\pi^2 m\frac{R}{kR^3},$$

$$F = (4\pi^2 m/k)R^{-2}.$$

Newton, of course, was able to derive an expression for the constant k from his law of universal gravitation.

15-31

The motion to be described is about the center of mass of the system and, therefore, the center of mass must first be located.

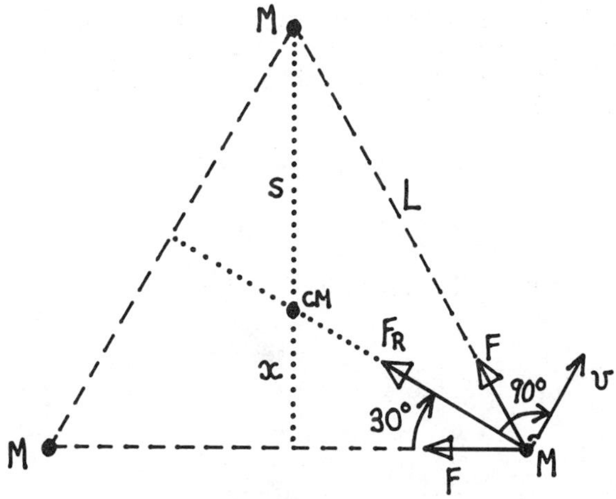

Clearly, it lies along the perpendicular bisectors drawn from the masses, at their intersection. The length of a perpendicular bisector is $L\sin 60° = 3^{\frac{1}{2}}L/2$. Hence, the center of mass is at a distance x along a bisector, as shown, where x is found from

$$(3M)x = M(3^{\frac{1}{2}}L/2) + 2M(0),$$

$$x = 3^{\frac{1}{2}}L/6$$

by the definition of center of mass. The distance s of any mass from the center of mass is

$$s = 3^{\frac{1}{2}}L/2 - x = 3^{\frac{1}{2}}L/3.$$

Now s is the radius of the circular orbit of each mass; for uniform circular motion,

$$F_R = Mv^2/s,$$

F_R the resultant force on each mass M. But,

$$F_R = 2F\cos 30° = 3^{\frac{1}{2}}F,$$

F the gravitational attraction between any pair of masses. By the law of gravitation,

$$F = GM^2/L^2$$

since L is the distance between any pair of masses. Putting the last four equations together yields

$$3^{\frac{1}{2}}GM^2/L^2 = Mv^2/(3^{\frac{1}{2}}L/3),$$

$$v = (GM/L)^{\frac{1}{2}} \quad \underline{\text{Ans}}.$$

15-36

(a) The escape speed at the earth's surface is $v_e = (2GM_e/R_e)^{\frac{1}{2}}$. Using $g = GM_e/R_e^2$ this becomes $v_e = (2gR_e)^{\frac{1}{2}}$. Since the speed of the rocket was $v = 2(gR_e)^{\frac{1}{2}} > v_e$, the rocket escapes.

(b) By conservation of energy,

$$\tfrac{1}{2}mv^2 - GM_em/R_e = \tfrac{1}{2}mV^2 - 0,$$

the potential energy being nearly zero when the rocket is very far from the earth, its speed out there being V. Solving for V gives

$$V^2 = v^2 - 2GM_e/R_e = 4gR_e - 2gR_e = 2gR_e,$$

$$V = (2gR_e)^{\frac{1}{2}} \quad \underline{\text{Ans}}.$$

15-37

Let i,f refer to the initial configuration (masses far apart) and to the configuration when the separation of the particles is d. Then,

$$\Delta U = U_f - U_i = -\frac{GMm}{d} - 0,$$

$$\Delta K = K_f - K_i = \tfrac{1}{2}mv^2 + \tfrac{1}{2}MV^2 - 0,$$

where v, V are the speeds of the particles relative to a fixed coordinate system when their separation is d. By the conservation of energy, $\Delta U + \Delta K = 0$ so that

$$\frac{GMm}{d} = \tfrac{1}{2}mv^2 + \tfrac{1}{2}MV^2.$$

The initial momentum was zero; thus, to conserve momentum

$$mv + M(-V) = 0,$$

$$mv = MV.$$

Using this, the energy equation becomes

$$\frac{GMm}{d} = \tfrac{1}{2}mv(v + V).$$

Let the relative speed of the two oncoming particles be u; using the momentum equation, u can be expressed in terms of v:

$$u = v + V = v + \frac{m}{M}v = \frac{m + M}{M}v.$$

Therefore,

$$\frac{GMm}{d} = \tfrac{1}{2}m\left(\frac{Mu}{m + M}\right)u = \tfrac{1}{2}\frac{Mm}{m + M}u^2$$

$$u = [2G(m + M)/d]^{\frac{1}{2}} \quad \underline{\text{Ans.}}$$

15-39

(a) Let M, R be the mass and radius of the planet (asteroid?). The escape speed v_e is

$$v_e = (2GM/R)^{\frac{1}{2}}.$$

Since

$$g = \frac{GM}{R^2},$$

the escape speed in terms of the surface gravity is

$$v_e = (2gR)^{\frac{1}{2}} = 1732 \text{ m/s} \quad \underline{\text{Ans}},$$

since $g = 3 \text{ m/s}^2$ and $R = 5 \times 10^5$ m.

(b) Let h be the height reached above the surface of the planet. The potential energy $U = -GMm/r$, r the distance to the center of the planet and m the mass of the particle. From (a),

$$M = gR^2/G,$$

so that

$$U = -G\,\frac{(gR^2/G)m}{(R + h)} = -mg\,\frac{R^2}{R + h}.$$

Hence, if v is the launch speed from h = 0 and V the speed at height h, conservation of energy gives

$$\tfrac{1}{2}mv^2 - mgR = \tfrac{1}{2}mV^2 - mg\,\frac{R^2}{R + h},$$

$$v^2 - V^2 = 2g(R - \frac{R^2}{R + h}),$$

$$v^2 - V^2 = 2gR(\frac{h}{R + h}) .$$

At the top of the path V = 0:

$$v^2 = 2gR\,\frac{h}{R + h}.$$

Substituting the numbers ($R = 5 \times 10^5$ m, etc.) gives

$$\frac{1}{3} = \frac{h}{R + h},$$

$$h = \tfrac{1}{2}R = 250 \text{ km} \quad \underline{\text{Ans}}.$$

(c) Apart from the numbers, the motion in (c) is just the reverse of that in (b); hence,

$$v^2 = 2gR\,\frac{h}{R + h}.$$

This time h = 1000 km = 2R, leading to

$$v^2 = \frac{4}{3}\,gR = \frac{4}{3}(3)(5 \times 10^5),$$

$$v = 1414 \text{ m/s} \quad \underline{\text{Ans}}.$$

<u>15-41</u>

Let m, M be the masses of the stars, D their initial separation and d their separation when their speeds are v, V. By conservation of energy,

$$-\frac{GMm}{D} = -\frac{GMm}{d} + \tfrac{1}{2}mv^2 + \tfrac{1}{2}MV^2.$$

Linear momentum is also conserved:

$$mv = MV. \qquad V = \frac{mv}{M}$$

Hence,

$$-\frac{GMm}{D} = -\frac{GMm}{d} + \tfrac{1}{2}mv^2 + \tfrac{1}{2}M\left(\frac{mv}{M}\right)^2,$$

$$GMm\left(\frac{1}{d} - \frac{1}{D}\right) = \tfrac{1}{2}mv^2\left(1 + \frac{m}{M}\right).$$

Since $m = M$,

$$v^2 = GM\left(\frac{1}{d} - \frac{1}{D}\right).$$

(a) Here $d = \frac{1}{2}D$, giving

$$v^2 = \frac{GM}{D} = \frac{(6.67 \times 10^{-11})(10^{30})}{(10^{10})},$$

$$v = V = 8.17 \times 10^4 \text{ m/s} = 81.7 \text{ km/s} \quad \underline{\text{Ans}}.$$

(b) As they collide $d = 2R = 2 \times 10^5$ m (the radius used is far too small for actual normal stars); thus,

$$v^2 = V^2 \cong \frac{GM}{d},$$

$$v = V = 1.83 \times 10^7 \text{ m/s} = 1.83 \times 10^4 \text{ km/s} \quad \underline{\text{Ans}}.$$

<u>15-45</u>

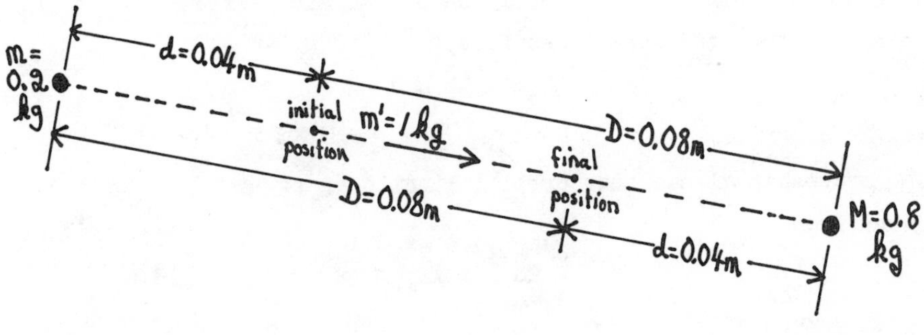

(a) The force F due to M and the force f due to m are

$$F = \frac{Gm'M}{D^2}; \quad f = \frac{Gm'm}{d^2},$$

and are oppositely directed. But $D = 2d$, $D^2 = 4d^2$ and $M = 4m$; hence $f = F$ and the resultant vector sum is zero Ans.

(b) The potential energies are scalars and simply add as scalars:

$$U = -\frac{Gmm'}{d} - \frac{GMm'}{D} = -Gm'\left(\frac{m}{d} + \frac{M}{D}\right),$$

$$U = -1.00 \times 10^{-9}\ \text{J} \quad \underline{\text{Ans.}}$$

(c) In the final position the distances d and D are interchanged, so that now

$$U = -Gm'\left(\frac{m}{D} + \frac{M}{d}\right) = -1.50 \times 10^{-9}\ \text{J}.$$

The work required is the difference in potential energies at the two positions:

$$W = (-1.50 \times 10^{-9}) - (-1.00 \times 10^{-9}),$$

$$W = -5 \times 10^{-10}\ \text{J} \quad \underline{\text{Ans.}}$$

15-46

(a) Let M_s, M_e, m be the masses of the sun, earth and spaceship. Also, let $d = R_e + h$ be the distance between the spaceship and the center of the earth while in orbit, and D the earth-sun distance If K is the total kinetic energy needed to escape from rest at height h above the earth's surface,

$$K - \frac{GM_e m}{d} - \frac{GM_s m}{D} = 0.$$

The spaceship already has orbital kinetic energy $\frac{1}{2}mv^2$, where v is given from

$$\frac{GM_e m}{d^2} = m\frac{v^2}{d},$$

so that

$$\tfrac{1}{2}mv^2 = \tfrac{1}{2}\frac{GM_e m}{d}.$$

Similarly, the spaceship possesses kinetic energy due to the earth's motion about the sun. Hence, the energy K' still needed is

$$K' = \frac{GM_e m}{d} + \frac{GM_s m}{D} - \tfrac{1}{2}\frac{GM_e m}{d} - \tfrac{1}{2}\frac{GM_s m}{D},$$

$$K' = Gm\left(\tfrac{1}{2}\frac{M_e}{d} + \tfrac{1}{2}\frac{M_s}{D}\right).$$

Numerically, $M_e = 6 \times 10^{24}$ kg, $M_s = 2 \times 10^{30}$ kg, $d = (6400 + 480)$ km $= 6.88 \times 10^6$ m, $D = 1.5 \times 10^{11}$ m; these give

$$K' = (0.265 + 4.046) \times 10^{14}\ \text{J} = 4.311 \times 10^{14}\ \text{J} \quad \underline{\text{Ans}}.$$

(b) The fraction demanded is

$$\frac{M_s/D}{M_e/d + M_s/D} = \frac{4.046}{4.311} = 94\ \% \quad \underline{\text{Ans}}.$$

<u>15-47</u>

Let R_e, R_m be the radii of the earth and moon and d the distance between their centers. The point considered lies at a distance R from the earth's center and r from the moon's center.

(a) The potential energies add as scalars:

$$U = -GM_e m/R - GM_m m/r \quad \underline{\text{Ans}}.$$

(b) The field will be zero for a point between earth and moon located by

$$GM_e m/R^2 = GM_m m/r^2 = GM_m m/(d - R)^2.$$

Since $M_e = 81\ M_m$, this gives $R = 0.9d$ <u>Ans</u>.

In the following, it is assumed that the points considered lie on the line segment joining the center of the earth with the center of the moon. At either of the locations of interest the forces due to earth and moon are in opposite directions.

(c) On the earth's surface $R = R_e$, $r = d - R_e$ so that

$$U = -GM_e m/R_e - GM_m m/(d - R_e),$$
$$g = GM_e/R_e^2 - GM_m/(d - R_e)^2,$$

the last towards the earth's center.

(d) At the moon's surface $r = R_m$, $R = d - R_m$ giving

$$U = -GM_e m/(d - R_m) - GM_m m/R_m,$$
$$g = GM_m/R_m^2 - GM_e/(d - R_m)^2,$$

the last towards the moon's center.

15-50

Let r be the radius of the orbit, R the radius of the earth; M,m are the masses of earth and satellite. The minimum energy K needed to take the satellite from the earth's surface, neglecting rotation, to a distance r is given by

$$K - GMm/R = -GMm/r,$$
$$K = GMm(\frac{1}{R} - \frac{1}{r}).$$

The kinetic energy K' required to put it into circular orbit once it is there is $\frac{1}{2}mv^2$, with

$$GMm/r^2 = mv^2/r,$$
$$K' = \frac{1}{2}GMm/r.$$

(a) $r = 1.25R$: this gives $K = 0.2GMm/R$, $K' = 0.4GMm/R$, so that $K = \frac{1}{2}K'$, and the answer is NO.

(b) $r = 1.5R$, giving $K = \frac{1}{3}\,GMm/R$, $K' = \frac{1}{3}\,GMm/R$. Thus, the energies are the same.

(c) $r = 1.75R$: $K = \frac{3}{7}\,GMm/R$, $K' = \frac{2}{7}\,GMm/R$; the answer is YES.

15-52

(a) Use the notation R, h, M, m for the radius of the earth, height of the circular orbit above the earth's surface, mass of earth and of satellite. For circular orbits,

$$\frac{GMm}{(R+h)^2} = m\,\frac{v^2}{R+h},$$

$$v = \left(\frac{GM}{R+h}\right)^{\frac{1}{2}} = 7.54\ \text{km/s} \quad \underline{\text{Ans}},$$

using $M = 5.98 \times 10^{24}$ kg, $R = 6.37 \times 10^6$ m.

(b) The period T is given by

$$T = \frac{2\pi(R+h)}{v} = \frac{2\pi\ (7.01 \times 10^6\ \text{m})}{7.543 \times 10^3\ \text{m/s}},$$

$$T = 5.84 \times 10^3\ \text{s} = 97.3\ \text{min} \quad \underline{\text{Ans}}.$$

(c) The mechanical energy E is

$$E = K + U = \tfrac{1}{2}mv^2 - \frac{GMm}{R+h} = -\tfrac{1}{2}\,\frac{GMm}{R+h}.$$

As the orbit lowers h diminishes and therefore E changes at a rate

$$\frac{dE}{dt} = \tfrac{1}{2}\,\frac{GMm}{(R+h)^2}\,\frac{dh}{dt}$$

assuming that the orbit remains almost circular. Numerically

$$\frac{GMm}{(R+h)^2} = 1.7857 \times 10^3\ \text{N},$$

and therefore

$$-1.4 \times 10^5 \text{ J/rev} = \tfrac{1}{2}(1.7857 \times 10^3 \text{ N})\frac{dh}{dt},$$

$$\frac{dh}{dt} = -156.8 \text{ m/rev},$$

the negative sign indicating that h is diminishing, just as the minus sign for dE/dt implies a loss of energy. After 1500 revolutions, the orbital altitude drops by (156.8)(1500) = 235 km. Thus, the new altitude is 640 - 235 = 405 km. The equations in (a) and (b) now give v = 7.68 km/s and T = 92.3 min.

(d) Let F be the force; the rate of energy loss (power) equals the product of torque and angular speed; the latter is $2\pi/T$ and therefore

$$\frac{dE}{dt} = F(R + h)\cdot\frac{2\pi}{T}.$$

The period is

$$T = 2\pi \frac{(R + h)^{3/2}}{(GM)^{\frac{1}{2}}}$$

(Kepler's third law), so that

$$F = \left(\frac{R + h}{GM}\right)^{\frac{1}{2}} \frac{dE}{dt}.$$

For the highest (initial) orbit, $dE/dt = 1.4 \times 10^5 \text{ J}/5.84 \times 10^3 \text{ s}$ = 24 J/s in absolute value; for this orbit $F = 3.18 \times 10^{-3}$ N.

(e) The resistive force exerts a torque on the satellite and its orbital angular momentum diminishes. If all influences originating outside the earth-satellite system (sun, moon, etc.) are ignored, the system is isolated and its total angular momentum must remain constant.

CHAPTER 16

16-2

Let p be the pressure inside the box. The air remaining in the box exerts a force pA against the lid from the inside, in the same direction as the force F required to pull off the lid. The force tending to keep the lid on is PA, where P is the outside air pressure. If F is the minimum force needed, then,

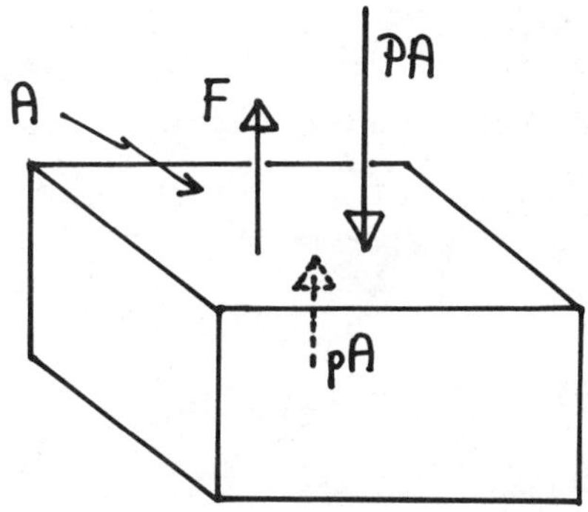

$$pA + F = PA,$$

$$p = P - \frac{F}{A} = 15 - \frac{108}{12},$$

$$p = 6 \text{ lb/in.}^2 \quad \underline{\text{Ans.}}$$

16-3

(a) Each horse must exert a force F equal, in magnitude, to the resultant of the forces due to the air inside and outside the hemisphere. The vertical components due to the air vanish; it is the horizontal components that must be calculated. Hence,

$$F = \int (dF_{out} - dF_{in})\cos\theta = \int (P_{out} - P_{in})\cos\theta \, dA,$$

$$F = \int P\cos\theta \, dA = P\int \cos\theta \, dA.$$

In spherical coordinates $dA = 2\pi R^2 \sin\theta d\theta$ and therefore

$$F = 2\pi PR^2 \int_0^{\pi/2} \sin\theta\cos\theta d\theta = \pi R^2 P.$$

(b) Since atmospheric pressure is 14.7 lb/in.^{2},

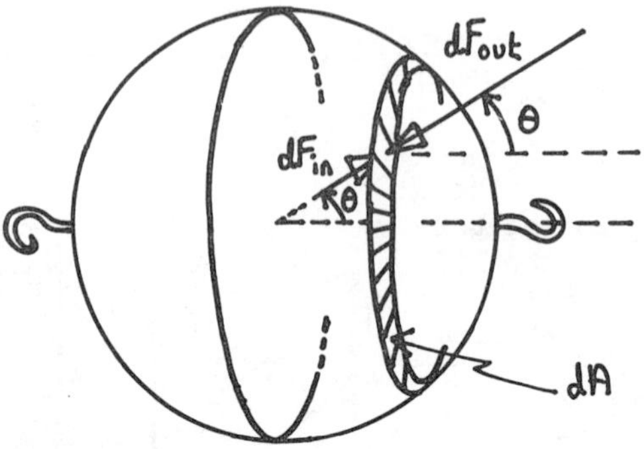

$$P = 14.7 - \frac{1}{10}(14.7) = 13.23 \text{ lb/in.}^2,$$

giving

$$F = \pi(1 \text{ ft})^2 \frac{13.23 \text{ lb}}{(1/12 \text{ ft})^2} = 6000 \text{ lb} \quad \underline{\text{Ans.}}$$

(c) Two teams of horses look more impressive than one, but one could be used if the other hemisphere is hooked to the side of a building.

16-12

(a) The force dF exerted by the water against the shaded area of the dam face is

$$dF = p\,dA = (\rho g x)(W dx).$$

The force F exerted against the entire face of the dam is

$$F = \int dF = \rho g W \int_0^D x\,dx = \tfrac{1}{2}\rho g W D^2 \quad \underline{\text{Ans.}}$$

(b) The horizontal component of the torque due to dF is $d\tau = dF(D - x)$; the total horizontal torque becomes

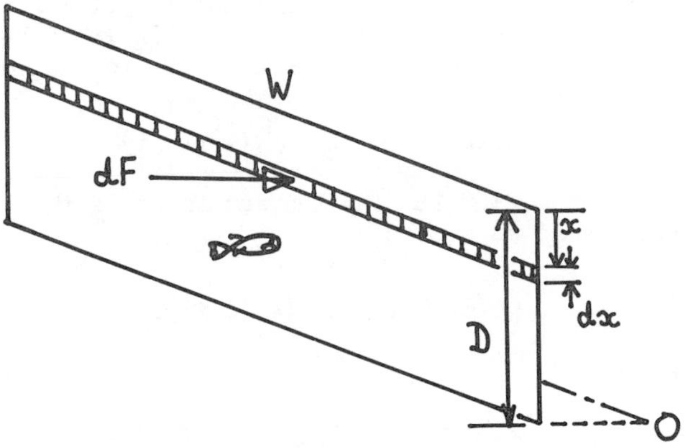

$$\tau = \int_0^D \rho g W x(D - x)dx = \frac{1}{6}\rho g W D^3 \quad \underline{\text{Ans}}.$$

(c) The line of action of F in (a) to yield the torque in (b) must be at a distance d above the bottom of the dam, where

$$(\tfrac{1}{2}\rho g W D^2)d = \frac{1}{6}\rho g W D^3,$$

$$d = \frac{D}{3} \quad \underline{\text{Ans}}.$$

16-13

Gravity takes a slab of fluid $\frac{1}{2}(h_2 - h_1)$ in thickness and lets it fall a vertical distance equal to its thickness. Thus, the work W done by gravity is

$$W = mg\cdot\tfrac{1}{2}(h_2 - h_1).$$

But the mass m of the slab equals the density ρ times the volume of the slab, and the volume is the base area A times its thickness. Hence,

$$W = \tfrac{1}{2}(h_2 - h_1)A\rho g\cdot\tfrac{1}{2}(h_2 - h_1) = \tfrac{1}{4}\rho A(h_2 - h_1)^2 g \quad \underline{\text{Ans}}.$$

16-18

(a) For the minimum area A of ice, let the ice sink 1 ft so that the top surface of the ice is at the water line. The buoyant force must support the weight of the car + ice. The ice weighs

$$(0.92)(62.4\ \text{lb/ft}^3)(A)(1\ \text{ft})$$

and the car weighs 2500 lb. The buoyant force is the weight of the water displaced and so equals

$$(62.4\ \text{lb/ft}^3)(A)(1\ \text{ft}).$$

For equilibrium,

$$62.4A = 2500 + (0.92)(62.4)A,$$

$$A = 501\ \text{ft}^2 \quad \underline{\text{Ans}}.$$

(b) Place the car in the center of the ice, otherwise the ice will tilt and the resulting buoyant force will diminish, since there is less ice in the water and therefore less water displaced.

16-23

Let r, R be the inner and outer radii; ρ, ρ' the densities of the liquid and sphere. The weight W of the sphere is

$$W = \rho' gV = \rho' g \cdot \frac{4\pi}{3}(R^3 - r^3).$$

The volume of liquid displaced is half the external volume of the sphere; the buoyant force, then, is

$$F = \rho g V_{ext} = \rho g \cdot \tfrac{1}{2}\frac{4\pi}{3} R^3.$$

These forces must balance for equilibrium; setting F = W yields

$$\rho'(R^3 - r^3) = \tfrac{1}{2}\rho R^3,$$

$$\rho' = \tfrac{1}{2}\rho \frac{1}{1 - (r/R)^3}.$$

Numerically, $\rho = (0.80)(1000\ \text{kg/m}^3) = 800\ \text{kg/m}^3$; $r/R = 8/9$. These give

$$\rho' = 1344 \text{ kg/m}^3 \quad \underline{\text{Ans.}}$$

16-26

(a)

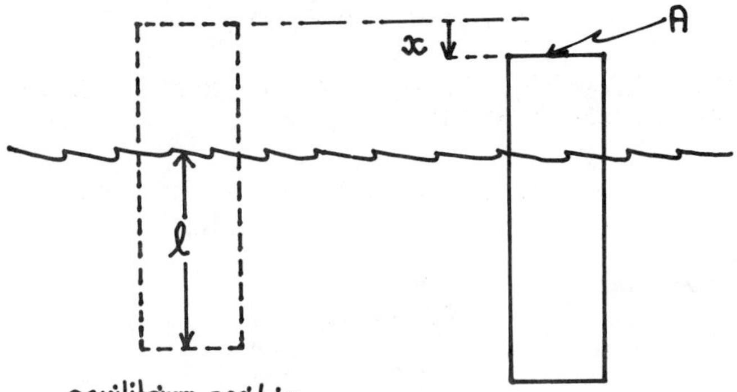

The forces acting on the log are its weight W and the buoyant force F(x), x measuring the vertical displacement from equilibrium. If a length l of the log is submerged when in equilibrium, then

$$W = \rho g A l,$$

ρ the density of water and A the cross-sectional area of the log. When the log is displaced downward a distance x, x positive down, Newton's second law gives

$$W - F = ma = \frac{W}{g} a,$$

$$\rho g A l - \rho g A(l + x) = \frac{\rho g A l}{g} \frac{d^2x}{dt^2},$$

$$\frac{d^2x}{dt^2} + \frac{g}{l} x = 0.$$

The last equation describes simple harmonic motion (e.g., of the simple pendulum).

(b) The period of oscillation is $T = 2\pi\left(\frac{l}{g}\right)^{\frac{1}{2}} = 3.17$ s $\underline{\text{Ans.}}$

16-27

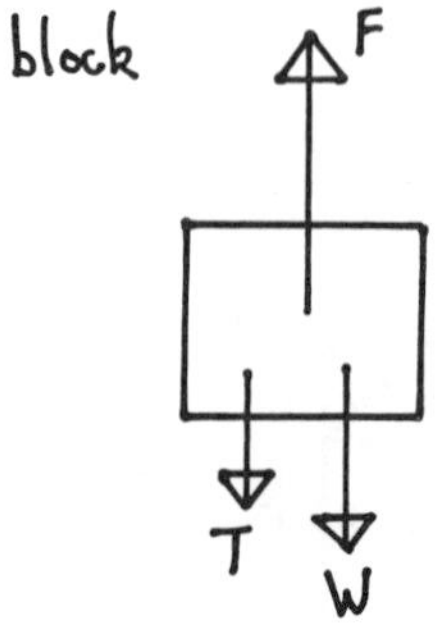

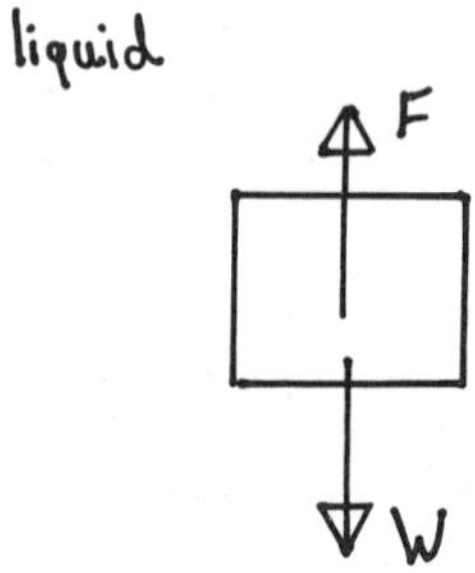

For the block, Newton's second law gives

$$F - W - T = ma = \frac{W}{g} a,$$
$$T = F - W(1 + \frac{a}{g}),$$

m,W the mass and weight of the block, F the buoyant force. Also, Newton's second law applied to an equal volume of liquid yields

$$F - \rho gV = (\rho V)a,$$

V the volume of block and liquid (which is of density ρ). The buoyant force F is the same for block and liquid since the buoyant force depends only on the volume of liquid displaced, and the volumes have been chosen equal above. The previous equation can be rewritten as

$$F = \rho gV(1 + \frac{a}{g}) = F_0(1 + \frac{a}{g}),$$

F_0 the buoyant force for zero acceleration, in which case the equation of motion of the block becomes

$$F_0 - W - T_0 = 0,$$
$$F_0 = W + T_0.$$

Finally, the second, fourth and sixth equations combined show that

$$T = (W + T_0)(1 + \frac{a}{g}) - W(1 + \frac{a}{g}),$$

$$T = T_0(1 + \frac{a}{g}).$$

16-30

The work W done by the pump on a mass m of water is

$$W = mgh + \tfrac{1}{2}mv^2.$$

Thus the power P supplied by the pump is

$$P = \frac{dW}{dt} = \frac{dm}{dt}(gh + \tfrac{1}{2}v^2).$$

But the mass flow rate is

$$\frac{dm}{dt} = Av\rho,$$

so that

$$P = Av\rho(gh + \tfrac{1}{2}v^2),$$

$$P = 65.8\ \text{W}\quad \underline{\text{Ans}}.$$

16-38

(a) Bernoulli's equation applied to points 1 and 2 gives

$$p_1 + \tfrac{1}{2}\rho V^2 + \rho gH = p_2 + \tfrac{1}{2}\rho v^2 + \rho g(H - h),$$

V the speed at which the liquid surface is falling, v the speed of efflux. If the tank is open, the pressure at the surface is just atmospheric pressure; also, the pressure at the hole is atmospheric pressure (definition of hole); that is, $p_1 = p_2 = p_{atm}$. Using this, Bernoulli's equation becomes

$$\tfrac{1}{2}\rho V^2 = \tfrac{1}{2}\rho v^2 - \rho gh.$$

The equation of continuity states that

$$AV = av.$$

Using this to eliminate V yields

$$\tfrac{1}{2}\rho(\frac{av}{A})^2 = \tfrac{1}{2}\rho v^2 - \rho gh,$$

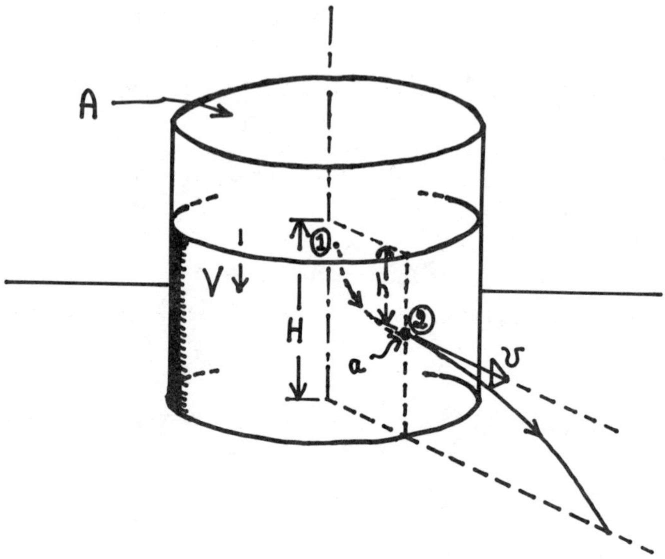

$$v^2 = \frac{2gh}{1 - (a/A)^2}.$$

If the area of the hole is much smaller than the area of the tank, that is a $\ll$ A, then approximately

$$v = (2gh)^{\frac{1}{2}}.$$

(b) The result above is identically equal to the speed of an object after falling freely from rest a vertical distance h. Thus, if the orifice is curved upward, the liquid would reach a height h above the hole, that is, would reach to the water line.

(c) Bernoulli's equation is the work-energy theorem restated in a manner convenient for the study of fluids. Friction was neglected in its derivation. As viscosity and turbulence imply frictional energy losses, Torricelli's law would not hold when these are present.

16-41

(a) The speed v at which the water leaves is, by Problem 16-38,

$$v = (2gh)^{\frac{1}{2}},$$

with h = 1 ft. Hence v = 8 ft/s. The volume flow rate is

$$R = av = (\tfrac{1}{144}\ \text{ft}^2)(8\ \text{ft/s}) = 0.0555\ \text{ft}^3/\text{s} \quad \underline{\text{Ans}}.$$

(b) Let x be vertical distance below the hole. The speed of the water after falling distance x is the free-fall speed V:

$$V^2 = v^2 + 2gx.$$

If A is the cross-sectional area of the stream at distance x, then

$$AV = av,$$

by the equation of continuity. Setting $A = \frac{1}{2}a$ gives $V = 2v$. Hence,

$$4v^2 = v^2 + 2gx,$$

$$x = 3v^2/2g = 3\ \text{ft} \quad \underline{\text{Ans}},$$

since v = 8 ft/s.

<u>16-42</u>

(a) Assume slow removal, so that $v_D = 0$ approximately. If the container is open, then $p_D = p_{atm}$. Also, $p_C = p_{atm}$ since the liquid emerges into the atmosphere at C. Thus, Bernoulli's equation applied between points D and C gives

$$p_{atm} = p_{atm} + \tfrac{1}{2}\rho v^2 - \rho g(h_2 + d),$$

v the speed of efflux at C. Hence,

$$v = [2g(h_2 + d)]^{\frac{1}{2}} \quad \underline{\text{Ans}}.$$

(b) If the tube has a uniform cross-sectional area, then $v_B = v_C = v$, by the equation of continuity. Applying Bernoulli's equation between points D and B gives

$$p_{atm} = p_B + \tfrac{1}{2}\rho v^2 + \rho g h_1,$$

$$p_B = p_{atm} - \rho g(h_2 + d + h_1) \quad \underline{\text{Ans}},$$

using (a).

(c) The minimum $p_B = 0$, in which event

$$p_{atm} = \rho g(h_2 + d + h_1).$$

For maximum h_1, arrange things so that $h_2 + d = 0$ to obtain

$$h_{1,max} = p_{atm}/\rho g = 39 \text{ ft} = 10.3 \text{ m} \quad \underline{\text{Ans}}.$$

<u>16-43</u>

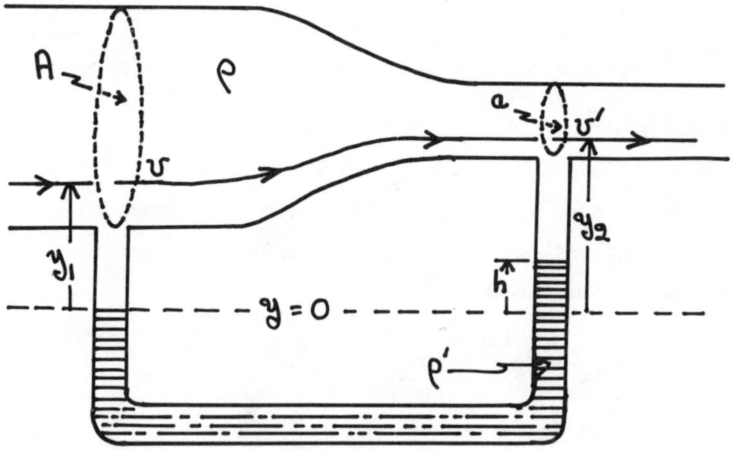

Apply Bernoulli's equation to points 1 and 2 along a streamline:

$$p_1 + \rho g y_1 + \tfrac{1}{2}\rho v^2 = p_2 + \rho g y_2 + \tfrac{1}{2}\rho v'^2.$$

By the continuity equation

$$Av = av',$$

so that

$$p_1 + \rho g y_1 + \tfrac{1}{2}\rho v^2 = p_2 + \rho g y_2 + \tfrac{1}{2}\rho(A^2/a^2)v^2.$$

Equate expressions for the pressure at $y = 0$ in the left and right hand tubes of the U-tube:

$$p_1 + \rho g y_1 = p_2 + \rho g(y_2 - h) + \rho' gh.$$

Solve each of the two preceding equations for $p_1 - p_2 + \rho g(y_1 - y_2)$ and equate the results; then solve for v^2 to get

$$v^2 = a^2 \frac{2(\rho' - \rho)gh}{\rho(A^2 - a^2)}.$$

16-49

Let P, V be the pressure and speed relative to the wing of air flowing across the upper wing surface, and p, v the same for air flowing below the lower wing surface. These air streams lie on different streamlines that, however, have a common origin just in front of the wing. Hence, if ρ is the density of air,

$$P + \tfrac{1}{2}\rho V^2 + \rho gY = p + \tfrac{1}{2}\rho v^2 + \rho gy.$$

The lift is $(p - P)A$, A the wing area; the lift per unit area is $(p - P)$ and, since $y \approx Y$, this is

$$(p - P) = \tfrac{1}{2}\rho(V^2 - v^2),$$

$$900 = \tfrac{1}{2}(1.3)(V^2 - 110^2),$$

$$V = 116.1 \text{ m/s} \quad \underline{\text{Ans}}.$$

CHAPTER 17

17-6

(a) At some instant t,

$$kx_1 - \omega t = kx_2 - \omega t + \frac{\pi}{3}.$$

Thus, the distance between the two points is

$$x_1 - x_2 = \frac{\pi}{3k} = \frac{\pi}{3(2\pi/\lambda)} = \frac{\lambda}{6}.$$

To find the wavelength λ, set

$$v = \nu\lambda,$$
$$350 = 500\lambda,$$
$$\lambda = 0.70 \text{ m}.$$

Hence,

$$x_1 - x_2 = \frac{0.7}{6} = 0.117 \text{ m} = 11.7 \text{ cm} \quad \underline{\text{Ans.}}$$

(b) In this case,

$$\Delta(\text{phase}) = (kx - \omega t_1) - (kx - \omega t_2),$$
$$\Delta(\text{phase}) = \omega(t_2 - t_1) = 2\pi\nu(t_2 - t_1),$$
$$\Delta(\text{phase}) = 2\pi(500)(0.001) = \pi \text{ rad} \quad \underline{\text{Ans.}}$$

17-10

(a) The wave speed v is given by

$$v = \nu\lambda = (25)(24) = 600 \text{ cm/s} \quad \underline{\text{Ans.}}$$

(b) The maximum particle displacement is the amplitude A, so that $A = 0.3$ cm. The equation for a wave moving in the -x-direction is

$$y = A\sin(kx + \omega t + \phi),$$

where ϕ is the phase angle. But $k = 2\pi/\lambda = 2\pi/24 = \pi/12$, and $\omega = 2\pi\nu = 50\pi$, and these lead to

$$y = 0.3\sin(\frac{\pi x}{12} + 50\pi t + \phi).$$

Since the displacement y at $x = 0$ when $t = 0$ is zero,

$$0 = 0.3\sin(0 + \phi),$$

$$\phi = 0.$$

Hence, the equation of the wave is

$$y = 0.3\sin(\frac{\pi}{12}x + 50\pi t) \quad \underline{\text{Ans}}.$$

17-16

The speed v of waves on the wire is given by $v = (T/\mu)^{\frac{1}{2}}$. The mass per unit length μ of the wire is just $(0.1 \text{ kg})/(10 \text{ m}) = 0.01 \text{ kg/m}$. Therefore,

$$v = (250/0.01)^{\frac{1}{2}} = 158.1 \text{ m/s}.$$

Let the later disturbance be generated at $t = 0$. The distance x' travelled by the disturbance in time t is $x' = vt$. But, in this same time the other disturbance, generated earlier, has travelled $x = v(t + 0.03)$. When the disturbances meet, then $x + x' = 10$. Hence, at the instant of the meeting,

$$v(t + 0.03) + vt = 10,$$

$$2t + 0.03 = \frac{10}{v},$$

$$t = 0.01662 \text{ s},$$

using the value of v from above. The waves, then, meet at a point 2.63 m from that end of the wire from which the later disturbance originated.

17-19

(a) If T is the tension in the rope,

$$v(y) = [T(y)/\mu]^{\frac{1}{2}}$$

where $\mu = m/L$. Consider a small portion of the rope a distance y from the lower end (which is at $y = 0$). The tension in this portion is the weight of that part of the rope hanging beneath it:

$$T = (\mu y)g.$$

Therefore,

$$v = [\frac{\mu g y}{\mu}]^{\frac{1}{2}} = (yg)^{\frac{1}{2}}.$$

(b) Since $v = dy/dt$, the result of (a) can be written as

$$y^{-\frac{1}{2}}dy = g^{\frac{1}{2}}dt.$$

Upon integrating, this last becomes

$$2y^{\frac{1}{2}} = g^{\frac{1}{2}}t + C,$$

C being the constant of integration. To evaluate C, let the wave start at the bottom ($y = 0$) at time $t = 0$:

$$2(0)^{\frac{1}{2}} = g^{\frac{1}{2}}(0) + C,$$

$$C = 0.$$

Hence,

$$2y^{\frac{1}{2}} = g^{\frac{1}{2}}t.$$

If the wave reaches the top ($y = L$) of the rope at time $t = t^*$,

$$2L^{\frac{1}{2}} = g^{\frac{1}{2}}t^*,$$

$$t^* = 2(\frac{L}{g})^{\frac{1}{2}}.$$

(c) Provided that the rope has a non-zero mass so that a tension will appear in the rope, the results in (a) and (b) are independent of the numerical value of the mass.

17-20

For a travelling wave, the particle displacement y can be written

$$y = A\sin(kx \pm \omega t).$$

The particle velocity u is

$$u = \frac{\partial y}{\partial t} = \pm A\omega\cos(kx \pm \omega t).$$

The maximum particle speed is just

$$u_{max} = A\omega.$$

The desired ratio, then, becomes

$$\frac{u_{max}}{v} = \frac{A\omega}{v} = \frac{2\pi A\nu}{v}.$$

In terms of the mass density μ of the string,

$$\frac{u_{max}}{v} = 2\pi A\nu\left(\frac{\mu}{T}\right)^{\frac{1}{2}}.$$

For strings of different materials stretched to the same tension T, this ratio (assuming the same frequency of excitation) is smaller for the lighter materials; i.e., smaller for nylon than for wire.

17-22

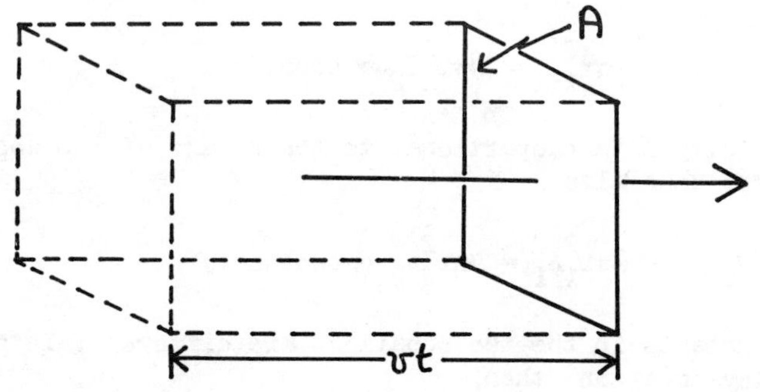

(a) Over a very short time interval t the wave can be considered plane-parallel. In this time t, the wave travels forward a distance vt and therefore energy contained in a cylinder of base area A and height vt crosses area A in time t. This energy is just the energy density u (energy per unit volume) times the cylinder volume Avt:

$$\text{energy} = u(Avt).$$

The energy crossing per unit area per unit time is

$$\frac{\text{energy}}{At} = I = uv.$$

(b) For a spherical wavefront a distance r from a source of power P,

$$I = \frac{P}{4\pi r^2}$$

so that

$$u = \frac{P}{4\pi r^2 v}.$$

With $P = 50{,}000$ W, $r = 4.8 \times 10^5$ m and $c = 3 \times 10^8$ m/s, this gives an energy density $u = 5.76 \times 10^{-17}$ J/m^3.

17-25

(a) Assuming no absorption, the power crossing a sphere centered at the source is independent of the sphere's radius. By definition of intensity this means that

$$4\pi r_1^2 I_1 = 4\pi r_2^2 I_2 = \text{constant}.$$

But the intensity I is proportional to the square of the amplitude A of the wave; thus also

$$4\pi r_1^2 A_1^2 = 4\pi r_2^2 A_2^2 = \text{constant},$$

where the constants in the two equations are different. In general (i.e., for any location) then,

$$A = \frac{\text{constant}}{r} = \frac{Y}{r}$$

say. Let the source be at $r = 0$, as implied above. The sinusoidal part of the particle displacement is

$$\sin(kr - \omega t) = \sin k(r - \frac{\omega}{k}t) = \sin k(r - vt),$$

giving

$$y = \frac{Y}{r}\sin k(r - vt),$$

when combined with the previous result.

(b) The sine factor in the above is dimensionless. Since both y and r have dimensions of length, Y must have dimensions of the square of length.

17-28

The difference in the lengths of the paths travelled by the direct wave from S and that reflected from height H must equal an integral number N of wavelengths for constructive interference:

$$L_H - d = N.\lambda.$$

However, waves reflected from the layer when at height H + h are out of phase with the direct wave, so that

$$L_{H+h} - d = N.\lambda + \tfrac{1}{2}\lambda,$$

that is, the waves are out of phase by half a wavelength, leading to destructive interference. Subtracting the two equations gives

$$L_{H+h} - L_H = \tfrac{1}{2}\lambda.$$

But the path lengths to the atmosphere and back are twice the hypotenuse of the right triangles; i.e.,

$$L_H = 2[(\tfrac{1}{2}d)^2 + H^2]^{\frac{1}{2}} = (d^2 + 4H^2)^{\frac{1}{2}},$$

$$L_{H+h} = [d^2 + 4(H + h)^2]^{\frac{1}{2}}.$$

Putting these two results into the previous equation and then multiplying by two yields

$$\lambda = 2[d^2 + 4(H + h)^2]^{\frac{1}{2}} - 2(d^2 + 4H^2)^{\frac{1}{2}}.$$

17-31

Waves having the same period have the same frequency. Therefore, with $\theta = kx - \omega t$, the waves can be written

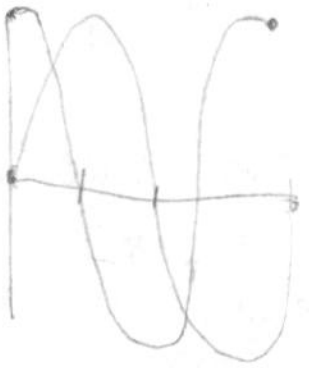

$$y_1 = A\sin\theta,$$
$$y_2 = \frac{A}{2}\sin(\theta + \frac{\pi}{2}) = \frac{A}{2}\cos\theta,$$
$$y_3 = \frac{A}{3}\sin(\theta + \pi) = -\frac{A}{3}\sin\theta.$$

The resultant $y_T = y_1 + y_2 + y_3$ becomes

$$y_T = \frac{2A}{3}\sin\theta + \frac{A}{2}\cos\theta = \frac{2A}{3}(\sin\theta + \frac{3}{4}\cos\theta),$$
$$y_T = \frac{2A}{3}[(\alpha\sin\beta)\sin\theta + (\alpha\cos\beta)\cos\theta],$$
$$y_T = \frac{2\alpha A}{3}\cos(\theta - \beta).$$

This requires that

$$\alpha\sin\beta = 1,$$
$$\alpha\cos\beta = \frac{3}{4}.$$

Solving these gives

$$\alpha = \frac{5}{4};\ \beta = \tan^{-1}(\frac{4}{3}).$$

Thus the amplitude of y_T is $2A\alpha/3 = 5A/6$. At $\theta = 0$, $y_T = \frac{5A}{6}\cos(-\beta)$ $= \frac{5A}{6}\cos\beta = \frac{5A}{6}\cos[\tan^{-1}(\frac{4}{3})] = \frac{A}{2}$. The resultant wave

$$y_T = \frac{5A}{6}\cos[\theta - \tan^{-1}(\frac{4}{3})]$$

is sketched on the following page.

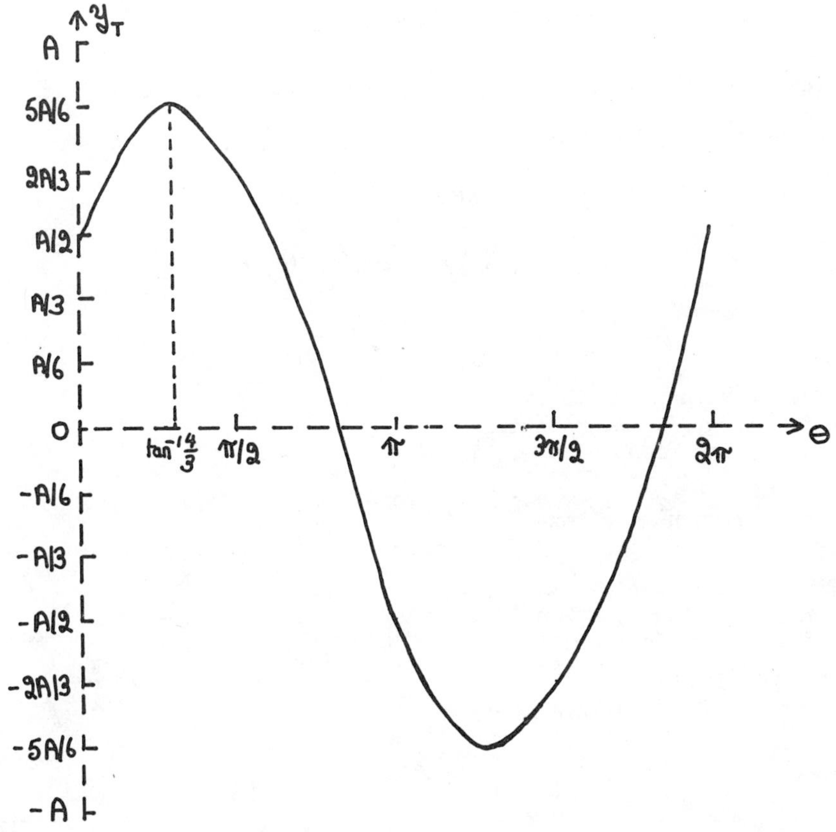

17-33

(a) If $y_1 = y_m \sin(kx - \omega t)$ and $y_2 = y_m \sin(kx + \omega t)$, then the sum $y = y_1 + y_2$ is given by

$$y = 2y_m \sin(kx)\cos(\omega t) = 0.5\sin\left(\frac{\pi x}{3}\right)\cos(40\pi t).$$

Therefore the amplitudes are $y_m = \frac{1}{2}(0.5) = 0.25$ cm. The wave speed v is

$$v = \frac{\omega}{k} = \frac{40\pi}{\pi/3} = 120 \text{ cm/s} \quad \underline{\text{Ans.}}$$

(b) The required distance is $\lambda/2$. But

$$\lambda = \frac{2\pi}{k} = \frac{2\pi}{\pi/3} = 6 \text{ cm},$$

and therefore the distance between adjacent nodes is 3 cm.

(c) The particle velocity u is

$$u = \frac{\partial y}{\partial t} = -\ 20\pi \sin\left(\frac{\pi x}{3}\right)\sin(40\pi t),$$

$$u = -\ 20\pi \sin\left(\frac{\pi}{2}\right)\sin(45\pi) = -\ 20\pi(+1)(0) = 0 \quad \underline{\text{Ans.}}$$

17-39

Associated with a length dx is kinetic energy dK and potential energy dU. The first of these is

$$dK = \tfrac{1}{2}(dm)u^2 = \tfrac{1}{2}(\mu dx)u^2$$

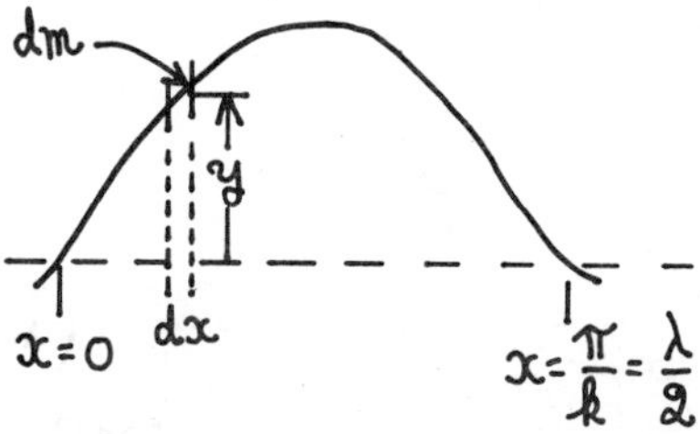

where u is the speed of the element. From the equation of a standing wave this speed is

$$u = \frac{\partial y}{\partial t} = -\ 2y_m\omega \sin(kx)\sin(\omega t)$$

so that

$$dK = 2\mu\omega^2 y_m^2 \sin^2 kx \cdot \sin^2 \omega t\ dx.$$

The potential energy is that of a simple harmonic oscillator of mass dm; if 'a' is the force constant,

$$dU = \tfrac{1}{2}ay^2,$$

where

$$a = (dm)\omega^2 = \mu\omega^2 dx.$$

Hence,

$$dU = \tfrac{1}{2}\mu\omega^2 y^2 dx = 2\mu\omega^2 y_m^2 \sin^2 kx \cdot \cos^2 \omega t\, dx.$$

Adding dU and dK to get the total energy dE,

$$dE = 2\mu\omega^2 y_m^2 \sin^2 kx \, dx,$$

and integrating over one loop, substituting $\omega = vk = 2\pi\nu$,

$$E = 2\mu\omega^2 y_m^2 \int_0^{\pi/k} \sin^2 kx \, dx = 2\pi^2 \mu y_m^2 \nu v \quad \underline{\text{Ans}}.$$

17-44

(a) From the tension T in the wires,

$$T = mg = (10 \text{ kg})(9.8 \text{ m/s}^2) = 98 \text{ N},$$

the wave speeds in the two wires can be found from

$$v_1 = (T/\mu_1)^{\frac{1}{2}}, \quad v_2 = (T/\mu_2)^{\frac{1}{2}}.$$

In terms of the density, the mass per unit length is

$$\mu = \rho AL/L = \rho A,$$

where A is the cross-sectional area of the wire and ρ is the density of the wire. Numerically,

$$\mu_1 = (2600 \text{ kg/m}^3)(10^{-6} \text{ m}^2) = 2.6 \times 10^{-3} \text{ kg/m},$$

$$\mu_2 = 7.8 \times 10^{-3} \text{ kg/m}.$$

These give for the wave speeds $v_1 = 194.1$ m/s and $v_2 = 112.1$ m/s. The distance between adjacent nodes is $\lambda/2$ so that if it is required that the joint be a node,

$$n_1\lambda_1/2 = l_1, \quad n_2\lambda_2/2 = l_2.$$

It is given that $l_1 = 0.6$ m and $l_2 = 0.866$ m and therefore, since $v = \lambda\nu$,

$$n_1 = 0.00618\nu_1, \quad n_2 = 0.0155\nu_2.$$

It is found by trial and error that the smallest integers n_1 and n_2 giving $\nu_1 = \nu_2$ are $n_1 = 2$ and $n_2 = 5$. The frequency is $2/0.00618 = 5/0.0155 = 323$ Hz <u>Ans.</u>

(b) There are $5 + 2 = 7$ loops or 8 nodes in all. If the two at the ends are not counted, the number of nodes is $8 - 2 = 6$ <u>Ans.</u>

18-8

The time t_1 required for a stone to fall to the bottom of a well of depth d is

$$t_1 = (2d/g)^{\frac{1}{2}},$$

and the time t_2 needed for sound, traveling at speed v, to cover this same distance d is

$$t_2 = d/v.$$

The total time t that elapses between dropping the stone and hearing the splash, then, is $t_1 + t_2$:

$$t = (2d/g)^{\frac{1}{2}} + d/v.$$

Rearranging and squaring the above equation gives a quadratic equation for d in terms of t:

$$gd^2 - d[(2v)(gt + v)] + (vt)^2 g = 0,$$

$$d = v[(\frac{v}{g} + t) \pm \sqrt{(\frac{v}{g})(\frac{v}{g} + 2t)}].$$

The negative sign is appropriate since t = 0 should indicate d = 0. With $g = 9.8\ m/s^2$, v = 331 m/s, a time of t = 3 s gives d = 40.6 m.

18-10

(a) Light travels almost instantaneously compared to sound: the time delays can be attributed to sound alone. Thus, the distance to each spectator is vt, v the speed of sound and t the time delay:

$$d_1 = (331\ m/s)(0.6\ s) = 198.6\ m \quad \underline{Ans.}$$
$$d_2 = (331\ m/s)(0.9\ s) = 297.9\ m \quad \underline{Ans.}$$

(b) Since the lines of sight meet at a right angle, the distance d the spectators are apart is

$$d = (d_1^2 + d_2^2)^{\frac{1}{2}},$$

$$d = (198.6^2 + 297.9^2)^{\frac{1}{2}} = 358 \text{ m} \quad \underline{\text{Ans.}}$$

18-18

(a) The path difference between the two waves going via SBD in the two positions must be half a wavelength:

$$\tfrac{1}{2}\lambda = 2(1.65 \text{ cm}),$$
$$\lambda = 0.066 \text{ m}.$$

The frequency ν must be

$$\nu = \frac{v}{\lambda} = \frac{331 \text{ m/s}}{0.066 \text{ m}} = 5015 \text{ Hz} \quad \underline{\text{Ans.}}$$

(b) Let A be the amplitude of the wave when at D that went via route SAD and B the amplitude at D of the wave that traversed SBD in either of the two positions. Since the intensity of a wave is proportional to the square of the resultant amplitude,

$$(A + B)^2 = 900,$$
$$(A - B)^2 = 100.$$

Thus,

$$\frac{B}{A} = 0.5 \quad \underline{\text{Ans.}}$$

(c) The waves going by SAD and SBD travel different distances and therefore lose different amounts of energy by, for example, gas friction with the walls of the tubes.

18-19

Assuming that the angle of incidence equals the angle of reflection,

$$\tan\theta = \frac{90}{50 - s} = \frac{10}{s},$$

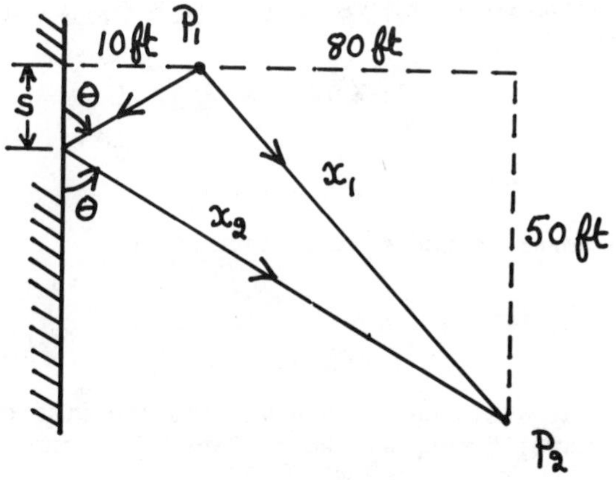

$$s = 5 \text{ ft.}$$

The total path lengths are

$$x_1 = (80^2 + 50^2)^{\frac{1}{2}} = 94.34 \text{ ft},$$

$$x_2 = (10^2 + 5^2)^{\frac{1}{2}} + (90^2 + 45^2)^{\frac{1}{2}} = 111.80 \text{ ft.}$$

The path difference is 17.46 ft. For constructive interference this must equal an odd number of half wavelengths, there being a phase change on reflection: that is,

$$(n + \tfrac{1}{2})\lambda = 17.46 \text{ ft.}$$

(i) $n = 0$: $\lambda = 2(17.46) = 34.92$ ft; $\nu = v/\lambda = (1100 \text{ ft/s})/(34.92 \text{ ft}) = 31.5$ Hz.

(ii) $n = 1$: $\lambda = 11.64$ ft; $\nu = 94.5$ Hz.

18-22

By equation 18-7,

$$I = 2\pi^2 y_m^2 \nu^2 \rho_0 v$$

so that the amplitude y_m of the particle displacements is

$$y_m = \frac{1}{\pi\nu}\left(\frac{I}{2\rho_0 v}\right)^{\frac{1}{2}}.$$

With $\nu = 300$ Hz, $I = 10^{-6}$ W/m^2, $\rho_0 = 1.22$ kg/m^3 and $v = 330$ m/s

$$y_m = 3.74 \times 10^{-8} \text{ m} \quad \underline{\text{Ans.}}$$

18-23

(a) If S is the power of the source, then

$$I = \frac{S}{4\pi r^2}$$

is the intensity at distance r since $4\pi r^2$ is the area of the spherical sound wave. Hence, at two different distances r_1 and r_2,

$$\frac{I_1}{I_2} = \frac{r_2^2}{r_1^2},$$

$$\frac{9.6 \times 10^{-4}}{I_2} = \frac{(30)^2}{(6.1)^2},$$

$$I_2 = 3.97 \times 10^{-5} \text{ W/m}^2 \quad \underline{\text{Ans.}}$$

(b) The displacement amplitude y_m is related to the intensity I by

$$y_m = \frac{1}{\pi\nu}\left(\frac{I}{2\rho_0 v}\right)^{\frac{1}{2}},$$

$$y_m = 1.74 \times 10^{-7} \text{ m} \quad \underline{\text{Ans,}}$$

at 6.1 m, using $\rho_0 = 1.22$ kg/m^3 for the density of air and $v = 330$ m/s as the speed of sound.

(c) The pressure amplitude p_m is given by

$$p_m = (2\rho_0 vI)^{\frac{1}{2}} = 0.88 \text{ Pa} \quad \underline{\text{Ans,}}$$

again at 6.1 m.

18-26

(a) A point source emits a spherical wave and therefore

$$I = \frac{S}{4\pi r^2} = \frac{10^{-6}}{4\pi(3)^2} = 8.842 \times 10^{-9}\ W/m^2 \quad \underline{Ans.}$$

(b) The reference intensity $I_0 = 10^{-12}$ W/m^2; hence, $I/I_0 = 8842$ and the sound level β becomes

$$\beta = 10\log(I/I_0) = 10\log 8842 = 39.5\ dB \quad \underline{Ans.}$$

18-29

Let d be the limiting distance for the whispered conversation to be audible with the reflector. If S is the power output of the conversation, then the intensity at the reflector is $S/4\pi d^2 = I_s$. Sound energy enters the reflector at the rate $I_s(\pi R^2)$, where R is the radius of the reflector. The intensity at the tube opening is $I_s\pi R^2/\pi r^2$, r the radius of the tube. But only 12% of this actually passes down the tube, so that the actual intensity I available to the ear is $0.12I_sR^2/r^2$ or

$$I = 0.12(S/4\pi d^2)(R^2/r^2) = 1200S/4\pi d^2,$$

since R = 50 cm and r = 0.5 cm. Now consider a whisper at a distance of 1 m. Its intensity I_w at the reflector is $S_w/4\pi(1)^2 = S_w/4\pi$ where S_w is the power of the whisper. The sound level of the whisper is

$$\beta_w = 10\log(S_w/4\pi I_0) = 20\ dB.$$

But the conversation is also whispered: $S = S_w$. Hence, the sound level of the conversation at the earpiece of the reflector is

$$\beta = 10\log(1200S_w/4\pi d^2 I_0) = 10\log(S_w/4\pi I_0) + 10\log(1200/d^2),$$

$$\beta = \beta_w + 10\log 1200 - 20\log d,$$

$$0 = 20 + 10\log(1200) - 20\log d,$$

$$\log d = 2.5396,$$

$$d = 346 \text{ m} \quad \underline{\text{Ans}}.$$

18-36

(a) In the fundamental mode,

$$L = \lambda_0/2 = v/2\nu_0.$$

If ℓ is the length by which the string is shortened,

$$L - \ell = v(2\cdot r\nu_0).$$

Eliminating ν_0 between these equations gives

$$\ell = L(1 - \frac{1}{r}) \quad \underline{\text{Ans}}.$$

(b) For L = 80 cm, successive sustitution of r = 6/5, 5/4, 4/3, 3/2 gives ℓ = 13.3, 16, 20, 26.7 cm.

18-41

(a) Displacement antinodes occur at each open end of the pipe. As the distance between adjacent antinodes is half a wavelength,

$$L = n\cdot\tfrac{1}{2}\lambda, \quad n = 1, 2, 3...$$

In terms of the frequency $\nu = v/\lambda$,

$$\nu = \frac{nv}{2L} = n\,\frac{1130}{2(1.5)} = n\,\frac{1130}{3} \text{ Hz}.$$

With the frequency limited to 1000 Hz to 2000 Hz, acceptable values of n and the associated frequencies are

$$n = 3, \quad \nu = 1130 \text{ Hz},$$

$$n = 4, \quad \nu = 1507 \text{ Hz},$$

$$n = 5, \quad \nu = 1883 \text{ Hz}.$$

(b)

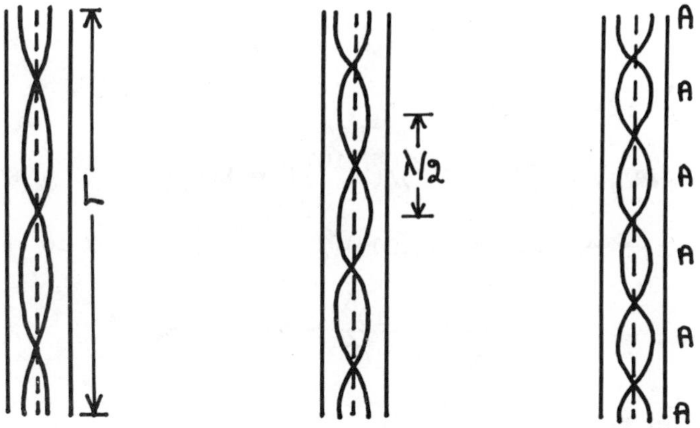

18-42

A displacement node exists at the bottom of the well and an antinode is present at the top. Since the frequency is the smallest allowable, the wavelength must be the longest permitted for resonance. The distance between a node and the adjacent antinode is one quarter of a wavelength; hence, if v is the speed of sound and D the depth of the well,

$$D = \tfrac{1}{4}\lambda = \frac{v}{4\nu}.$$

The speed of sound is given by

$$v = \left[\frac{c_p}{c_v} P/\rho\right]^{\frac{1}{2}} = \left[\frac{(1.4)(9.5 \times 10^4)}{1.1}\right]^{\frac{1}{2}} = 347.7 \text{ m/s}.$$

Hence,

$$D = \frac{347.7}{4(7)} = 12.4 \text{ m} \quad \underline{\text{Ans}}.$$

18-44

(a) For spherically symmetric pulsations, the center of the star must be a displacement node.

(b) In the fundamental mode of oscillation, antinodes exist at the surface and nowhere else. Since the distance between antinodes is one-half the wavelength and $\lambda = v/\nu$, v the sound speed,

$$\tfrac{1}{2}v/\nu = 2R,$$

$$\frac{1}{\nu} = T = 4R/v \quad \underline{\text{Ans.}}$$

(c) The speed of sound is

$$v = (\gamma P/\rho)^{\frac{1}{2}} = [(\tfrac{4}{3})(10^{22})/(10^{10})]^{\frac{1}{2}} = 1.1547 \times 10^6 \text{ m/s}.$$

The radius R of the star is $(0.009)(7 \times 10^8 \text{ m}) = 6.3 \times 10^6$ m. Thus,

$$T = \frac{(4)(6.3 \times 10^6)}{1.1547 \times 10^6} = 22 \text{ s} \quad \underline{\text{Ans.}}$$

<u>18-45</u>

Possible frequencies are

$$\frac{v}{2L}, \frac{v}{L}, \frac{3v}{2L} \text{ etc.,}$$

among which are 880, 1320 Hz as consecutive frequencies. But 880/1320 = 2/3 and therefore

$$880 = \frac{v}{L},\ 1320 = \frac{3}{2}\frac{v}{L}.$$

Either of these gives

$$v = 880L = (880)(0.316) = 278 \text{ m/s}.$$

But

$$v = (T/\mu)^{\frac{1}{2}}$$

and $\mu = 0.00065$ kg/m so that

$$T = \mu v^2 = (6.5 \times 10^{-4})(278)^2 = 50 \text{ N} \quad \underline{\text{Ans.}}$$

<u>18-46</u>

In the fundamental mode $\lambda = 2L = v/\nu$ and therefore

$$\nu = \frac{1}{2L}(T/\mu)^{\frac{1}{2}}.$$

To obtain six beats change the frequency of one wire by 6 Hz. Take differentials of the equation above and then divide the result by the equation itself to obtain

$$\frac{\Delta\nu}{\nu} = \frac{1}{2}\frac{\Delta T}{T}.$$

Since $\Delta\nu/\nu = 0.01$, $\Delta T/T = 2\%$.

18-51

The frequency of each bleep as it strikes the cave wall is

$$\nu' = \nu\left(\frac{v}{v - v_s}\right),$$

where ν is the frequency of the bleep as produced by the bat. Also ν' is the frequency of the reflected bleep. The bat now "sees" sound of frequency ν' emitted from a source at which she is approaching at speed v_0. Hence, the frequency of this reflected wave as perceived by the approaching bat is

$$\nu'' = \nu'\left[\frac{v + v_0}{v}\right].$$

Eliminating ν' and noting that $v_0 = v_s$,

$$\nu'' = \nu\left(\frac{v + v_s}{v - v_s}\right) = \nu\left(\frac{1 + v_s/v}{1 - v_s/v}\right).$$

The ratio $v_s/v = 1/40$; $\nu = 39,000$ Hz so that

$$\nu'' = (39,000)\frac{41/40}{39/40} = 41,000 \text{ Hz} \quad \underline{\text{Ans}}.$$

18-55

(a) The frequency the uncle hears is

$$\nu' = \nu\left(\frac{v}{v + v_s}\right) = (500)\frac{331}{331 + 10} = 485 \text{ Hz} \quad \underline{\text{Ans}}.$$

(b) Since the relative velocity between the girl and the train is zero, she hears the rest frequency of the whistle: to wit, 500 Hz.

(c) The velocities relative to the air enter into the Doppler shift equations. The train moves at 20 m/s and the uncle 10 m/s relative to the air, both to the east (same sign). Since they are moving apart,

$$\nu' = (500)\frac{331 + 10}{331 + 20} = 486 \text{ Hz} \quad \underline{\text{Ans}}.$$

(d) The girl is still at rest relative to the train and hears 500 Hz, as before.

18-57

Since the plane sees an approaching source the frequency ν' at which it receives the signal is

$$\nu' = \nu(1 + \frac{u}{c})$$

where ν is the rest frequency of the source. The receiver also sees an approaching source (the plane) and it receives the approaching reflected microwaves as at a frequency ν'' given by

$$\nu'' = \nu'(1 + \frac{u}{c}) = \nu(1 + \frac{u}{c})^2.$$

Since $\frac{u}{c} \ll 1$ for today's planes,

$$\nu'' = \nu(1 + 2\,\frac{u}{c}),$$

$$\frac{\nu'' - \nu}{\nu} = 2\,\frac{u}{c}.$$

But $\nu = c/\lambda$ and therefore

$$u = \frac{\lambda}{2}(\nu'' - \nu) = \tfrac{1}{2}(0.1 \text{ m})(990 \text{ s}^{-1}) = 49.5 \text{ m/s} \quad \underline{\text{Ans}}.$$

18-61

(a) Let u be the speed of the jet. If θ is the half-angle of the cone,

$$\sin\theta = v/u = v/1.5v = 2/3,$$

$$\theta = 42^\circ \quad \underline{\text{Ans}}.$$

(b) The shock travels with the plane at a speed (1.5)(331 m/s) = 496.5 m/s. It must cover a distance L = $h\cot\theta$ = (5000 m)(1.118) = 5590 m to reach the ground observer. Clearly it will take a time t = (5590 m)/(496.5 m/s) = 11 s <u>Ans.</u>

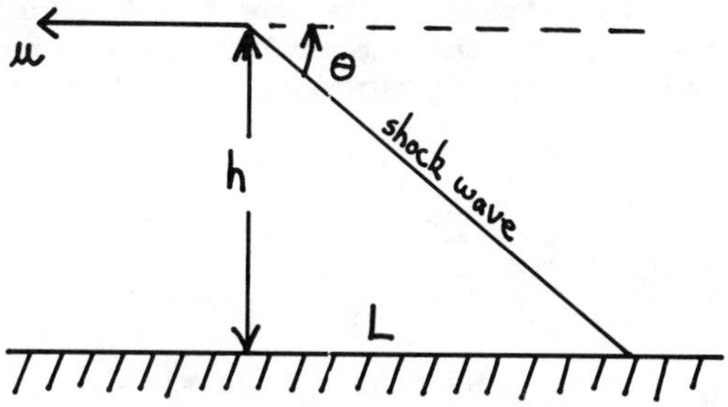

CHAPTER 19

<u>19-4</u>

(a) The cooling of an object can occur by radiation, conduction, convection, evaporation etc., and these will be affected by the nature of the material, its surface area and temperature, the orientation of the object, the temperature of the surroundings, the presence or absence of air currents, etc. The units in the equation

$$\frac{d\Delta T}{dt} = -K(\Delta T)$$

are

$$\text{kelvins/seconds} = K(\text{kelvins});$$

therefore, the units of K must be $(\text{seconds})^{-1}$, or dimensions of 1/time.

(b) Rearrange Newton's law of cooling to read

$$\frac{d\Delta T}{\Delta T} = -K\,dt,$$

and integrate, to get

$$\Delta T = Ce^{-Kt}$$

where C is a constant (i.e., independent of T and t). To evaluate C, set $\Delta T(t = 0) = \Delta T_0$. This gives $\Delta T_0 = C$. Hence,

$$\Delta T = \Delta T_0 e^{-Kt}.$$

<u>19-9</u>

(a) Since $T_F = 32 + \frac{9}{5}T_C$, setting $T_F = T_C = T$, say, gives

$$T = 32 + \frac{9}{5}T,$$

$$T = -40° \quad \underline{\text{Ans.}}$$

(b) The relation between the Kelvin and Fahrenheit scales is

$$T_C = T_K - 273.15 = \frac{5}{9}(T_F - 32).$$

If $T_K = T_F = T$,

$$T - 273.15 = \frac{5}{9}T - 17.78,$$

$$T = 575 \quad \underline{\text{Ans}}.$$

(c) The Kelvin and Celsius scales are identical except for the choice of a zero poi..t. Therefore they can never be equal, but always differ by the difference in their zero points.

19-16

Let Δd be the change in diameter on heating. It is desired that

$$d_s + \Delta d_s = d_b + \Delta d_b,$$

when the ring just slides onto the rod; d_s and d_b are the diameters at 25°C. Since

$$\Delta d = \alpha d \Delta T,$$

this becomes

$$d_s(1 + \alpha_s \Delta T) = d_b(1 + \alpha_b \Delta T).$$

Numerically, this is

$$3(1 + 11 \times 10^{-6}\Delta T) = 2.992(1 + 19 \times 10^{-6}\Delta T),$$

$$\Delta T = 335\ \text{C}^\circ.$$

Hence, the common temperature must be $25 + 335 = 360°C$ Ans.

19-22

The area of the plate is $A = ab$ before heating. After the rise in temperature, the area will be

$$A + \Delta A = (a + \Delta a)(b + \Delta b).$$

This may be written

$$A + \Delta A = a(1 + \frac{\Delta a}{a})\cdot b(1 + \frac{\Delta b}{b}) = ab(1 + \frac{\Delta a}{a} + \frac{\Delta b}{b} + \frac{\Delta a \Delta b}{ab}).$$

If $\Delta a/a$, $\Delta b/b \ll 1$, then the last term above is an order of magnitude smaller than either $\Delta a/a$ or $\Delta b/b$, and two orders of magnitude smaller than 1. Under these conditions, the last term can be dropped to give

$$A + \Delta A \approx ab(1 + \frac{\Delta a}{a} + \frac{\Delta b}{b}).$$

But $\Delta a = \alpha a \Delta T$ and $\Delta b = \alpha b \Delta T$ so that

$$A + \Delta A = A(1 + 2\alpha\Delta T),$$

$$\Delta A = 2\alpha A \Delta T \quad \underline{\text{Ans}}.$$

19-25

The change in radius is

$$\Delta R = \alpha R \Delta T = (23 \times 10^{-6})(10)(100) = 0.023 \text{ cm}.$$

The new radius is $R + \Delta R = 10.023$ cm so that the volume V' at the higher temperature is

$$V' = \frac{4}{3}\pi (R + \Delta R)^3 = 4217.76 \text{ cm}^3.$$

The volume at 0°C is

$$V = \frac{4}{3}\pi R^3 = 4188.79 \text{ cm}^3.$$

The change in volume is just

$$V' - V = \Delta V = 28.97 \text{ cm}^3 \quad \underline{\text{Ans}}.$$

19-28

Let A be the cross-sectional area of the glass tube. Then,

$$\Delta V = \beta V\Delta T = \beta(Ah)\Delta T,$$

h the original height. But

$$V = Ah,$$
$$\Delta V = A\Delta h,$$

if $\Delta A = 0$ (compare 3α for glass with β for mercury). Hence,

$$A\Delta h = \beta Ah\Delta T,$$
$$\Delta h = \beta h\Delta T.$$

19-31

The capacity V of the cup increases by $\Delta V = 3\alpha V\Delta T$, α the linear coefficient for aluminum. The volume of the mercury, which at 12°C equals V, the capacity of the cup, increases by $\Delta V = \beta V\Delta T$. Hence, mercury will spill if $\beta V\Delta T > 3\alpha V\Delta T$. Since in fact $\beta > 3\alpha$ some of the mercury will spill, the amount being

$$\text{spill} = (\beta - 3\alpha)V\Delta T,$$

$$\text{spill} = (1.8 \times 10^{-4} - 6.9 \times 10^{-5})(100)(6),$$

$$\text{spill} = 0.0666\ \text{cm}^3 = 66.6\ \text{mm}^3 \quad \underline{\text{Ans}},$$

since 0.1 liter = 100 cm^3.

19-32

The period P of a pendulum is $P = 2\pi(L/g)^{\frac{1}{2}}$. Hence, the period at the higher temperature is $P + \Delta P$, where

$$\Delta P = \frac{\partial P}{\partial L}(\Delta L) = \tfrac{1}{2}\alpha P\Delta T,$$

since $\Delta L = \alpha L\Delta T$. Numerically,

$$\Delta P = \tfrac{1}{2}(7 \times 10^{-7})(0.5)(10) = 1.75 \times 10^{-6}\ \text{s}.$$

Since there are 86,400 s in one day, the clock makes 5,184,000 oscillations in 30 days, leading to a total correction of $(5.184 \times 10^6)(1.75 \times 10^{-6}) = 9.07$ s, the clock running slow.

19-34

Let x be the length of each side of the cube, and let the cube be submerged to a depth y. The weight of the cube is $x^3\rho_a g$. Before heating, the buoyant force is $x^2 y\rho_m g$ and therefore

$$x^2 y\rho_m g = x^3\rho_a g,$$

$$y\rho_m = x\rho_a.$$

All quantities in the last equation change as the temperature rises, so that

$$y\Delta\rho_m + \rho_m\Delta y = x\Delta\rho_a + \rho_a\Delta x.$$

But the mass $M = x^3\rho_a$ of the cube remains constant:

$$\Delta M = 0 = 3x^2\rho_a\Delta x + x^3\Delta\rho_a,$$

$$x\Delta\rho_a = -3\rho_a\Delta x.$$

Therefore,

$$y\Delta\rho_m + \rho_m\Delta y = -2\rho_a\Delta x.$$

If β is the coefficient of volume expansion of mercury, then $\Delta\rho_m = -\beta\rho_m\Delta T$ (problem 19-27). Substitute this into the equation above and solve for Δy:

$$\frac{\Delta y}{\Delta T} = (\beta y - 2\frac{\rho_a}{\rho_m}\alpha x) = (\beta - 2\alpha)x\frac{\rho_a}{\rho_m}.$$

Putting in the numbers ($\rho_a = 2.7$ g/cm^3, $\rho_m = 13.6$ g/cm^3) gives $\Delta y = 0.266$ mm Ans.

19-35

(a) Let 2y be the horizontal span and x the sag of the cable. For a parabola, the length L of the cable is

$$L = (4x^2 + y^2)^{\frac{1}{2}} + \frac{y^2}{2x} \ln\left[\frac{2x + (4x^2 + y^2)^{\frac{1}{2}}}{y}\right].$$

The half-span $y = 4200/2 = 2100$ ft and is assumed to remain unchanged. At 50°F, $x = 470$ ft, so that the length of the cable at 50°F is given by the equation above with $y = 2100$, $x = 470$. This turns out to be $L = 4336$ ft. Hence,

$$\Delta L = \alpha L \Delta T = (6.5 \times 10^{-6})(4336)(80),$$

$$\Delta L = 2.25 \text{ ft} \quad \underline{\text{Ans}}.$$

(b) The change in the sag x cannot be computed from $\Delta x = \alpha x \Delta T$ because of the constraint on y. Instead, write,

$$\Delta x = \frac{\Delta x}{\Delta L}(\Delta L) \approx \frac{\Delta L}{dL/dx}.$$

Use the first equation to compute dL/dx:

$$\frac{dL}{dx} = \frac{4x}{(4x^2 + y^2)^{\frac{1}{2}}} - \frac{y^2}{2x^2} \ln[\;] + \frac{y^2}{2x} \frac{2 + 4x(4x^2 + y^2)^{-\frac{1}{2}}}{2x + (4x^2 + y^2)^{\frac{1}{2}}},$$

where

$$[\;] = \frac{2x + (4x^2 + y^2)^{\frac{1}{2}}}{y}.$$

Numerically $dL/dx = 0.56436$ so that

$$\Delta x = 2.25/0.56436 = 4.0 \text{ ft} \quad \underline{\text{Ans}}.$$

CHAPTER 20

20-4

Heat exchanges with the glass itself are ignored. The heat that is required to raise the temperature of the ice from -15°C to 0°C is

$$Q_1 = mc\Delta T = (100\ g)(\tfrac{1}{2}\ cal/g{\cdot}C°)(15\ C°) = 750\ cal.$$

Melting the ice at 0°C requires

$$Q_2 = (100\ g)(80\ cal/g) = 8000\ cal$$

of heat. Thus, 8750 cal are needed to convert 100 g of ice at -15°C to liquid water at 0°C. However, if the 200 g of water originally in the glass at 25°C were cooled to 0°C, they would liberate

$$Q_3 = (200\ g)(1\ cal/g{\cdot}C°)(25\ C°) = 5000\ cal$$

of heat. Now this is greater than Q_1 but less than $Q_1 + Q_2$. Hence, the equilibrium temperature is 0°C, and not all of the ice melts.

20-7

The sphere, mass M, will pass through the ring, mass m, when the sphere's diameter D equals the inner diameter d of the ring. Let T be the equilibrium temperature; T may be found from

$$mc_c(T) = Mc_a(100 - T),$$

$$\frac{M}{m} = \frac{c_cT}{c_a(100 - T)},$$

$$\frac{M}{m} = \frac{0.0923T}{21.5 - 0.215T},$$

after substituting numerical values for the specific heat c_c of copper and c_a of aluminum. The diameters of the ring and sphere at temperature T will be, with d, D the initial diameters,

$$d + \Delta d = d + \alpha_c d(T) = d(1 + \alpha_c T),$$

$$d + \Delta d = (1.0)(1 + 17 \times 10^{-6} T);$$

$$D + \Delta D = D[1 - \alpha_a (100 - T)],$$

$$D + \Delta D = (1.002)[1 - 23 \times 10^{-6}(100 - T)].$$

Setting $d + \Delta d = D + \Delta D$, so that the sphere just passes through the ring, gives

$$1 + 17 \times 10^{-6} T = 1.002 - 23.046 \times 10^{-4} + 23.046 \times 10^{-6} T,$$

$$T = 50.38°C.$$

Substituting this into the third equation, for M/m, yields

$$\frac{M}{m} = 0.436 \quad \underline{\text{Ans.}}$$

<u>20-17</u>

Consider a section of ice with a cross-sectional area of 1 cm^2. Since its thickness is 5 cm, the rate dQ/dt at which heat flows through it to the outside is

$$\frac{dQ}{dt} = kA\frac{dT}{dx} = (0.004)(1)\frac{0 - (-10)}{5} = 0.008 \text{ cal/s}.$$

Therefore it takes (80 cal/g)/(0.008 cal/s) = 10,000 seconds to freeze 1 g of water. In 10,000 s, then, the layer of ice will grow by x cm, where x is found from

$$(0.92 \text{ g/cm}^3)(1 \text{ cm}^2)(x \text{ cm}) = 1 \text{ g},$$

$$x = 1.087 \text{ cm},$$

each 10,000 s (since 1 cm^3 of water has a mass of 1 g). Hence, the hourly growth rate dx/dt is

$$\frac{dx}{dt} = \frac{3600 \text{ s/h}}{10,000 \text{ s}}(1.087 \text{ cm}) = 0.39 \text{ cm/h} \quad \underline{\text{Ans.}}$$

20-19

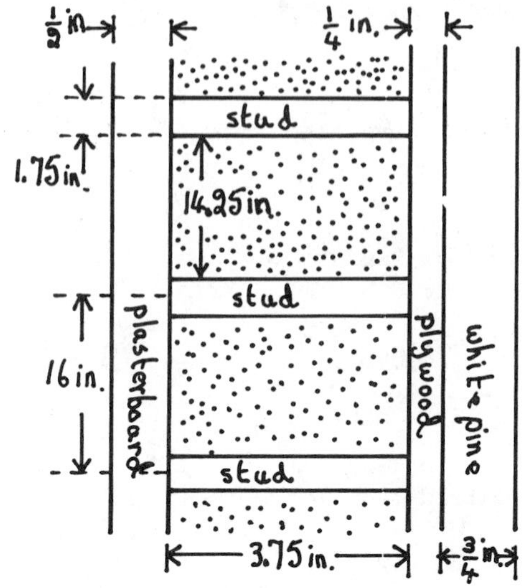

(a) The R value is defined by $R = L/k$. For the foam, if $R = 5.9$ for $L = 1$ in. thickness, then $R = (5.9)(3.75) = 22.125$ for the actual 3.75 in. thickness. Similarly, the R value of the wood studs is $(1.3)(3.75) = 4.875$ for each stud. The rate of heat flow dQ/dt through a multilayered section is given by

$$\frac{dQ}{dt} = \frac{A(T_2 - T_1)}{\Sigma R}.$$

For flow through the studs, the area is

$$A = (16)(\frac{1.75}{12}\ \text{ft})(12\ \text{ft}) = 28\ \text{ft}^2.$$

Hence, the rate H_s of heat flow through the studded portion is

$$H_s = \frac{(28)(30)}{0.47 + 4.875 + 0.30 + 0.98} = 126.79\ \text{Btu/h},$$

the units of R being $ft^2 \cdot F^\circ \cdot h/Btu$. The foam sections each have a width of $16 - 1.75 = 14.25$ in. $= 1.1875$ ft between adjacent studs, and there are $16 - 1 = 15$ such sections. They occupy a total area

$$A = (15)(1.1875\ \text{ft})(12\ \text{ft}) = 213.75\ \text{ft}^2.$$

(Note that $213.75 + 28 = 241.75 \approx 240\ \text{ft}^2$, so that the dimensions have been "rounded-off" a bit.) The heat flow rate H_f through the total foam section is

$$H_f = \frac{(213.75)(30)}{0.47 + 22.125 + 0.3 + 0.98} = 268.59\ \text{Btu/h}.$$

The total rate of heat flow is

$$H = H_s + H_f = 126.79 + 268.59 = 395.4\ \text{Btu/h} \quad \underline{\text{Ans}}.$$

(b) The R value for the whole wall is defined from

$$H = \frac{A_{tot}\Delta T}{R},$$

$$395.4 = \frac{(240)(30)}{R},$$

$$R = 18.2 \quad \underline{\text{Ans}}.$$

(c) From (a), the ratio is $(28\ \text{ft}^2)/(240\ \text{ft}^2) = 11.67\%$.

(d) Also from (a),

$$\frac{H_s}{H} = \frac{126.79}{395.4} = 32\% \quad \underline{\text{Ans}}.$$

<u>20-25</u>

The work W done by friction is

$$W = \Delta K = 0 - \tfrac{1}{2}m_1v_1^2 = -\tfrac{1}{2}(50)(5.38)^2 = -723.61\ \text{J}.$$

This implies that there is available 723.61 J = 172.86 cal of energy available to melt ice. Since it takes 80 cal to melt 1 g of ice, the mass of ice melted will be (172.86 cal)/(80 cal/g) = 2.16 g <u>Ans</u>.

20-28

In one-half liter there are 500 g = 0.5 kg of water. To begin boiling, this water must be brought to 100°C starting from 59°F = 15°C, that is, heated by 85 C°. The heat required is

$$Q = mc\Delta T = (500\ g)(1\ cal/g\cdot C°)(85\ C°) = 42,500\ cal.$$

If any additional heat is supplied, the water will begin to boil. Each shake produces

$$E = mg\Delta h = (0.5)(9.8)(0.3048) = 1.49352\ J$$

of energy or 0.3568 cal of heat. Therefore 42,500/0.3568 = 119,114 shakes are needed. At the rate of 30 shakes/min = 0.5 shakes/s, this will take 238,228 s or 238,228/86,400 = 2.76 days.

20-34

The internal energy change $\Delta U = 0$ over a complete cycle. Hence, by the first law,

$$\Delta U = Q - W$$

over the cycle,

$$Q = W.$$

Now W is the work done by the gas and this equals the area inside the curve of pressure p versus volume V. This curve is a triangle and therefore

$$W = -\tfrac{1}{2}(\Delta p)(\Delta V) = -\tfrac{1}{2}(20\ N/m^2)(3\ m^3) = -30\ J,$$

the negative sign since the path is traversed counterclockwise. It follows that the heat added is Q = -30 J = -7.17 cal (i.e., heat flowed out of the gas).

20-35

The first law of thermodynamics is

$$Q = \Delta U + W.$$

Over a complete cycle $\Delta U = 0$ and therefore

$$W = Q = -(100\ g)(80\ cal/g) = -8000\ cal,$$

negative since heat left the gas to melt ice. W is the work done by the gas and therefore the work done on the gas is +8000 cal.

20-36

(a) The change in internal energy $\Delta U_{fi} = U_f - U_i$ between points i and f is the same regardless of path. Along path iaf, Q = 50 cal and W = 20 cal; hence $\Delta U_{fi} = Q - W = 50 - 20 = 30$ cal. Along path ibf

$$\Delta U_{fi} = Q - W,$$

$$30 = 36 - W,$$

$$W = 6 \text{ cal} \quad \underline{\text{Ans.}}$$

(b) From (a), $\Delta U_{if} = U_i - U_f = -30$ cal. Hence, $\Delta U_{if} = -30 = Q - (-13)$, giving Q = -43 cal Ans.

(c) If $U_i = 10$ cal, then since from (a), $U_f - U_i = 30$, $U_f = 40$ cal.

(d) If $U_b = 22$ cal, $\Delta U_{bi} = U_b - U_i = 22 - 10 = 12$ cal. For the process ib, then, 12 = Q - W. Now, W = 6 cal for ibf, but W = 0 along bf since bf occurs at constant volume; hence W = 6 cal for ib. This gives 12 = Q - 6, or Q = 18 cal Ans. Along bf, $U_b = 22$ and $U_f = 40$ so that $U_f - U_b = 18$ cal. But W = 0 along bf as noted above, so that Q = 18 cal here also.

20-37

(a) The actual time that the ball is in contact with the floor is very short so that unless the floor is extraordinarily hot no heat will flow to the ball: Q = 0.

(b) By assumption $\Delta U > 0$ so that, by the first law $Q = \Delta U + W$, Q = 0 gives $W < 0$ indicating that work has been done on the ball.

(c) The change in internal energy ΔU equals, in magnitude, the loss of mechanical energy which, if M is the mass of the ball, is

$$\Delta U = Mg\Delta h = M(9.8 \text{ m/s}^2)(9.5 \text{ m}) = 93M \text{ J}$$

or 93 J/kg.

(d) The rise in temperature is given by

$$\Delta T = \frac{\Delta U}{Mc} = (93M)/(M)(120)(4.186) = 0.185\ C^\circ \quad \underline{\text{Ans.}}$$

since c = 120 cal/kg·C° and 4.186 is the mechanical equivalent of heat.

<u>20-38</u>

(a) Let m, ρ be the mass and density of the steam and V' its volume, V the volume of the chamber and M, A, v the mass, area and speed of the piston. Although V' ≠ V, dV'/dt ≈ dV/dt if the water level rises , due to steam condensation, slowly compared to the fall of the piston. Under this assumption, and disregarding a minus sign,

$$\frac{dV'}{dt} = \frac{dV}{dt} = Av = \frac{d}{dt}\left(\frac{m}{\rho}\right) = \frac{1}{\rho}\frac{dm}{dt},$$

$$\frac{dm}{dt} = \rho Av = (6 \times 10^{-4})(2)(0.3) = 3.6 \times 10^{-4}\ \text{g/s} \quad \underline{\text{Ans.}}$$

(b) Since 540 cal are liberated for each gram of steam that condenses,

$$\frac{dQ}{dt} = -\left(\frac{dm}{dt}\right)(540) = -(3.6 \times 10^{-4})(540),$$

$$\frac{dQ}{dt} = -0.1944\ \text{cal/s} = -0.8138\ \text{J/s} \quad \underline{\text{Ans.}}$$

(c) Differentiate the first law to obtain

$$\frac{dQ}{dt} = \frac{dU}{dt} + \frac{dW}{dt}.$$

To evaluate dW/dt, note that

$$\frac{dW}{dt} = p\frac{dV}{dt} = p(-Av).$$

The pressure p is given by

$$p = p_{atm} + \frac{Mg}{A},$$

so that

$$\frac{dW}{dt} = -(p_{atm}A + Mg)v.$$

Numerically, $p_{atm}A = (1.013 \times 10^5\ N/m^2)(2 \times 10^{-4}\ m^2) = 20.26$ N and $Mg = 19.6$ N; since $v = 0.003$ m/s

$$\frac{dW}{dt} = -0.1196\ J/s.$$

Combining this with (b) gives

$$\frac{dU}{dt} = -0.8138 - (-0.1196),$$

$$\frac{dU}{dt} = -0.6942\ J/s \quad \underline{Ans.}$$

21-4

For a fixed amount of gas undergoing a change from p_1, V_1, T_1 to p_2, V_2, T_2, the ideal gas law $pV = nRT$ applied to each state gives

$$p_1V_1 = nRT_1, \quad p_2V_2 = nRT_2.$$

Dividing these equations yields

$$\frac{p_1V_1}{p_2V_2} = \frac{T_1}{T_2}.$$

Let point 1 apply at the surface and 2 indicate the bottom of the lake. Then $p_1 = p_{atm} = 1.013 \times 10^5$ Pa, and

$$p_2 = p_{atm} + \rho gh = p_{atm} + (1000\ \text{kg/m}^3)(9.8\ \text{m/s}^2)(40\ \text{m}),$$

$$p_2 = (1.013 + 3.92) \times 10^5 = 4.933 \times 10^5\ \text{Pa}.$$

The temperatures are $T_1 = 20 + 273 = 293$ K, $T_2 = 4 + 273 = 277$ K. Hence,

$$\frac{1.013 \times 10^5}{4.933 \times 10^5} \cdot \frac{V_1}{20} = \frac{293}{277},$$

$$V_1 = 103\ \text{cm}^3 \quad \underline{\text{Ans}},$$

(cm^3, since the units of V_2 are cm^3).

21-10

The initial pressure is $p_i = 14.7 + 15 = 29.7\ \text{lb/in}^2$. Since the expansion is isothermal (T constant),

$$p_iV_i = p_fV_f,$$

$$(29.7)(5) = (14.7)V_f,$$

$$V_f = 10.10\ \text{ft}^3.$$

The work W_e done during this expansion is

$$W_e = \int_i^f p\,dV = nRT\cdot\ln(V_f/V_i) = p_iV_i\ln(V_f/V_i) = (29.7)(5)\ln(\frac{10.1}{5}),$$

$$W_e = 104.44\ \text{ft}^3\cdot\text{lb/in}^2.$$

For the compression at constant pressure,

$$W_c = p\Delta V = p_f(V_i - V_f) = (14.7)(5 - 10.1) = -74.97\ \text{ft}^3\cdot\text{lb/in}^2.$$

Therefore the total work $W = W_e + W_c$ is

$$W = 29.47\ \text{ft}^3\cdot\text{lb}/(\frac{1}{144}\ \text{ft}^2),$$

$$W = 4244\ \text{ft}\cdot\text{lb}\quad \underline{\text{Ans}}.$$

<u>21-11</u>

The work done at constant pressure is

$$W = p\Delta V = p_0(V_2 - V_1).$$

But, from the equation of state,

$$V = (AT - BT^2)/p$$

so that

$$W = (AT_2 - BT_2^2) - (AT_1 - BT_1^2),$$

$$W = A(T_2 - T_1) - B(T_2^2 - T_1^2)\quad \underline{\text{Ans}}.$$

21-13

The momentum imparted to the wall on each collision is

$$\Delta p = m\Delta v_p = 2mv_p = 2mv\cos 45°,$$

where v_p is the component of the velocity perpendicular to the wall. If N is the number of collisions per unit time, then the pressure P on the wall is

$$P = \frac{F}{A} = \frac{N\Delta p}{A} = \frac{2Nmv\cos 45°}{A},$$

$$P = [(2)(10^{23})(3.3 \times 10^{-24})(10^5)\cos 45°]/(2),$$

$$P = 2.33 \times 10^4 \text{ dyne/cm}^2 = 2330 \text{ Pa} \quad \underline{\text{Ans.}}$$

21-23

(a) The escape speed and root-mean-square speeds are

$$v_e = \left(\frac{2GM}{R}\right)^{\frac{1}{2}} = (2gR)^{\frac{1}{2}},$$

$$v_{rms} = \left(\frac{3kT}{m}\right)^{\frac{1}{2}}.$$

If $v_e = v_{rms}$, then

$$T = \frac{2}{3}\frac{Rgm}{k}.$$

For the earth, $R = 6.4 \times 10^6$ m, $g = 9.8$ m/s^2. H_2 has a molecular weight of 2 so that $m = 2(1.66 \times 10^{-27}$ kg); the equation above then gives T = 10,000 K. O_2 has a molecular weight of 32, or 16 times greater than H_2, giving T = 160,000 K for O_2.
(b) On the moon $g = 0.16(9.8$ m/s$^2)$ and $R = 1.74 \times 10^6$ m. The equation for T then gives T = 440 K for H_2 and 7000 K for O_2.
(c) Although 1000 K $\ll$ 10,000 K, there are enough molecules in the "tail" of the Maxwell-Boltzmann speed distribution to ensure depletion of molecular hydrogen over the 3 or 4 billion years since the atmosphere was formed. Most of the O_2, fortunately, is

retained.

21-24

(a) The molecular weight M of water (H_2O) is 2(1) + 16 = 18. The number n of molecules per gram is

$$n = N_0/M = 6.02 \times 10^{23}/18 = 3.3444 \times 10^{22}.$$

Thus,

$$\epsilon = \frac{540}{n} = (540 \text{ cal/g})/(3.3444 \times 10^{22} \text{ g}^{-1}),$$

$$\epsilon = 161 \times 10^{-22} \text{ cal} = 6.76 \times 10^{-20} \text{ J} \quad \underline{\text{Ans.}}$$

(b) The average translational kinetic energy of a molecule is

$$K = \frac{3}{2}kT = \frac{3}{2}(1.38 \times 10^{-23})(300) = 6.21 \times 10^{-21} \text{ J}.$$

Evidently, $\epsilon/K = 10.9$ Ans.

21-27

The work W done by the gas is

$$W = \int_i^f p \cdot dV = \int_i^f \frac{nRT}{V}\,dV = nRT\int_i^f \frac{dV}{V},$$

$$W = nRT \cdot \ln(V_f/V_i),$$

since, for an isothermal process, T is constant and can be drawn out from the integral. The change in internal energy is

$$\Delta U = \frac{3}{2}nR\,\Delta T = 0$$

for an isothermal process. Then, by the first law $\Delta U = Q - W$,

$$0 = Q - W,$$

$$Q = W = RT \cdot \ln(V_f/V_i),$$

for n = 1, i.e., one mole.

21-38

(a) Use the adiabatic equation

$$p_1V_1^\gamma = p_2V_2^\gamma,$$

with p in atmospheres and V in liters:

$$(1)(4)^{3/2} = p_2(1)^{3/2},$$

$$p_2 = 4^{3/2} = 8 \text{ atm} \quad \underline{\text{Ans.}}$$

(b) For an ideal gas pV = nRT. Combine this with the adiabatic equation to eliminate p:

$$\left(\frac{nRT_1}{V_1}\right)V_1^\gamma = \left(\frac{nRT_2}{V_2}\right)V_2^\gamma$$

$$T_1V_1^{\gamma-1} = T_2V_2^{\gamma-1},$$

$$(300)(4)^{\frac{1}{2}} = T_2(1)^{\frac{1}{2}},$$

$$T_2 = 600 \text{ K} \quad \underline{\text{Ans.}}$$

21-40

(a) Process 1-2: since the volume does not change, $W = \int p\,dV = 0$; $\Delta U = \frac{3}{2}nR\Delta T = \frac{3}{2}(1)(8.314)(600 - 300) = 3741.3$ J. From the first law, $Q = \Delta U + W = 3741.3 + 0 = 3741.3$ J.

Process 2-3: this is adiabatic, meaning that Q = 0. $\Delta U = \frac{3}{2}R\Delta T = \frac{3}{2}(8.314)(455 - 600) = -1808.295$ J. The first law gives $W = -\Delta U = +1808.295$ J.

Process 3-1: the pressure is constant and therefore $W = p\Delta V = nR\Delta T$ (by the ideal gas law pV = nRT); hence $W = (1)(8.314)(300 - 455) = -1288.67$ J. $\Delta U = \frac{3}{2}(8.314)(300 - 455) = -1933.005$ J. The first law then gives $Q = \Delta U + W = -3221.675$ J.

Whole cycle: adding the results above gives $\Delta U = 3741.3 - 1808.295 - 1933.005 = 0$ J (as expected); $Q = 3741.3 + 0 - 3221.675 =$

519.625 J; W = 0 + 1808.295 - 1288.67 = 519.625 J = Q, as expected from the first law applied to the whole cycle.

(b) The volume at point 1 is given from the ideal gas law pV = nRT:

$$(1.013 \times 10^5)(V_1) = (1)(8.314)(300),$$

$$V_1 = 0.02462 \text{ m}^3.$$

At point 2, $V_2 = V_1 = 0.02462 \text{ m}^3$. Also,

$$\frac{p_1V_1}{p_2V_2} = \frac{T_1}{T_2},$$

$$\frac{(1)}{p_2} = \frac{300}{600},$$

$$p_2 = 2 \text{ atm}.$$

At point 3, $p_3 = p_1 = 1$ atm, so that

$$\frac{p_1V_1}{p_3V_3} = \frac{T_1}{T_3},$$

$$\frac{0.02462}{V_3} = \frac{300}{455},$$

$$V_3 = 0.03734 \text{ m}^3.$$

As a check, since 2-3 is an adiabat, the values found above should satisfy

$$p_2V_2^\gamma = p_3V_3^\gamma,$$

$$(2)(0.02462)^{5/3} = (1)(0.03734)^{5/3},$$

$$(2)(0.00208) = 0.00417,$$

equal, that is, within roundoff error.

<u>21-41</u>

The constant pressure expansion is easily plotted as a straight

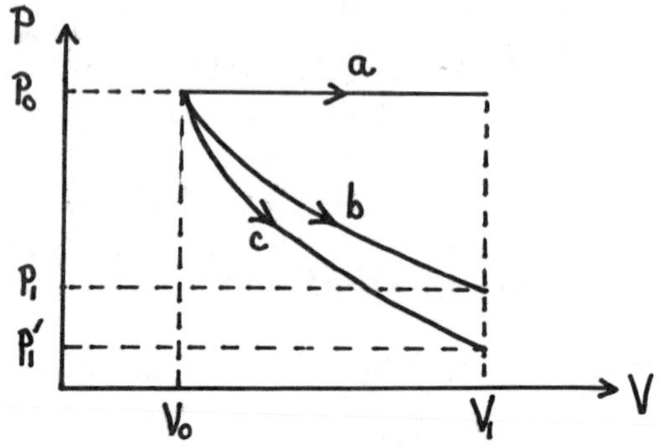

line parallel to the V-axis. For the isothermal process

$$p_0V_0 = nRT = p_1V_1,$$

so that

$$p_1 = p_0(V_0/V_1),$$

the graph of the isotherm being a hyperbola. For the adiabatic process $(Q = 0)$,

$$p_1' = p_0(V_0/V_1)^{\gamma},$$

since pV^{γ} = constant. But $\gamma > 1$ and $V_1 > V_0$ so that $p_1 > p_1'$. Hence the curves are positioned as shown. Since the area under each is the work W done, these works can be ordered immediately: $W_a > W_b > W_c > 0$. For an ideal gas ΔU is proportional to ΔT. Hence, $\Delta U_b = 0$. Since $pV = nRT$ the expansion at constant pressure is accompanied by an increase in temperature giving $\Delta U_a > 0$. For the adiabatic process the first law $Q = \Delta U + W$ shows that as $Q_c = 0$ and $W_c > 0$, then $\Delta U_c < 0$. Therefore, for the constant pressure process the change in internal energy and the work done must be greater than for the other processes and for the adiabatic process they must be least. The first law then implies that the same statement may be made also for Q.

21-47

Equation 21-3 is $p = \frac{1}{3}\rho\overline{v^2}$. From equipartition, $\frac{1}{2}m\overline{v^2} = \frac{3}{2}kT$. Hence,

$$p = \frac{1}{3}\rho\left(\frac{3kT}{m}\right) = \frac{\rho}{m}\cdot kT.$$

Since ρ = mass of gas/volume of gas, and m is the mass of each particle of the gas, then ρ/m = number of particles/volume of gas = N/V, so that

$$p = \frac{N}{V}kT.$$

Hence, if two gasses a and b have the same temperature and pressure, this last equation gives

$$\frac{N_a}{N_b} = \frac{V_a}{V_b}.$$

Finally, this implies that if $V_a = V_b$, then $N_a = N_b$.

21-48

Let V be the volume of the room and n the number of molecules per unit volume in the air inside. Since the pressure remains unchanged

$$p_0 = n_1kT_1 = n_2kT_2.$$

The internal energy $U = \frac{5}{2}NkT = \frac{5}{2}(nV)kT$; this includes rotation and translation. Hence,

$$U_2 = \frac{5}{2}(n_2V)kT_2 = \frac{5}{2}V\left(\frac{n_1T_1}{T_2}\right)kT_2 = \frac{5}{2}(n_1V)kT_1 = U_1,$$

as asserted. Of course, although the internal energy remains unchanged, the temperature of the remaining air has been increased, so the room feels warmer.

21-51

(a) The gas does not pass through equilibrium states during a free expansion, so that the adiabatic equation

$$pV^{\gamma} = \text{constant}$$

is not observed. However, the equation of state $pV = nRT$ holds at the initial and final configurations (but not during the expansion) and therefore

$$\frac{p_0V_0}{p_1V_1} = \frac{T_0}{T_1}.$$

But $Q = 0$, $W = 0$ so that $\Delta U = 0$ by the first law. Hence $\Delta T = 0$ also, or $T_1 = T_0$. The equation above now gives

$$\frac{p_0V_0}{p_1(3V_0)} = 1,$$

$$p_1 = p_0/3 \quad \underline{\text{Ans}}.$$

(b) The compression being slow, equilibium configurations are implied. Hence,

$$p_1V_1^\gamma = p_2V_2^\gamma,$$

$$(p_0/3)(3V_0)^\gamma = (3^{1/3}p_0)V_0^\gamma,$$

$$3^{\gamma-1} = 3^{1/3},$$

$$\gamma - 1 = \frac{1}{3},$$

$$\gamma = \frac{4}{3},$$

indicating a polyatomic gas.

(c) The final kinetic energy (translational) $K_f = K_2 = \frac{3}{2}kT_2$; similarly $K_1 = \frac{3}{2}kT_0$. Therefore $K_f/K_1 = T_2/T_0$. But

$$\frac{T_2}{T_0} = \frac{p_2V_2}{p_0V_0} = (3^{1/3}p_0)(V_0)/p_0V_0 = 3^{1/3}.$$

Hence,

$$K_f/K_1 = 3^{1/3} = 1.44 \quad \underline{\text{Ans}}.$$

21-57

The number n of jelly beans per unit volume is $n = 15/1000\ cm^3 = 0.015\ cm^{-3}$, and the diameter d of each bean is d = 1.0 cm. Hence, assuming a Maxwellian distribution of speeds (which, with only 15 beans, can be an approximation only), the mean free path l will be

$$l = (2^{\frac{1}{2}}\pi d^2 n)^{-1} = 15.0\ \text{cm}\quad \underline{\text{Ans.}}$$

21-59

(a) The mass of a hydrogen molecule is $m = (2)(1.66 \times 10^{-27}\ kg) = 3.32 \times 10^{-27}$ kg. Since T = 4000 K and $k = 1.38 \times 10^{-23}$ J/K, the root-mean-square speed v_{rms} will be

$$v_{rms} = (3kT/m)^{\frac{1}{2}} = 7.06 \times 10^3\ \text{m/s}\quad \underline{\text{Ans.}}$$

(b) When the hydrogen molecule and argon atoms touch, their distance of closest approach, center to center, will be

$$d = r_H + r_A = \tfrac{1}{2}(1 + 3) \times 10^{-8} = 2 \times 10^{-8}\ \text{cm}\quad \underline{\text{Ans,}}$$

assuming they can be pictured as rigid spheres.

(c) By definition of mean free path, the collision frequency f is

$$f = v/l = 2^{\frac{1}{2}}\pi n_A d^2 v_{rms} = 5.02 \times 10^{10}\ s^{-1}\quad \underline{\text{Ans,}}$$

using $v_{rms} = 7.06 \times 10^5$ cm/s from (a), $d = 2 \times 10^{-8}$ cm/s from (b) and $n_A = 4 \times 10^{19}\ cm^{-3}$.

21-60

In the laboratory frame, the velocities of N particles are distributed uniformly on the surface of a sphere in a v_x, v_y, v_z coordinate system. The center of the sphere is at the origin and the radius of the sphere is v. There are

$$\Delta N(\theta) = (\frac{N}{4\pi}\cdot 2\pi\sin\theta)\Delta\theta = \frac{N}{2}\sin\theta\Delta\theta$$

particles with velocities between θ and $\theta + \Delta\theta$, as shown. As seen from one of the particles (one with $v_y = -v$, $v_x = v_z = 0$) the distribution is exactly the same, except that the origin is moved to 0', a "distance" v from 0 along the negative v_y-axis. The relative speed v_r of a particle with a laboratory velocity $\vec{v}_A$ may be obtained by applying the law of cosines to triangle 0A0':

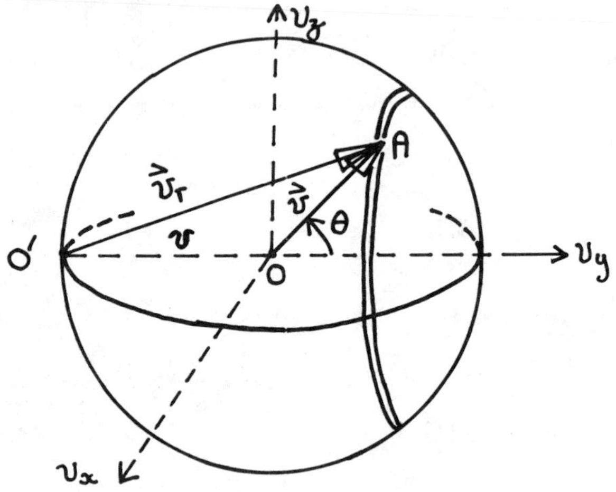

$$v_r^2 = v^2 + v^2 + 2v\cdot v\cos\theta = 2v^2(1 + \cos\theta),$$

$$v_r = 2^{\frac{1}{2}}v(1 + \cos\theta)^{\frac{1}{2}}.$$

The average value of v_r is then

$$\bar{v}_r = \frac{1}{N}\int_0^{\pi} v_r(\theta)\Delta N = \frac{1}{N}\int_0^{\pi} 2^{\frac{1}{2}}v(1 + \cos\theta)^{\frac{1}{2}}\cdot\frac{N}{2}\cdot\sin\theta\, d\theta,$$

$$\bar{v}_r = 2^{-\frac{1}{2}}v\, \frac{(1 + \cos\theta)^{3/2}}{3/2}\Big]_0^{\pi} = \frac{4}{3}v.$$

21-63

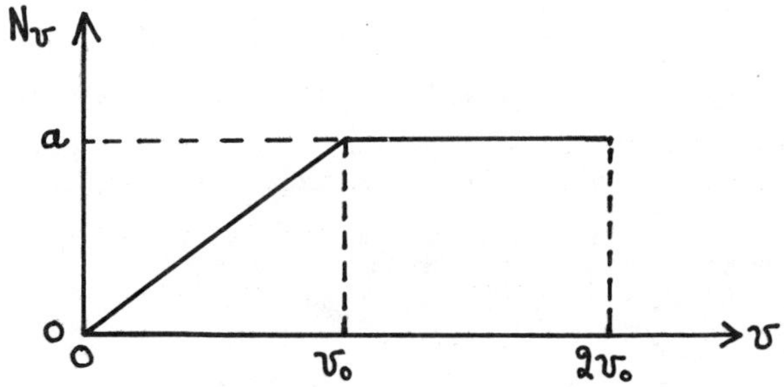

(a) The area under the curve equals the total number N of particles:

$$N = \tfrac{1}{2}av_0 + av_0 = \frac{3}{2}av_0,$$

$$a = 2N/3v_0 \quad \underline{\text{Ans}}.$$

(b) The area under the curve between $1.5v_0$ and $2v_0$ equals the number of particles with speeds between these limits; hence, this number is

$$\tfrac{1}{2}av_0 = \tfrac{1}{2}(2N/3v_0)v_0 = \frac{N}{3} \quad \underline{\text{Ans}}.$$

(c) The distribution function is $N_v = \frac{a}{v_0}v$ for $0 \le v \le v_0$ and $N_v = a$ for $v_0 \le v \le 2v_0$. Therefore, by definition,

$$\overline{v} = \frac{1}{N}\int vN_v dv = \frac{1}{N}\int_0^{v_0} v\left(\frac{a}{v_0}v\right) dv + \frac{1}{N}\int_{v_0}^{2v_0} vadv,$$

$$\overline{v} = \frac{1}{N}(av_0^2/3) + \frac{1}{N}(3av_0^2/2) = \frac{1}{N}(11av_0^2/6),$$

$$\overline{v} = \frac{11}{9}v_0 \quad \underline{\text{Ans}}.$$

21-64

(a) The speed distribution is not Maxwellian and therefore the basic definition of average speed must be used. This gives

$$\overline{v} = \frac{1}{10}(2 + 3 + 4 + 5 + 6 + 7 + 8 + 9 + 10 + 11),$$

$$\overline{v} = 6.5\ \text{km/s}\quad \underline{\text{Ans}}.$$

(b) The root-mean-square speed is the square-root of the average of the squared speeds; that is,

$$\overline{v^2} = \frac{1}{10}(2^2 + 3^2 + 4^2 + 5^2 + 6^2 + 7^2 + 8^2 + 9^2 + 10^2 + 11^2),$$

$$\overline{v^2} = 50.5;$$

hence,

$$v_{rms} = (\overline{v^2})^{\frac{1}{2}} = 7.1\ \text{km/s}\quad \underline{\text{Ans}}.$$

CHAPTER 22

22-9

The first law of thermodynamics applied to the engine and to the refrigerator gives

$$Q_1 = W + Q_2,$$
$$W + Q_4 = Q_3.$$

Eliminating W yields

$$Q_1 - Q_2 = Q_3 - Q_4.$$

This equation can be rewritten as follows:

$$\frac{Q_1 - Q_2}{Q_3 - Q_4} = 1,$$

$$\frac{Q_1 - Q_2}{Q_3 - Q_4}\cdot\frac{Q_3}{Q_1} = \frac{Q_3}{Q_1},$$

$$\left(\frac{Q_1 - Q_2}{Q_1}\right)/\left(\frac{Q_3 - Q_4}{Q_3}\right) = \frac{Q_3}{Q_1},$$

$$[1 - (Q_2/Q_1)]/[1 - (Q_4/Q_3)] = \frac{Q_3}{Q_1}.$$

But for Carnot engines and refrigerators, these heat ratios are equal to the absolute temperature ratios, so that

$$\frac{Q_3}{Q_1} = [1 - (T_2/T_1)]/[1 - (T_4/T_3)].$$

22-12

Along the path ab,

$$Q_{ab} = nC_v\Delta T = n(\tfrac{3}{2}R)\Delta T,$$

where $\Delta T = T_b - T_a$. But

$$pV = nRT,$$

$$(\Delta p)V_0 = nR\Delta T$$

since along ab the volume is constant and equals $V_0 = 1\ \text{m}^3$. Hence,

$$Q_{ab} = \tfrac{3}{2}(\Delta p)V_0 = \tfrac{3}{2}(p_b - p_a)V_0.$$

Now $p_a = p_c$, and

$$p_bV_b^\gamma = p_cV_c^\gamma,$$

$$(10)(V_0)^{5/3} = p_c(8V_0)^{5/3},$$

$$p_a = p_c = \tfrac{5}{16}\ \text{atm}.$$

Hence,

$$Q_{ab} = \tfrac{3}{2}(10 - \tfrac{5}{16})(1)\ \text{atm}\cdot\text{m}^3 = \tfrac{465}{32}(101300)\ \text{J},$$

$$Q_{ab} = 1.472 \times 10^6\ \text{J}.$$

Since the path bc is an adiabat, $Q_{bc} = 0$. Along ca,

$$Q_{ca} = nC_p\Delta T = n(\tfrac{5}{2}R)(T_a - T_c).$$

But for this constant pressure process,

$$p\Delta V = nR\Delta T,$$

so that

$$Q_{ca} = \tfrac{5}{2}(\Delta V)p = \tfrac{5}{2}(V_a - V_c)p_c,$$

$$Q_{ca} = \frac{5}{2}(1 - 8)V_0p_c = -\frac{35}{2}(1)(\frac{5}{16}) \text{ atm}\cdot\text{m}^3,$$

$$Q_{ca} = -5.540 \times 10^5 \text{ J}.$$

From the above results, the following conclusions can be drawn.

(a) The heat added = Q_{ab} = 1.472×10^6 J <u>Ans.</u>

(b) The heat removed = Q_{ca} = -5.540×10^5 J <u>Ans.</u>

(c) Over a complete cycle $\Delta U = 0$. The first law then predicts that $W = Q = Q_{ab} + Q_{bc} + Q_{ca} = 1.472 \times 10^6 + 0 - 5.540 \times 10^5 =$ 9.18×10^5 J <u>Ans.</u>

(d) Although the heats are not added or removed at a constant temperature, an efficiency can still be calculated from

$$e = 1 - (Q_C/Q_H) = 1 - (5.540 \times 10^5)/(14.72 \times 10^5),$$

$$e = 0.624 \quad \underline{\text{Ans.}}$$

<u>22-17</u>

(a)

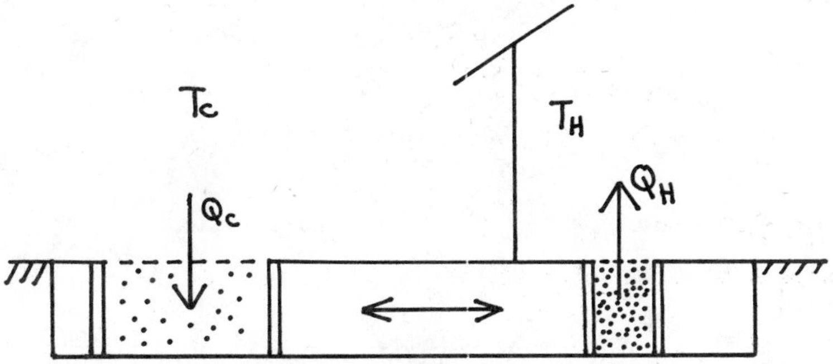

(b) First let the gas in the pump expand so that the gas temperature is decreased and lower than the outside temperature T_C. Therefore a quantity of heat Q_C is extracted from the outside atmosphere at T_C. Then move the piston into the house and compress the gas until its temperature is higher than inside the house, which is at T_H. A larger quantity of heat Q_H is delivered to the inside of the house. Thus, in principle there is no significant difference between the heat pump and the refrigerator. In practical use, in a refrigerator the quantity of heat Q_C is extracted at T_C through the vaporization of a liquid instead of the expansion of a gas as in the heat pump.

(c) By the first law, $Q_H = Q_C + W$, the change in internal energy being zero over one cycle.

(d) The heat pump can be reversed by reversing the expansion and compression stages.

(e) The advantages of the pump are that Q_H, the heat delivered, is greater than W, the energy paid for (to the power company), and that the pump can be operated to heat and cool.

<u>22-19</u>

(a) Apply the ideal gas law to points 1 and 2. Since $p_2 = 3p_1$ and $V_1 = V_2$, this gives

$$p_1V_1/p_2V_2 = T_1/T_2,$$

$$p_1V_1/(3p_1)V_1 = T_1/T_2,$$

$$T_2 = 3T_1 \quad \underline{\text{Ans}}.$$

Points 2 and 3 are connected by the adiabatic relation

$$p_2V_2^\gamma = p_3V_3^\gamma,$$

$$(3p_1)(V_1)^\gamma = p_3(4V_1)^\gamma,$$

$$p_3 = (3/4^\gamma)p_1 \quad \underline{\text{Ans}}.$$

Also,

$$p_1V_1/p_3V_3 = T_1/T_3,$$

$$p_1V_1/(3/4^\gamma)p_1(4V_1) = T_1/T_3,$$

$$T_3 = (3/4^{\gamma-1})T_1 \quad \underline{\text{Ans}}.$$

For point 4,

$$p_1 V_1^\gamma = p_4 V_4^\gamma,$$

$$p_1 V_1^\gamma = p_4 (4V_1)^\gamma,$$

$$p_4 = p_1/4^\gamma \quad \underline{\text{Ans}}.$$

Finally,

$$p_1 V_1/p_4 V_4 = T_1/T_4,$$

$$p_1 V_1/(4^{-\gamma} p_1)(4V_1) = T_1/T_4,$$

$$T_4 = T_1/4^{\gamma-1} \quad \underline{\text{Ans}}.$$

(b) From (a), $T_1 < T_2$ and $T_4 < T_1$. Hence, heat Q_{12} enters along path 1-2 and heat $Q_{34} > 0$ leaves along path 3-4. No heat enters or leaves along adiabats. Although the paths along which heat is transferred are not isotherms, the efficiency e can still be defined as

$$e = 1 - Q_{34}/Q_{12}.$$

The paths 1-2 and 3-4 are constant volume processes; hence,

$$Q_{12} = nC_v(T_2 - T_1).$$

But $C_p - C_v = R$ and $C_p/C_v = \gamma$ so that $C_v = R(\gamma - 1)$, giving

$$Q_{12} = nR(\gamma - 1)\cdot 2T_1.$$

Similarly,

$$Q_{34} = nR(\gamma - 1)(2T_1/4^{\gamma-1}),$$

since $Q_{34} = nC_v(T_3 - T_4)$ and using the results from (a) for the temperatures in terms of T_1. The efficiency now becomes

$$e = 1 - \frac{nR(\gamma - 1)(2T_1/4^{\gamma-1})}{nR(\gamma - 1)\cdot 2T_1} = 1 - 4^{1-\gamma} \quad \underline{\text{Ans}}.$$

<u>22-24</u>

(a) The work $W_{ac} = W_{ab} + W_{bc} = W_{ab}$, $W_{bc} = 0$ since bc is a constant volume process. The path ab takes place at constant pressure and therefore

$$W_{ac} = W_{ab} = p\Delta V = p_0(4V_0 - V_0),$$
$$W_{ac} = 3p_0V_0 \quad \underline{\text{Ans}}.$$

(b) The change in internal energy $\Delta U_{bc} = \frac{3}{2}R(T_c - T_b)$ as $n = 1$. For the temperatures the ideal gas law gives

$$p_bV_b/p_aV_a = T_b/T_a,$$
$$p_0(4V_0)/p_0V_0 = T_b/T_0,$$
$$T_b = 4T_0;$$
$$p_cV_c/p_aV_a = T_c/T_a,$$
$$(2p_0)(4V_0)/p_0V_0 = T_c/T_0,$$
$$T_c = 8T_0.$$

Hence,

$$\Delta U_{bc} = \frac{3}{2}R(8T_0 - 4T_0) = 6RT_0.$$

But $p_0V_0 = RT_0$, so that

$$\Delta U_{bc} = 6p_0V_0 \quad \underline{\text{Ans}}.$$

The change in entropy ΔS_{bc} is given by

$$\Delta S_{bc} = \int_b^c \frac{dQ}{T} = \int_b^c \frac{nC_v dT}{T} = \frac{3}{2}R\int_{4T_0}^{8T_0} \frac{dT}{T} = \frac{3}{2}R\cdot\ln 2 = \frac{3}{2}(p_0V_0/T_0)\ln 2 \quad \underline{\text{Ans}}.$$

(c) The internal energy and entropy being state functions, their change over the complete cycle (presumed reversible as it is shown on a p-V diagram) each are zero.

22-26

If the heating of the water and the cooling of the aluminum were carried out reversibly, the entropy change of each could be found as follows:

$$\Delta S = \int \frac{dQ}{T} = \int_{T_i}^{T_f} mc_p dT/T = mc_p \int_{T_i}^{T_f} dT/T = mc_p \cdot \ln(T_f/T_i).$$

To find the equilibrium temperature T, set the heat lost by the aluminum equal to the heat gained by the water. Each of these can be written $mc_p\Delta T$, so that

$$(200)(0.215)(100 - T) = (50)(1)(T - 20),$$

$$T = 57^{\circ}C = 330\ K.$$

The total entropy change is

$$\Delta S = \Delta S_{water} + \Delta S_{Al},$$

$$\Delta S = (50)(1)\ln\frac{330}{293} + (200)(0.215)\ln\frac{330}{373} = 5.946 - 5.267,$$

$$\Delta S = 0.68\ cal/K \quad \underline{Ans.}$$

22-28

The equilibrium temperature can safely be taken as the temperature of the lake, $15^{\circ}C = 288$ K. By Problem 22-26,

$$\Delta S_{ice} = (10)(0.5)\ln\frac{273}{263} = 0.1866\ cal/K,$$

$$\Delta S_{water} = (10)(1)\ln\frac{288}{273} = 0.5349\ cal/K,$$

the 'water' being the liquid formed by the melted ice. During the melting,

$$\Delta S_{melting} = \frac{Q}{T} = \frac{(10)(80)}{273} = 2.9304\ cal/K$$

the melting taking place at the constant temperature of 273 K. Turning now to the lake, although its temperature does not change it loses heat ΔQ required to warm the ice to $0^\circ C$, melt the ice and then bring the water formed to the lake temperature. This heat is

$$\Delta Q = (10)(0.5)(10) + (10)(80) + (10)(1)(15) = 1000 \text{ cal.}$$

Hence,

$$\Delta S_{lake} = \frac{-\Delta Q}{T} = -\frac{1000}{288} = -3.4722 \text{ cal/K}.$$

The total entropy change is the sum of the four listed above, i.e.,

$$\Delta S = 0.18 \text{ cal/K} \quad \underline{\text{Ans}}.$$

<u>22-30</u>

(a) If T is constant, then since $pV = nRT$, the work W will be

$$W = \int p dV = \int \frac{nRT}{V} dV = nRT \int \frac{dV}{V} = nRT \cdot \ln(V_2/V_1),$$

$$W = (4)(8.314)(400)\ln 2 = 9221 \text{ J} \quad \underline{\text{Ans}}.$$

(b) With T constant,

$$\Delta S = \frac{Q}{T} = \frac{W}{T} = \frac{9221}{400} = 23 \text{ J/K} \quad \underline{\text{Ans}},$$

since at constant T, $\Delta U = 0$ and $Q = W$ by the first law.

(c) For an adiabatic process $Q = 0$ so that $\Delta S = 0$ also.

<u>22-31</u>

(a) By definition of heat reservoirs, their temperatures do not change although they may gain or lose heat. The temperatures are $T_H = 400$ K and $T_C = 300$ K. The heat leaves the hot reservoir; thus,

$$\Delta S_H = \frac{Q}{T} = \frac{-1200}{400} = -3 \text{ cal/K},$$

$$\Delta S_C = \frac{Q}{T} = \frac{+1200}{300} = +4 \text{ cal/K}.$$

Evidently the entropy change of the system is 4 - 3 = 1 cal/K.

(b) If the rod simply transmits the heat, absorbing none itself, then its entropy change is zero, as assumed above.

22-34

(a) In equilibrium, $\frac{1}{2}(1773 + 227) = 1000$ g of ice are present, a gain of 1000 - 227 = 773 g over the amount present initially. This indicates that (773 g)(80 cal/g) = 61840 cal of heat is liberated to the environment. The entropy change of the ice+water is Q/T = (-61840 cal)/(273 K) = -226.52 cal/K.

(b) The entropy change can be calculated by substituting the process used in (a), i.e., gradual infinitesimal increments in temperature, but in reverse from (a). In this case heat flows into the system, the entropy change being +226.52 cal/K.

(c) In (a), the entropy change of the environment will be +226.52 cal/K for, to keep the process reversible, the temperature of the environment can only differ from the temperature of the system by a very small amount. In (b), the entropy change of the environment will be greater than -226.52 cal/K, so that the total change of system+environment entropy will exceed zero, as required by the second law.

22-36

Clearly the equilibrium temperature $T = \frac{1}{2}(T_H + T_C)$ as the objects are indentical. By Problem 22-29, the entropy changes are

$$\Delta S_H = mc\cdot\ln(T/T_H),$$

$$\Delta S_C = mc\cdot\ln(T/T_C),$$

so that $\Delta S = \Delta S_H + \Delta S_C$ becomes

$$\Delta S = mc[\ln(T/T_H) + \ln(T/T_C)] = mc\cdot\ln[(T/T_H)(T/T_C],$$

$$\Delta S = mc\cdot\ln(T^2/T_HT_C).$$

But the heat Q transferred is $Q = mc\Delta T = mc(T - T_C)$. Using this to eliminate mc yields

$$\Delta S = \frac{Q}{T - T_C}\cdot\ln(T^2/T_HT_C).$$

Finally, substitute $T = \frac{1}{2}(T_H + T_C)$ to obtain

$$\Delta S = \frac{2Q}{T_H - T_C}\cdot \ln\left[\frac{(T_H + T_C)^2}{4T_HT_C}\right].$$

CHAPTER 23

23-10

The third charge must lie between the two charges +q, +4q (i.e., on the line joining them), for if it did not it would feel a force of repulsion if it was positive and attraction if negative. In fact, the charge must be negative: each of the given charges +q, +4q, could not be in equilibrium otherwise, since each would feel repulsive forces from the other two charges. So let the third charge be -Q, Q > 0, located a distance x from the +q charge; then, for this charge to be in equilibrium,

$$\frac{1}{4\pi\epsilon_0}\frac{qQ}{x^2} = \frac{1}{4\pi\epsilon_0}\frac{Q(4q)}{(\ell - x)^2},$$

$$x = \frac{\ell}{3} \quad \underline{\text{Ans.}}$$

To evaluate Q, require that the +q charge, say, also be in equilibrium:

$$\frac{1}{4\pi\epsilon_0}\frac{qQ}{x^2} = \frac{1}{4\pi\epsilon_0}\frac{q(4q)}{\ell^2},$$

$$Q = 4q\left(\frac{x}{\ell}\right)^2 = \frac{4}{9}q \quad \underline{\text{Ans.}}$$

The same result is obtained when equilibrium of the +4q charge is examined. The equilibrium is unstable, for a displacement of -Q along the line joining +q, +4q results in a net force away from the position of equilibrium.

23-11

(a) Draw a free-body diagram of one of the balls; the electrical force F is of repulsion. Assuming equilibrium,

$$T\cos\theta - mg = 0,$$

$$T\sin\theta = F.$$

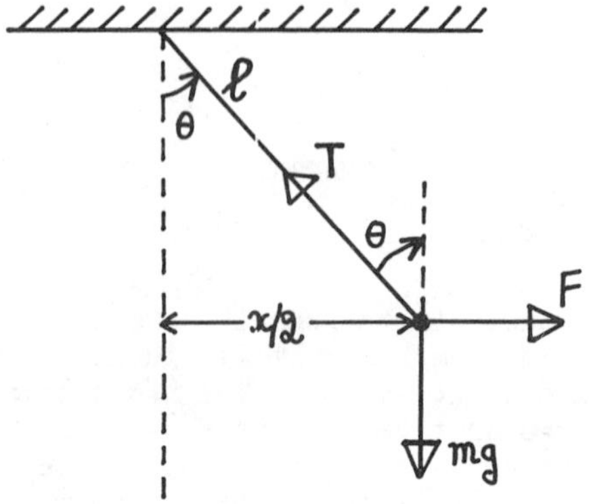

Therefore

$$\tan\theta = \frac{F}{mg}.$$

Now set $\tan\theta \approx \sin\theta = \frac{x/2}{\ell}$ and $F = (1/4\pi\epsilon_0)q^2/x^2$ to obtain

$$\frac{x/2}{\ell} = \frac{1}{4\pi\epsilon_0}\frac{q^2}{x^2}/mg,$$

$$x = \left[\frac{1}{2\pi\epsilon_0}\frac{q^2\ell}{mg}\right]^{1/3}.$$

(b) Setting $x = 5$ cm $= 0.05$ m, $\ell = 120$ cm $= 1.2$ m, $m = 10$ g $= 0.01$ kg, and $g = 9.8$ m/s^2, $1/4\pi\epsilon_0 = 9 \times 10^9$ N$\cdot$m^2/C^2 gives $q = 2.4 \times 10^{-8}$ C. Since repulsive forces are involved, the charges on the balls must either be both positive or both negative.

23-13

Assume that the spheres are small enough so that the charges are always uniformly distributed over their surfaces, allowing them to be treated as point charges (radii $\ll 0.5$ m). Let the initial charges be Q_1, $-Q_2$ with $Q_1, Q_2 > 0$. Then, before being connected by the wire,

$$F = \frac{1}{4\pi\epsilon_0}\frac{Q_1Q_2}{r^2},$$

where r = 0.5 m. Thus,

$$Q_1Q_2 = (0.108)(0.5)^2/(9 \times 10^9) = 3 \times 10^{-12}.$$

When the spheres are connected with the wire they and the wire form a single conductor carrying a charge $(Q_1 - Q_2)$. When equilibrium is reached, the wire will carry virtually no charge since its surface area is very small compared with the areas of the spheres; these spheres, being identical, will each carry half the net charge: $\frac{1}{2}(Q_1 - Q_2)$. Then,

$$\frac{1}{4\pi\epsilon_0} \frac{[\frac{1}{2}(Q_1 - Q_2)]^2}{r^2} = 0.036,$$

$$Q_1 - Q_2 = \pm 2 \times 10^{-6} \text{ C}.$$

But it has been determined above that $Q_1 = 3 \times 10^{-12}/Q_2$. Combining this with the previous equation gives

$$Q_2^2 \pm (2 \times 10^{-6})Q_2 - 3 \times 10^{-12} = 0,$$

$$Q_2 = -3 \times 10^{-6} \text{ C}, \; Q_1 = 1 \times 10^{-6} \text{ C, upper sign,}$$

$$Q_2 = 3 \times 10^{-6} \text{ C}, \; Q_1 = -1 \times 10^{-6} \text{ C, lower sign.}$$

Hence the initial charges on the spheres are 3×10^{-6} C, 1×10^{-6}C but of opposite sign.

23-15

(a) It is desired that

$$GM_eM_m/r^2 = (1/4\pi\epsilon_0)q^2/r^2,$$

$$q = (G4\pi\epsilon_0 M_e M_m)^{\frac{1}{2}}.$$

Since the mass of the moon $M_m \approx \frac{1}{81} M_e$, this gives

$$q = \left(\frac{6.67 \times 10^{-11}}{9 \times 10^{9}}\right)^{\frac{1}{2}}(5.98 \times 10^{24})/9,$$

$$q = 5.72 \times 10^{13} \text{ C} \quad \underline{\text{Ans.}}$$

(b) The distance between earth and moon does not enter numerically because the gravitational and electrostatic forces are both inverse square in the distance so that it drops out.

(c) Each hydrogen atom contributes 1.6×10^{-19} C of positive charge. Hence $(5.72 \times 10^{13})/(1.6 \times 10^{-19}) = 3.575 \times 10^{32}$ atoms are needed. In mass this is $(3.575 \times 10^{32})(1.66 \times 10^{-27}$ kg$) = 5.935 \times 10^{5}$ kg. On earth, this is equivalent to 13.09×10^{5} lb = 650 tons.

23-23

(a)

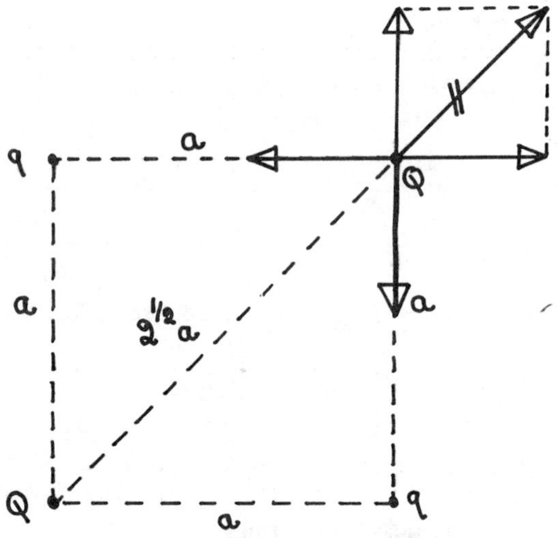

The force between the two Q charges must be of repulsion since like sign charges repel. To put Q in equilibrium, q must be opposite in sign to Q to yield attractive forces to balance the repulsion. Also,

$$\frac{1}{4\pi\epsilon_0}\frac{Q^2}{(2^{\frac{1}{2}}a)^2}\cos 45^\circ + \frac{1}{4\pi\epsilon_0}\frac{qQ}{a^2} = 0,$$

in order that the vector sum of the three forces is zero. This gives

$$Q = -2 \cdot 2^{\frac{1}{2}} q \quad \underline{\text{Ans.}}$$

(b) The magnitude of the attractive forces on q are the same as the attractive forces on Q. However, the magnitude of the repulsive force on q is proportional to q^2, whereas the repulsive force on Q is proportional to Q^2. Hence, if q is chosen to put Q in equilibrium, then q cannot be in equilibrium since the same attractive force cannot balance two different repulsive forces.

23-24

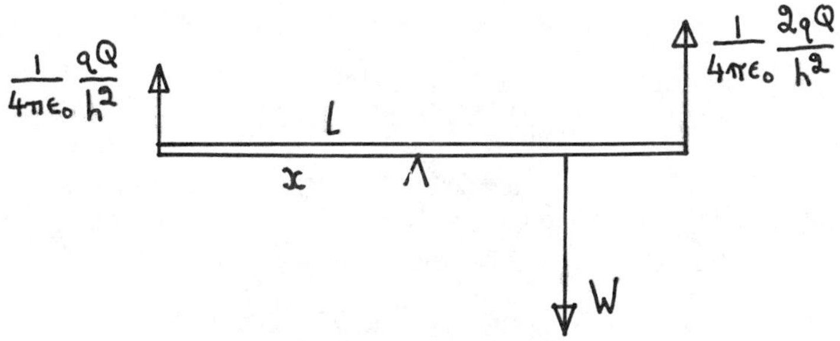

The forces on the rod are the weight W and the two electrostatic forces

$$\frac{1}{4\pi\epsilon_0}\frac{qQ}{h^2}, \quad \frac{1}{4\pi\epsilon_0}\frac{2qQ}{h^2}$$

on the left and right ends, respectively. It is assumed that no force acts at the bearing. For equilibrium the forces must sum to zero, so that

$$W = 3\frac{1}{4\pi\epsilon_0}\frac{qQ}{h^2}.$$

Taking moments about the left end:

$$W\cdot(x) = \frac{1}{4\pi\epsilon_0}\frac{2qQ}{h^2}\ell .$$

(a) Dividing these two equations gives

$$\frac{1}{x} = \frac{3}{2\ell},$$

$$x = \frac{2\ell}{3} \quad \underline{\text{Ans.}}$$

(b) From the first equation,

$$h = \left(\frac{3}{4\pi\epsilon_0}\frac{qQ}{W}\right)^{\frac{1}{2}} \quad \underline{\text{Ans.}}$$

23-25

For the mass m of the electron substitute the reduced mass μ,

$$\mu = \frac{mM}{m + M},$$

where M is the mass of the proton. With this done the proton can be considered at rest throughout the motion. By the work-energy theorem,

$$W = \Delta K = \tfrac{1}{2}\mu(2v_0)^2 - \tfrac{1}{2}\mu v_0^2 = \tfrac{3}{2}\mu v_0^2.$$

By analogy with a corresponding gravitational problem, for which $W = GmM/r$, set

$$W = \frac{1}{4\pi\epsilon_0}\frac{e^2}{r},$$

r the desired distance. Hence,

$$\tfrac{3}{2}\mu v_0^2 = \frac{1}{4\pi\epsilon_0}\frac{e^2}{r}.$$

Since $M \approx 2000m$, $\mu \approx m$ and therefore

$$r = \frac{2}{3}\frac{1}{4\pi\epsilon_0}(e^2/mv_0^2).$$

Substitution of numerical values gives $r = 1.65 \times 10^{-9}$ m <u>Ans</u>.

CHAPTER 24

24-3

(a) Since $\vec{F} = q\vec{E}$, the electric field E is

$$E = \frac{F}{q} = \frac{3 \times 10^{-6}}{2 \times 10^{-9}} = 1500 \text{ N/C} \quad \underline{\text{Ans.}}$$

(b) In part (a) the charge of the particle was negative (the sign was disregarded since only the strength of the field was required); hence, $\vec{F}$ is directed opposite to $\vec{E}$ and therefore $\vec{E}$ points upward. If a proton is placed in this field, the force it feels will be directed parallel to $\vec{E}$ (i.e., upward) since the charge (e) of the proton is positive. The magnitude of the force will be

$$F_e = eE = (1.6 \times 10^{-19} \text{ C})(1500 \text{ N/C}) = 2.4 \times 10^{-16} \text{ N} \quad \underline{\text{Ans.}}$$

(c) The gravitational force on the proton is

$$F_g = mg = (1.67 \times 10^{-27} \text{ kg})(9.8 \text{ m/s}^2) = 1.6366 \times 10^{-26} \text{ N} \quad \underline{\text{Ans.}}$$

(d) The ratio of these forces is

$$F_e/F_g = \frac{2.4 \times 10^{-16}}{1.6366 \times 10^{-26}} = 1.47 \times 10^{10} \quad \underline{\text{Ans.}}$$

24-6

Very close to the surface of the disc, at distances much less than the radius R, the lines must resemble those from an infinite plane of charge: that is, they leave the disc at right angles to its surface. Since the net charge on the disc is not zero, very far away the lines will appear to emanate from a point charge, equal to the total charge on the disc, located at the 'center of gravity' of the charge distribution, i.e., from the center of the disc. In the sketch the disc is assumed to be charged positively and only lines in the upper half plane are shown.

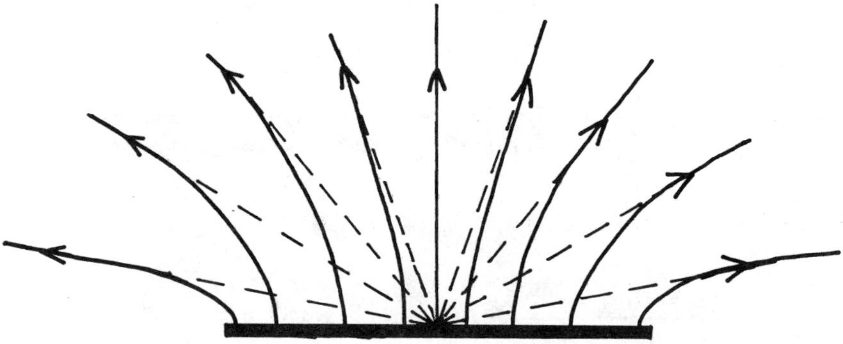

<u>24-18</u>

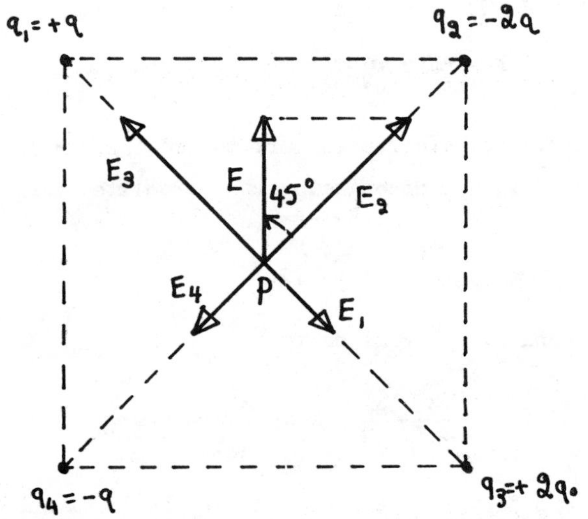

Since the point P is equidistant ($2^{\frac{1}{2}}a/2$) from each charge, the strengths of the electric fields due to the individual charges will be in the same proportion as the charges themselves (signs disregarded). The directions will be toward the negative charges and away from the positive. Thus, if

$$E_1 = \frac{1}{4\pi\epsilon_0}\frac{q}{(2^{\frac{1}{2}}a/2)^2},$$

then

$$E_2 = 2E_1,\ E_3 = 2E_1,\ E_4 = E_1.$$

From the diagram and the rules of vector addition, it is clear that if $\vec{E} = \vec{E}_1 + \vec{E}_2 + \vec{E}_3 + \vec{E}_4$, $\vec{E}$ will point vertically upward and have a magnitude

$$E = 2E_1 \cos 45°,$$

$$E = 2\frac{(9 \times 10^9)(10^{-8})}{[2^{\frac{1}{2}}(0.05)/2]^2}(2^{\frac{1}{2}}/2) = 1.02 \times 10^5 \text{ N/C} \quad \underline{\text{Ans}}.$$

24-24

The equilibrium position of the electron is at the center of the ring. The force on the electron a distance x away from this point along the ring's axis is

$$F = eE = \frac{1}{4\pi\epsilon_0}\frac{eqx}{(a^2 + x^2)^{3/2}},$$

and is attractive, being directed back toward the ring's center. If $x \ll a$, the x^2 in the denominator can be deleted to give

$$F = \frac{1}{4\pi\epsilon_0}\frac{eq}{a^3}x.$$

By Newton's second law, $\vec{F} = m\vec{a}$ so that

$$\frac{1}{4\pi\epsilon_0}\frac{eq}{a^3}x = -m\frac{d^2x}{dt^2}.$$

This has the same form as the equation for simple harmonic motion,

$$\frac{d^2x}{dt^2} + \omega^2 x = 0.$$

Comparing the last two equations gives

$$\omega = \left[\frac{eq}{4\pi\epsilon_0 ma^3}\right]^{\frac{1}{2}}$$

24-25

(a)

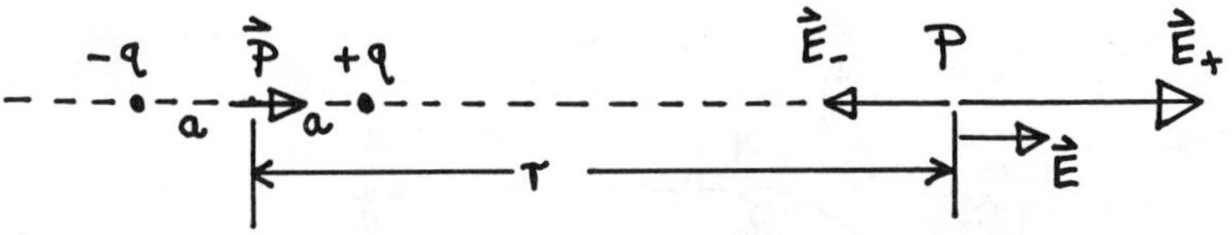

Let the point P be a distance r from the center of the dipole, along its axis, so that P is a distance r-a from the positive charge and r+a from the negative. The fields from these charges are oppositely directed, so that if $\vec{E} = \vec{E}_+ + \vec{E}_-$, then

$$E = E_+ - E_-,$$

$$E = \frac{1}{4\pi\epsilon_0}\frac{q}{(r-a)^2} - \frac{1}{4\pi\epsilon_0}\frac{q}{(r+a)^2},$$

$$E = \frac{q}{4\pi\epsilon_0}\left[\frac{1}{(r-a)^2} - \frac{1}{(r+a)^2}\right] = \frac{q}{4\pi\epsilon_0}\left[\frac{1}{r^2(1-a/r)^2} - \frac{1}{r^2(1+a/r)^2}\right],$$

$$E = \frac{q}{4\pi\epsilon_0 r^2}\left[\frac{1}{(1-a/r)^2} - \frac{1}{(1+a/r)^2}\right],$$

$$E \approx \frac{q}{4\pi\epsilon_0 r^2}[(1 + 2a/r) - (1 - 2a/r)] = \frac{q}{4\pi\epsilon_0 r^2}(4a/r),$$

$$E = \frac{2aq}{2\pi\epsilon_0 r^3} = \frac{1}{2\pi\epsilon_0}\frac{p}{r^3},$$

if $a/r \ll 1$ and $p = 2aq$.

(b) Since $\vec{p}$ points from -q to +q, then $\vec{E}$ is parallel to $\vec{p}$.

24-27

Due to the element of charge λdz located at Q as shown, the contribution to the electric field at P is

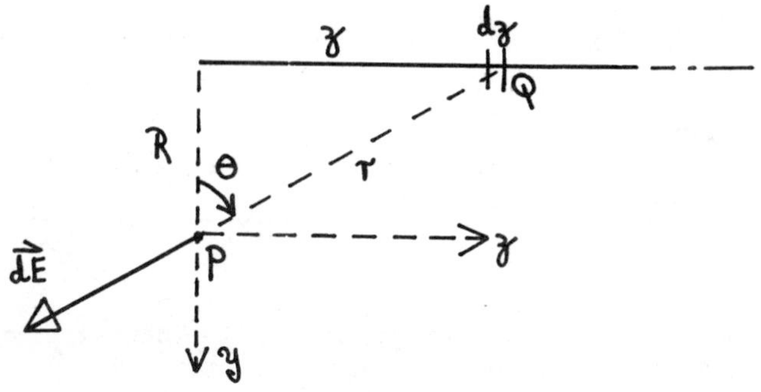

$$dE = \frac{1}{4\pi\epsilon_0}\frac{\lambda dz}{r^2},$$

along QP extended. Taking components, $dE_y = (dE)\cos\theta$, $dE_z = -(dE)\sin\theta$; also $\cos\theta = R/r$, $\sin\theta = z/r$, $r^2 = R^2 + z^2$. Therefore,

$$E_y = \int_0^\infty \frac{\lambda}{4\pi\epsilon_0}\frac{1}{R^2 + z^2}\frac{R}{(R^2 + z^2)^{\frac{1}{2}}}\,dz = \frac{\lambda}{4\pi\epsilon_0 R};$$

$$E_z = -\int_0^\infty \frac{\lambda}{4\pi\epsilon_0}\frac{1}{(R^2 + z^2)}\frac{z}{(R^2 + z^2)^{\frac{1}{2}}}\,dz = -\frac{\lambda}{4\pi\epsilon_0 R}.$$

In magnitude, then, $E_y = E_z$ and therefore $\theta = 45°$, where θ is the angle made by $\vec{E} = \vec{E}_y + \vec{E}_z$ with the rod.

24-29

The distances of P to the positive charges are r-a and r+a, and the distance to the two negative charges is r. Taking account of the directions of the three contributions to the field at P, and carrying out the approximation, for large distances, used in Problem 24-25 to the third term in the series yields

$$E = \frac{1}{4\pi\epsilon_0}\left[\frac{q}{(r - a)^2} + \frac{q}{(r + a)^2} - \frac{2q}{r^2}\right],$$

$$E = \frac{q}{4\pi\epsilon_0 r^2}\left[\frac{1}{(1 - a/r)^2} + \frac{1}{(1 + a/r)^2} - 2\right],$$

$$E \approx \frac{q}{4\pi\epsilon_0 r^2}[(1 + 2a/r + 3a^2/r^2) + (1 - 2a/r + 3a^2/r^2) - 2],$$

$$E = \frac{q}{4\pi\epsilon_0 r^2}(6a^2/r^2) = \frac{3Q}{4\pi\epsilon_0 r^4},$$

if $Q = 2qa^2$.

24-33

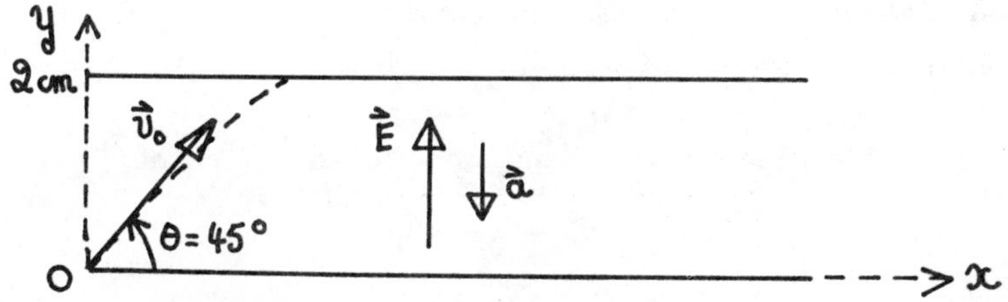

Since the charge on the electron is negative the force it feels will be directed downwards, opposite to the electric field. The acceleration is

$$a = \frac{F}{m} = \left(\frac{e}{m}\right)E,$$

$$a = (1.76 \times 10^{11}\ \text{C/kg})(2000\ \text{N/C}) = 3.52 \times 10^{14}\ \text{m/s}^2.$$

Clearly gravity can be ignored. To see if the electron strikes the top plate solve for t, the time required from projection for the electron to reach a height y = 2 cm = 0.02 m. Since the acceleration is constant,

$$y = \tfrac{1}{2}at^2 + v_{0y}t.$$

But $v_{0y} = v_0 \sin 45° = 3 \cdot 2^{\frac{1}{2}} \times 10^6$ m/s; set y = 0.02 m to obtain

$$0.02 = \tfrac{1}{2}(-3.52 \times 10^{14})t^2 + (3 \cdot 2^{\frac{1}{2}} \times 10^6)t,$$

$$t = 6.428 \times 10^{-9}\text{s},$$

choosing the smaller solution of the quadratic equation (the other applies to the descending part of the trajectory if the motion is not interrupted by the upper plate). The horizontal distance x traveled in this time is

$$x = v_{0x}t = (3 \cdot 2^{\frac{1}{2}} \times 10^6 \text{ m/s})(6.428 \times 10^{-9} \text{ s}) = 0.0273 \text{ m},$$

$$x = 2.73 \text{ cm}.$$

Since 2.7 cm < 10 cm, (a) the electron will strike the upper plate (b) at a distance 2.73 cm from the left edge.

24-34

(a) Equating the electrostatic and gravitational forces,

$$qE = mg,$$

$$qE = [\tfrac{4}{3}\pi\, r^3 \rho]g,$$

$$q(1.92 \times 10^5) = [\tfrac{4}{3}\pi(1.64 \times 10^{-6})^3(851)](9.8),$$

$$q = 8.0 \times 10^{-19} \text{ C}.$$

Since $e = 1.6 \times 10^{-19}$ C,

$$q = \frac{8.0 \times 10^{-19}}{1.6 \times 10^{-19}}e = 5e \quad \underline{\text{Ans.}}$$

(b) Individual electrons cannot be seen; also, the required electric field would be very small.

24-38

The period of a simple pendulum (small conducting sphere with a radius much less than the length of the string) is

$$T = 2\pi(l/g_e)^{\frac{1}{2}},$$

where g_e is the acceleration due to the "effective" gravitational field. This will be

$$g_e = g \pm a,$$

g being the acceleration due to gravity proper and a the acceleration imparted by the electric force. This latter is qE/m since $F_e = qE = ma$. Therefore

$$g_e = g \pm qE/m.$$

If a is directed down (like g) then $g_e = g + qE/m$; this will be the case with the lower plate charged negatively (the electric field is then down and the sphere, being charged positively, experiences an electrical force in the same direction as E). With the lower plate charged positively, a is directed upward, which is opposite to g. Hence gravity is reduced effectively and $g_e = g - qE/m$. If E is large enough ($qE/m > g$), "gravity is reversed" and the pendulum oscillates about the topmost point of its circular trajectory. Of course, with $qE/m = g$, gravity is "neutralised" and the period is infinite (no tendency to oscillate). To summarize:

(a) lower plate positive:

$$T = 2\pi\left(\frac{l}{g - qE/m}\right)^{\frac{1}{2}}, \quad qE/m < g;$$

$$T = \infty, \quad qE/m = g;$$

$$T = 2\pi\left(\frac{l}{qE/m - g}\right)^{\frac{1}{2}}, \quad g < qE/m.$$

(b) Lower plate negative:

$$T = 2\pi\left(\frac{l}{g + qE/m}\right)^{\frac{1}{2}}.$$

<u>24-40</u>

The torque on a dipole is $\tau = pE\sin\theta$ where θ is the displacement from equilibrium. But $\tau = I\alpha = -pE\sin\theta$, the minus sign added now since the torque tends to return the dipole to its equilibrium, $\theta = 0$, orientation. For small displacements, $\sin\theta \approx \theta$ and hence $d^2\theta/dt^2 + (pE/I)\theta = 0$, so that $\nu = \omega/2\pi = (pE/I)^{\frac{1}{2}}/2\pi$ <u>Ans.</u>

25-9

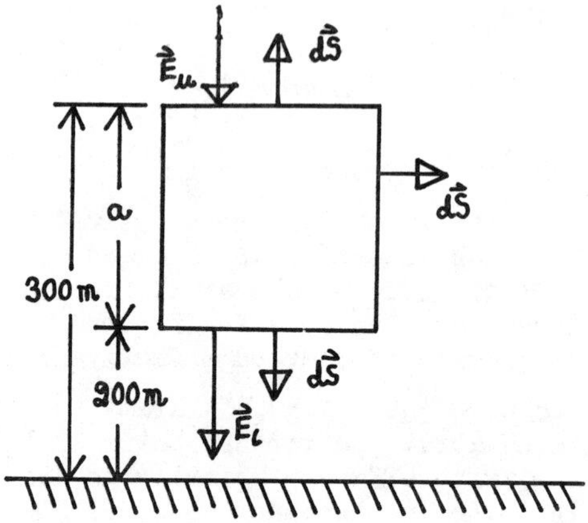

Along the vertical sides of the cube $\vec{E}$ is perpendicular to $d\vec{S}$ so that $\vec{E}\cdot d\vec{S} = (E)(dS)\cos 90^\circ = 0$. Along the upper face of the cube, the electric field and $d\vec{S}$ point in opposite directions. Since the field has the same value over the horizontal upper face,

$$\Phi_u = E_u S\cos 180^\circ = -E_u S = -E_u a^2.$$

Along the bottom face the electric field and $d\vec{S}$ point in the same direction, so that

$$\Phi_1 = E_1 S\cos 0^\circ = E_1 a^2.$$

Hence, by Gauss's law,

$$\Phi = E_1 a^2 - E_u a^2 = (E_1 - E_u)a^2 = q/\epsilon_0,$$

$$q = \epsilon_0(E_1 - E_u)a^2 = (8.85 \times 10^{-12})(100 - 60)(100)^2,$$

$$q = 3.54 \times 10^{-6}\ \text{C}$$ Ans.

25-10

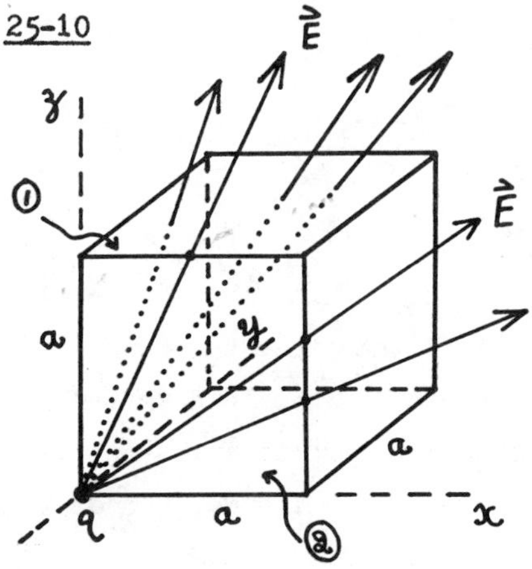

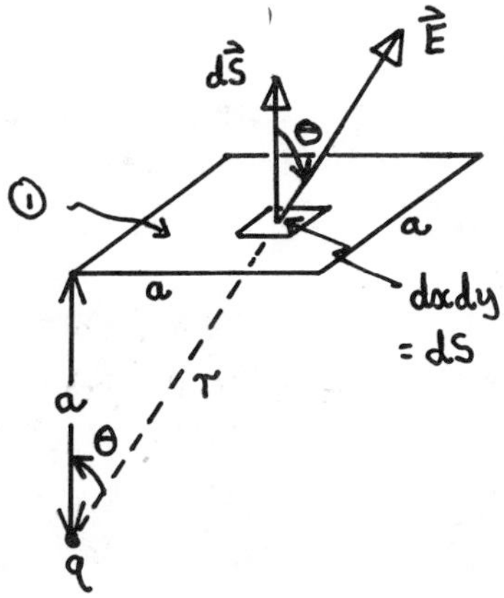

Clearly the flux through each of the three faces (such as number 2) that meet at the charge is zero since $\vec{E} \perp d\vec{S}$ on these surfaces. The flux through the other surfaces (represented by surface 1) are identical and can be computed as follows. The electric field strength is

$$E = \frac{1}{4\pi\epsilon_0}\frac{q}{r^2} = \frac{1}{4\pi\epsilon_0}\frac{q}{a^2 + x^2 + y^2};$$

also,

$$\cos\theta = \frac{a}{r}, \quad dS = dx\,dy\,.$$

Therefore $\Phi_E = \int \vec{E}\cdot d\vec{S}$ is simply

$$\Phi_E = \frac{aq}{4\pi\epsilon_0}\int_0^a\int_0^a \frac{dx\,dy}{(a^2 + x^2 + y^2)^{3/2}},$$

$$\Phi_E = \frac{aq}{4\pi\epsilon_0}\int_0^a \frac{x\,dy}{(a^2 + y^2)(a^2 + x^2 + y^2)^{\frac{1}{2}}}\Bigg|_0^a,$$

$$\Phi_E = \frac{a^2 q}{4\pi\epsilon_0}\int_0^a \frac{dy}{(a^2 + y^2)(2a^2 + y^2)^{\frac{1}{2}}} = \frac{q}{4\pi\epsilon_0}\tan^{-1}\left[\frac{y}{(2a^2 + y^2)^{\frac{1}{2}}}\right]_0^a = \frac{q}{24\epsilon_0}.$$

Hence, the flux through the three faces not touching the charge at 0 is $q/8\epsilon_0$, and as the flux through each of the other three faces is nil, the net flux through the cube is $q/8\epsilon_0$ also. This agrees with Gauss's law since the cube subtends at 0 a solid angle equal $\frac{1}{8}$th that subtended by the cube if the charge lay inside. Hence the flux should be $\frac{1}{8}(q/\epsilon_0)$, as obtained by direct calculation.

<u>25-22</u>

The field due to q is

$$E_q = \frac{1}{4\pi\epsilon_0}\frac{q}{r^2}.$$

The field due to the sphere, for $a < r < b$, is found from Gauss's law by constructing a spherical Gaussian surface centered at $r = 0$. Then,

$$\epsilon_0 E\cdot 4\pi r^2 = q_{enc} = \int \rho\, dV = \int_a^r \frac{A}{r}\, 4\pi r^2 dr = 2\pi A(r^2 - a^2),$$

so that

$$E = \frac{A}{2\epsilon_0}\left(1 - \frac{a^2}{r^2}\right)$$

radially in or out. In order that $\vec{E}_q + \vec{E}$ be independent of r, it must be true that

$$\frac{Aa^2}{2\epsilon_0} = \frac{q}{4\pi\epsilon_0},$$

$$A = \frac{q}{2\pi a^2} \quad \underline{\text{Ans}},$$

A and q having the same sign.

25-23

(a) From equation 25-9,

$$E = \frac{1}{4\pi\epsilon_0} \frac{qr}{R^3}.$$

But the charge density is $\rho = q/(\frac{4}{3}\pi R^3)$ so that $E = \rho r/3\epsilon_0$. If the charge is positive, $\vec{E}$ is radially out. Since $\vec{r}$ is radially out also

$$\vec{E} = \frac{\rho}{3\epsilon_0}\vec{r} \quad \underline{\text{Ans}}.$$

(b) Assume the charge is positive. If A be any point in the cavity, then from superposition it is expected that, at A,

(Field $\vec{E}'$ due to sphere of charge density ρ centered at O' and of radius R')

+ (Field $\vec{E}_0$ due to the given distribution of charge)

= (Field $\vec{E}$ due to sphere of charge density ρ centered at O and of radius R);

here O' is the center of the cavity of radius R', O and R being the center and radius of the sphere in which the cavity is located.

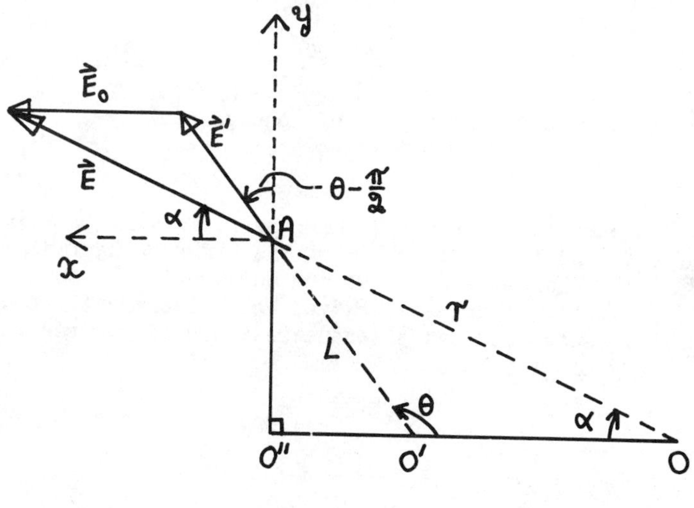

Therefore

$$\vec{E}_0 = \vec{E} - \vec{E}'.$$

Let 00' = a, 0A = r, 0'A = L; also, locate a point 0" such that A0" is perpendicular to 00'. Choose the x-axis parallel to 00' and the y-axis perpendicular to it in the plane 00'A. Then,

$$E_x = E\cos\alpha, \; E_y = E\sin\alpha,$$

$$E'_x = E'\sin(\theta - \tfrac{\pi}{2}), \; E'_y = E'\cos(\theta - \tfrac{\pi}{2}).$$

Then, from (a),

$$E_{0y} = E_y - E'_y = \frac{\rho r}{3\epsilon_0}\sin\alpha - \frac{\rho L}{3\epsilon_0}\sin\theta = \frac{\rho}{3\epsilon_0}(r\cdot\sin\alpha - L\cdot\sin\theta) = 0$$

by the sine rule applied to triangle A0'0. Turning to the x-component,

$$E_{0x} = E_x - E'_x = E\cos\alpha - E'\sin(\theta - \tfrac{\pi}{2}) = \frac{\rho}{3\epsilon_0}(r\cdot\cos\alpha + L\cdot\cos\theta),$$

$$E_{0x} = \frac{\rho}{3\epsilon_0}[r\cdot\cos\alpha - L\cdot\cos(\pi - \theta] = \frac{\rho}{3\epsilon_0}(00" - 0'0") = \frac{\rho}{3\epsilon_0}\cdot 00',$$

$$E_{0x} = \frac{\rho}{3\epsilon_0}\, a.$$

Thus, vectorially,

$$\vec{E}_0 = \frac{\rho}{3\epsilon_0}\vec{a}$$

with $\vec{a}$ being a vector from 0 to 0', charge positive.

<u>25-31</u>

(a) By symmetry it is expected that the field will be azimuthally symmetric around the cylinder axis, the lines being radially inward if the charge is negative and outward if the charge is positive. Positive charge is assumed here. Construct a coaxial, cylindrical Gaussian surface (ends are included) of radius r and length L. By Gauss's law,

$$\epsilon_0\oint\vec{E}\cdot d\vec{S} = q_{enc}$$

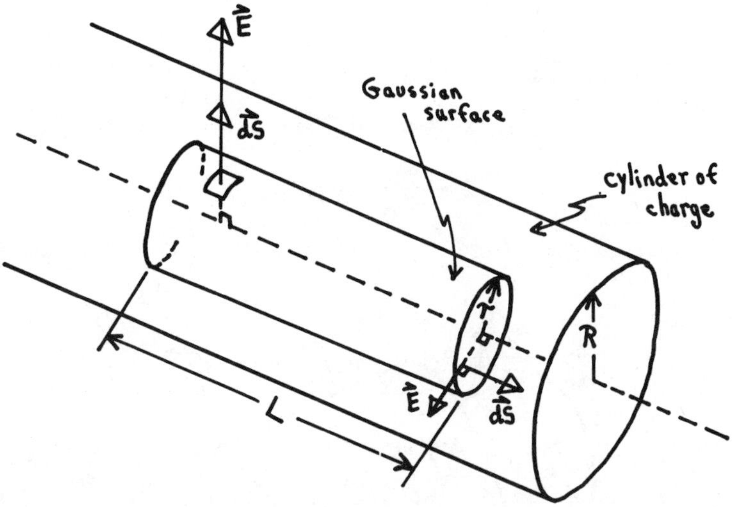

where q_{enc} is the charge inside the Gaussian surface over which the integral is taken. For convenience, the integral is divided into two parts, over the curved portion of the cylindrical surface and over the two end caps:

$$\epsilon_0 \int_{\substack{\text{curved}\\\text{surface}}} \vec{E}\cdot\vec{dS} + \epsilon_0 \int_{\text{ends}} \vec{E}\cdot\vec{dS} = q_{enc}\,.$$

Now $\vec{E}\cdot\vec{dS} = E(dS)\cos\sphericalangle(\vec{E},\vec{dS})$. On the flat ends the angle between $\vec{E}$ and $\vec{dS}$ is 90°, the cosine of which is zero. Hence, the integral over the ends (indeed, over each end) is nil. On the curved surface, however, the angle is zero (or 180° if the charge is negative), the cosine of which is +1 (-1). Also, since all points on the curved surface are at the same distance from the axis of the rod, E is independent of location on the curved surface, that is, is constant. Applying these considerations, over the curved part,

$$\int \vec{E}\cdot\vec{dS} = \int E\, dS = E\int dS = E\ 2\pi rL.$$

Since the entire Gaussian surface is inside the rod, the charge enclosed, which equals the product of charge density and volume of Gaussian surface occupied by charge, becomes

$$q_{enc} = (\rho)(\pi r^2 L),$$

so that Gauss's law gives

$$\epsilon_0 E(2\pi rL) = \rho\pi r^2 L,$$

$$E = \frac{\rho r}{2\epsilon_0} \quad \underline{\text{Ans.}}$$

(b)

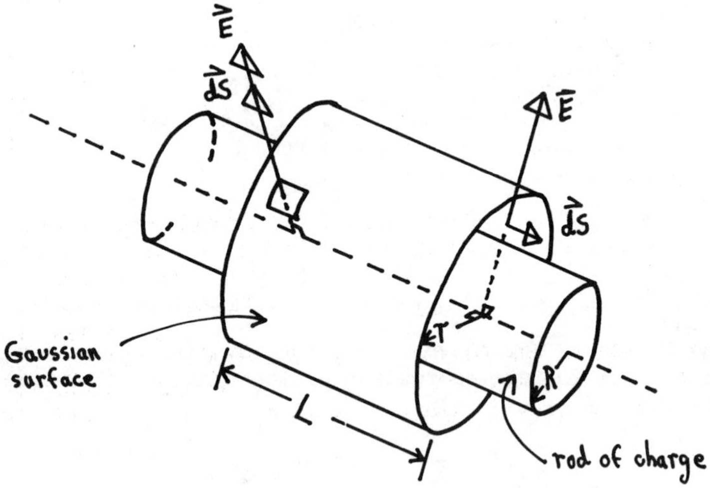

For $R < r$, draw the Gaussian surface outside the rod; now only part of the volume enclosed by the Gaussian surface is occupied by charge. Hence, by Gauss's law,

$$\epsilon_0 E(2\pi rL) = \rho\pi R^2 L,$$

$$E = \frac{\rho R^2}{2\epsilon_0 r} \quad \underline{\text{Ans}},$$

since $\pi R^2 L$ is the volume of the Gaussian surface that contains charge.

25-33

The restriction on points to be considered is tantamount to regarding the plates as infinite in two dimensions (thickness unaffected). The field lines, by symmetry, must pass, perpendicular to the sheets, from the positive to the negative charges. Under electrostatic conditions no lines can penetrate into the conducting plates, and since there is no fringing of the lines around the edges, there being no edges in effect, the field must be zero at all points not between the plates: i.e., (a) $E = 0$ and (c) $E = 0$ Ans.

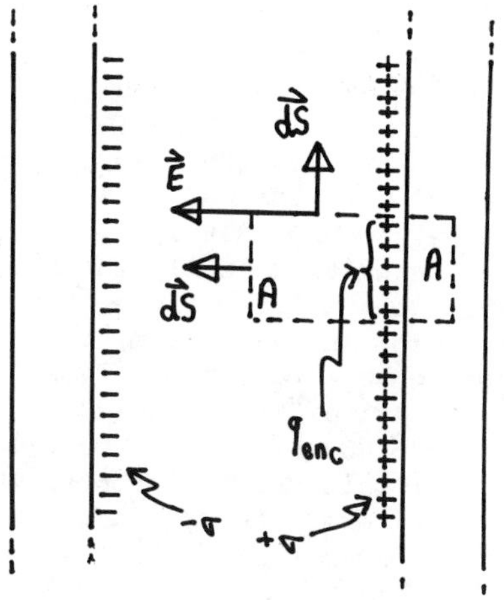

(b) To calculate E at points between the sheets, construct a right rectangular cylindrical "pillbox" Gaussian surface (any shape cross-sectional area A), with end caps parallel to the plates, as shown. The flux, $\int \vec{E}\cdot d\vec{S}$, over that part of the Gaussian surface lying inside the plate is zero as $E = 0$ there. The flux is also zero over the curved part of the surface found outside the plate; this is because here $\vec{E} \perp d\vec{S}$. Over the left end cap, however, $\vec{E} \parallel d\vec{S}$, and as all points on the cap are at the same distance from the plate Gauss's law yields

$$\epsilon_0 \oint \vec{E}\cdot d\vec{S} = \epsilon_0 \int_{\substack{\text{left}\\ \text{cap}}} \vec{E}\cdot d\vec{S} = \epsilon_0 \int E\, dS = \epsilon_0 E \int dS = \epsilon_0 EA = q_{enc}.$$

The charge enclosed by the Gaussian surface resides on that part of the plate lying inside the surface: this is q_{enc} = (charge per area) X (area) = σA. Therefore

$$\epsilon_0 EA = \sigma A,$$

$$E = \sigma/\epsilon_0 \quad \underline{\text{Ans}},$$

directed to the left. It may appear as though the presence of the negative plate was ignored, but this is not so for, if that plate was absent, the positive charge would then be found equally distributed on both sides of the plate, making $E = \frac{1}{2}\sigma/\epsilon_0$ at all points not inside the plate.

25-35

Focus on a single nonconducting sheet. To determine $\vec{E}$ construct a "pillbox" Gaussian surface, like that in Problem 25-33, oriented with the end caps parallel to the sheet. With the sheet being essentially infinite, $\vec{E}$ will be a uniform field directed normal to, and away from, the sheet (charge positive). Over the curved part of the Gaussian surface $\vec{E} \perp d\vec{S}$ and therefore $\int \vec{E}\cdot d\vec{S} = 0$ over this curved surface. The flux over each end cap is EA, so that by Gauss's law,

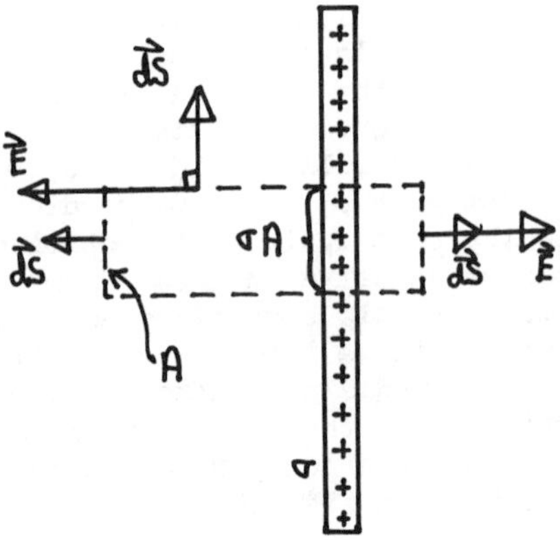

$$\epsilon_0(2EA) = q_{enc} = \sigma A,$$

$$E = \sigma/2\epsilon_0.$$

With both sheets present the total electric field $\vec{E} = \vec{E}_r + \vec{E}_l$, $\vec{E}_r$ and $\vec{E}_l$ due to the right and left sheets. These last are equal in magnitude (both being $\sigma/2\epsilon_0$). However, it is clear (b) that between the sheets their directions are opposite giving $\vec{E} = 0$ there. At all points not between the sheets the fields are parallel so that at these points

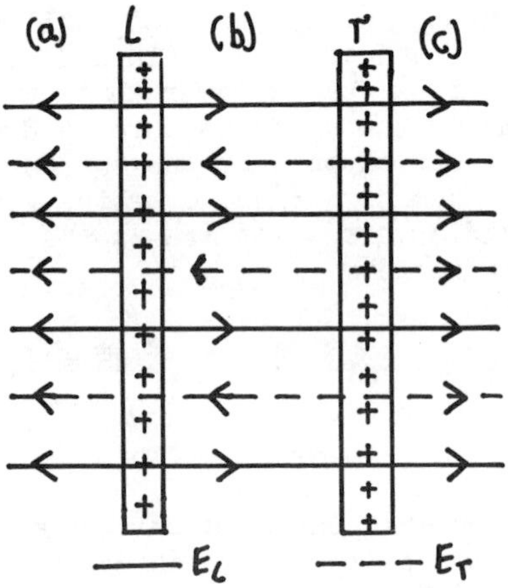

$$E = \sigma/2\epsilon_0 + \sigma/2\epsilon_0 = \sigma/\epsilon_0,$$

to the left in region (a) and to the right in region (c).

25-36

(a) Consider the slab to be of very large area and the charge positive. By symmetry, the field inside the slab will be directed at right angles to the slab face, but it will vary in magnitude with distance x from the median plane of the slab. Construct a "pillbox" Gaussian surface oriented as shown. The flux over the

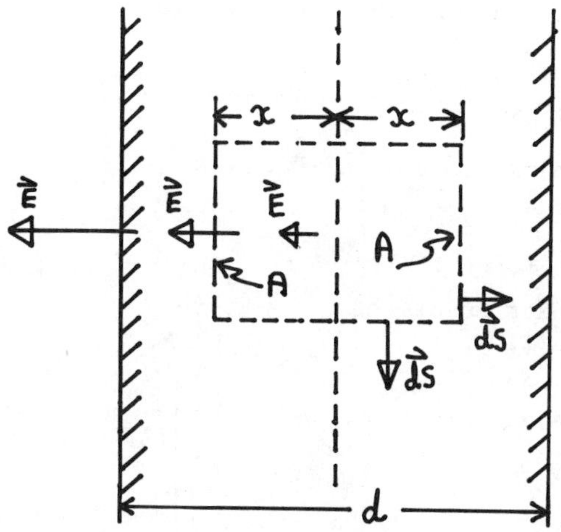

curved part of the surface is zero, and since each end cap is at the same distance from the median plane,

$$\Phi_E = \oint \vec{E}\cdot d\vec{S} = 2E(x)A,$$

where E(x) is the electric field strength at a distance x from the median plane and A is the area of each end cap. The charge enclosed by the surface is the volume charge density ρ of the slab multiplied by the volume enclosed by the Gaussian surface since the entire volume contains charge; hence,

$$q_{enc} = \rho(2xA).$$

Thus, by Gauss's law,

$$\epsilon_0(2EA) = \rho(2xA),$$

$$E = \rho x/\epsilon_0 \quad \underline{\text{Ans}}.$$

(b) Since the field outside the slab is directed at right angles to it the analysis for the electric field outside a single, non-conducting sheet, given in Problem 25-35, is valid here, so that $E = \sigma/2\epsilon_0$. Now σ is the charge per unit area of the slab, that is, the charge contained in a cylinder extending across the full width d of the slab and having a cross-sectional area of 1 m^2. The volume of the cylinder is d(1 m^2). Hence, expressions for the charge contained in this cylinder expressed in terms of σ on the one hand

and ρ on the other are (and they must be equal),

$$\sigma(1) = \rho[d(1)],$$

so that $\sigma = \rho d$. Therefore $E = \rho d/2\epsilon_0$.

25-38

Draw a free-body diagram of the sphere. For equilibrium,

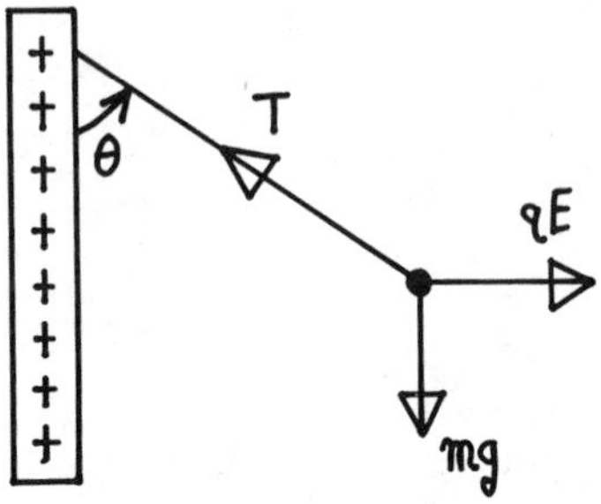

$$T\cos\theta = mg,$$

$$T\sin\theta = qE.$$

Eliminating T gives

$$\tan\theta = \frac{qE}{mg}.$$

But, from Problem 25-35, $E = \sigma/2\epsilon_0$. Putting this into the above equation and solving for σ gives

$$\sigma = (2\epsilon_0 mg\tan\theta)/q,$$

$$\sigma = (2)(8.854 \times 10^{-12})(10^{-6})(9.8\ )(3^{\frac{1}{2}}/3)/(2 \times 10^{-8}),$$

$$\sigma = 5.0 \times 10^{-9}\ C/m^2 \quad \underline{\text{Ans}}.$$

25-39

Assume that no charge is located in the immediate neighbourhood of P. Imagine a spherical Gaussian surface of very small radius with its center at P. Let the test charge +q be displaced from P to any point on the Gaussian surface. For stable equilibrium the charge must experience a force that tends to push it back to P: i.e., the electrostatic field E must have an inward directed normal component everywhere on the Gaussian surface. But, by Gauss's law, this means that net negative charge, responsible for setting up E, must reside inside the Gaussian surface. This is contrary to assumption. A similar argument may be made for negative test charges.

CHAPTER 26

26-1

The energy E released in the flash is

$$E = QV = (30\ C)(10^9\ V) = 3 \times 10^{10}\ J.$$

Hence, the mass M of ice that can be melted is given by

$$M = \frac{3 \times 10^{10}\ J}{3.3 \times 10^5\ J/kg} = 9.09 \times 10^4\ kg \quad \underline{Ans.}$$

26-8

Let the point P at which the potential is desired be at a distance r = a from the center of the sphere of charge, r < R. Then,

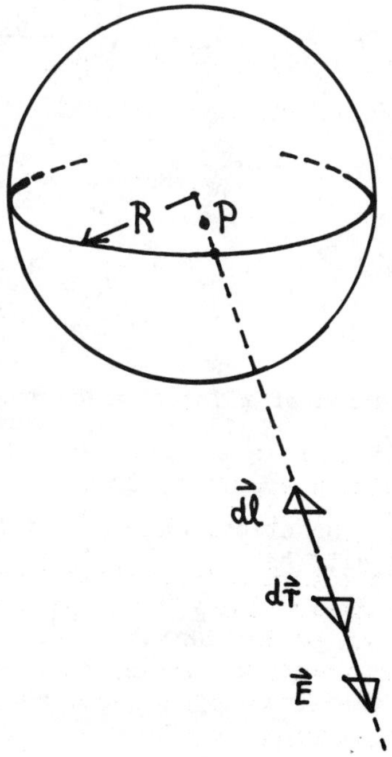

$$V_P - V_\infty = -\int_\infty^P \vec{E}\cdot\vec{dl};$$

choosing $V_\infty = 0$ gives

$$V_P = -\int_\infty^P \vec{E}\cdot\vec{dl}.$$

For convenience, integrate along a straight line from infinity to P. Now $\vec{dl}$ points in the direction of integration, opposite to $\vec{dr}$ which points outward from the sphere, since the center of the sphere is at r = 0. Hence,

$$\vec{E}\cdot\vec{dl} = E\,dl\cos\pi = -E\,dl = -E(-dr) = E\,dr,$$

assuming the charge is positive so that the electric field points outward. Substituting the last result into the integral for V_P gives

$$V_P = -\int_{r=\infty}^{r=a} E\,dr.$$

From Chapter 25 it is known that E(r) has different analytical forms for r < R, R < r:

$$E(r) = \frac{1}{4\pi\epsilon_0}\frac{q}{r^2},\ R \le r;\quad E(r) = \frac{1}{4\pi\epsilon_0}\frac{q}{R^3}r,\ r \le R.$$

For this reason, the integral must be divided into two parts, one for points exterior, and one for points interior, to the sphere of charge:

$$V_P = -\int_{\infty}^{R}\frac{q}{4\pi\epsilon_0}\frac{dr}{r^2} - \int_{R}^{a}\frac{q}{4\pi\epsilon_0 R^3}r\,dr,$$

$$V_P = -\frac{q}{4\pi\epsilon_0}[(-\frac{1}{R} + 0) + \frac{1}{R^3}(\tfrac{1}{2}a^2 - \tfrac{1}{2}R^2)],$$

$$V_P = \frac{q(3R^2 - a^2)}{8\pi\epsilon_0 R^3}.$$

Substitution of r for a gives the desired result.

26-15

(a) For a spherical charge distribution the potential at the surface will be

$$V = \frac{q}{4\pi\epsilon_0 R} = (9 \times 10^9)\frac{3 \times 10^{-11}}{R} = 500,$$

$$R = 5.4 \times 10^{-4}\ \text{m} = 0.54\ \text{mm}\ \underline{\text{Ans}}.$$

(b) The radius of the combined drops is found from

$$\frac{4}{3}\pi r^3 = 2(\frac{4}{3}\pi R^3),$$

$$r = 2^{1/3}R.$$

The total charge is $2q = 6 \times 10^{-11}$ C. If the drops act as conductors (due to impurities, say) so that the charge becomes uniformly distributed throughout the new drop, the surface potential will become

$$V' = \frac{2q}{4\pi\epsilon_0 2^{1/3}R} = 794 \text{ V} \quad \underline{\text{Ans}}.$$

26-16

Let Q be the total charge on the shell.

(a) For $r > r_2$, the potential is identical with that of a point charge Q at the center of the actual distribution: that is, $V = Q/4\pi\epsilon_0 r$.

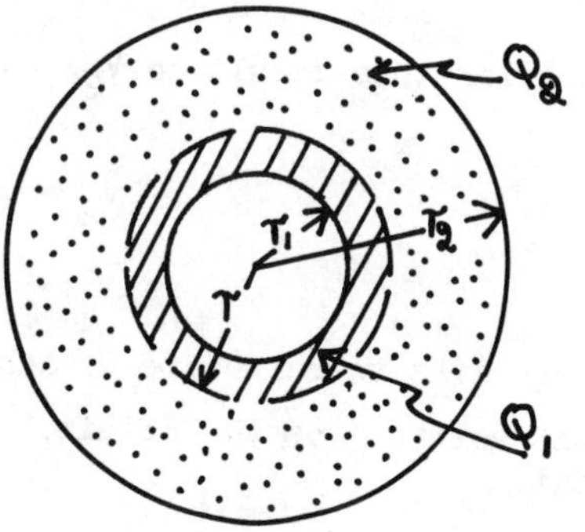

(b) In this region, the charge distribution can be considered as made up of two parts: charge Q_1 interior to r and Q_2 exterior to r. The potential V at r is the sum $V_1 + V_2$ of the potentials due to each of these distributions. The point considered is on the surface of Q_1 so $V_1 = Q_1/4\pi\epsilon_0 r$. To find V_2 it is necessary to calculate the potential V' in the cavity of a thick shell of charge. Consider such a shell carrying charge q and of inner radius a and outer radius b. The shell can be divided into many very thin shells, such as the one shown of

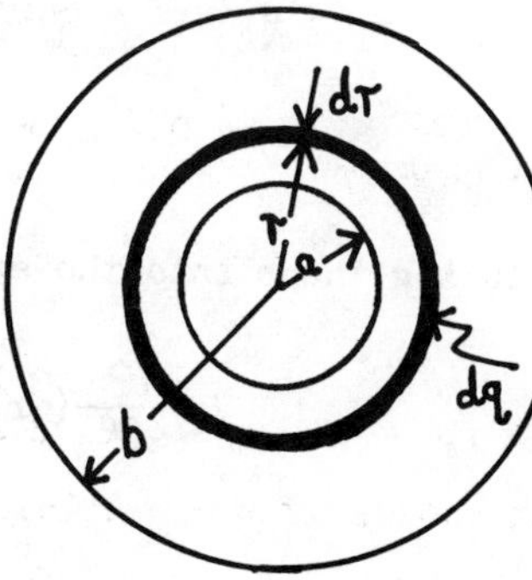

radius r and thickness dr. If the charge on the shell is dq, the potential dV' due to the shell is

$$dV' = dq/4\pi\epsilon_0 r.$$

But

$$dq = [q/(\tfrac{4}{3}\pi b^3 - \tfrac{4}{3}\pi a^3)](4\pi r^2 dr)$$

the term in square brackets being the charge per unit volume on the thick shell and $4\pi r^2 dr$ being the volume of the thin shell of radius r. Hence,

$$dV' = \frac{q\ rdr}{\tfrac{4}{3}\pi\epsilon_0 (b^3 - a^3)}$$

and

$$V' = \int_a^b dV' = \frac{3q}{8\pi\epsilon_0}\frac{b^2 - a^2}{b^3 - a^3}.$$

Applying this result to V_2, put $b = r_2$, $a = r$, $q = Q_2$ to get

$$V_2 = \frac{3Q_2}{8\pi\epsilon_0}\frac{r_2^2 - r^2}{r_2^3 - r^3}.$$

It remains to calculate Q_1 and Q_2 in terms of Q. The charge per unit volume ρ on the original thick shell is $Q/(\tfrac{4}{3}\pi r_2^3 - \tfrac{4}{3}\pi r_1^3)$, so that

$$Q_1 = \rho(\tfrac{4}{3}\pi r^3 - \tfrac{4}{3}\pi r_1^3) = Q(r^3 - r_1^3)/(r_2^3 - r_1^3),$$

$$Q_2 = Q - Q_1 = Q(r_2^3 - r^3)/(r_2^3 - r_1^3).$$

Substituting these into the expressions for V_1 and V_2 and adding gives

$$V = \frac{\rho}{3\epsilon_0}(\tfrac{3}{2}r_2^2 - \tfrac{1}{2}r^2 - \frac{r_1^3}{r})$$ <u>Ans.</u>

where $\rho = Q/[\frac{4}{3}\pi(r_2^3 - r_1^3)]$.

(c) For $r < r_1$, i.e., inside the cavity, use the result from (b), but now set $r = r_1$ (the inner boundary of the original shell), so that

$$V = \frac{\rho}{2\epsilon_0}(r_2^2 - r_1^2) \quad \underline{\text{Ans}}.$$

(d) The appropriate potentials must agree at the boundaries of the regions in which they are valid: that is, $V_a = V_b$ at $r = r_2$ and $V_b = V_c$ at $r = r_1$.

26-22

Put the origin at the middle charge. Then,

$$V = \frac{1}{4\pi\epsilon_0}\frac{q}{r - a} + \frac{1}{4\pi\epsilon_0}\frac{q}{r} + \frac{1}{4\pi\epsilon_0}\frac{-q}{r + a} = \frac{q}{4\pi\epsilon_0}\frac{r^2 + 2ar - a^2}{(r^2 - a^2)r}.$$

If $a \ll r$ this becomes, upon neglecting the a^2 terms,

$$V = \frac{q}{4\pi\epsilon_0}\left(\frac{r + 2a}{r^2}\right) = \frac{1}{4\pi\epsilon_0}\left(\frac{q}{r} + \frac{2aq}{r^2}\right) \quad \underline{\text{Ans}}.$$

The first term is the potential due to a point charge (the middle charge) and the second the potential due to a dipole (the outer charges) at large distances (compared to the dipole charge separation) along the dipole axis.

26-23

(a) The location r_p of the center of positive charge along the line of symmetry is given by the "center of mass" formula,

$$r_p = \frac{2e(d\cos 52°) + 8e(0)}{2e + 8e},$$

$$r_p = \frac{1}{5}(0.96 \times 10^{-10})\cos 52° = 0.11821 \times 10^{-10}\ \text{m} \quad \underline{\text{Ans}}.$$

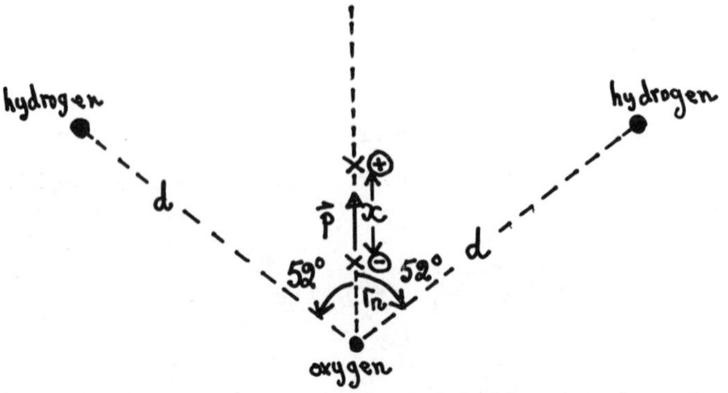

(b) Let the distance between the centers of positive and negative charge be x. Then p = xq = x(10e), so that

$$x = \frac{p}{10e} = \frac{6.2 \times 10^{-30}}{10(1.6 \times 10^{-19})} = 0.03875 \times 10^{-10} \text{ m.}$$

Since p points from the negative to the positive charge centers, the distance r_n of the center of negative charge from the oxygen nucleus is $r_n = (0.11821 - 0.03875) \times 10^{-10} = 0.079 \times 10^{-10}$ m <u>Ans.</u> This calculation assumes that the two hydrogen atoms share their electrons with the oxygen atom, the electrons of which remain close to their parent nucleus.

<u>26-26</u>

Since one particle is to be kept fixed the reduced mass need not be introduced. With the second particle released from rest at P, thereby having zero initial kinetic energy, conservation of kinetic plus potential energy gives

$$\frac{1}{4\pi\epsilon_0}\frac{q^2}{r_1} + 0 = \frac{1}{4\pi\epsilon_0}\frac{q^2}{r_2} + \tfrac{1}{2}mv^2,$$

$$v^2 = \frac{2}{m}\frac{q^2}{4\pi\epsilon_0}\left(\frac{1}{r_1} - \frac{1}{r_2}\right),$$

v being the speed at r_2. Putting in the numbers,

$$v^2 = (2/2 \times 10^{-5})(9 \times 10^9)(3.1 \times 10^{-6})^2\left(\frac{1}{9} - \frac{1}{25}\right)10^4,$$

$$v = 2.48 \times 10^3 \text{ m/s} \quad \underline{\text{Ans.}}$$

26-33

Let v be the initial speed. The potential energy of the moving electron is zero at infinity and its kinetic energy, in turn, is zero at the instant the electron is at rest midway between the two fixed electrons. Conservation of kinetic plus potential energy gives

$$\tfrac{1}{2}mv^2 + 0 = 0 + 2\cdot\frac{1}{4\pi\epsilon_0}\frac{(-e)^2}{\tfrac{1}{2}D},$$

where $D = 0.02$ m is the distance between the fixed electrons and $-e = -1.6 \times 10^{-19}$ C is the charge of an electron. Since $m = 9.11 \times 10^{-31}$ kg, the equation yields $v = 318$ m/s Ans.

26-37

Let W_e = work to be done against the electrostatic forces; this is

$$W_e = (-q)(V_2 - V_1) = -q\left(\frac{1}{4\pi\epsilon_0}\frac{Q}{r_2} - \frac{1}{4\pi\epsilon_0}\frac{Q}{r_1}\right) = \frac{Qq}{4\pi\epsilon_0}\left(\frac{1}{r_1} - \frac{1}{r_2}\right).$$

The initial kinetic energy of the electron can supply some of this work, however. For uniform circular motion,

$$\frac{1}{4\pi\epsilon_0}\frac{Qq}{r^2} = m\frac{v^2}{r},$$

so that

$$K = \tfrac{1}{2}mv^2 = \frac{1}{2}\frac{1}{4\pi\epsilon_0}\frac{Qq}{r}.$$

Therefore, the kinetic energy available for work is

$$K_1 - K_2 = \frac{1}{2}\frac{Qq}{4\pi\epsilon_0}\left(\frac{1}{r_1} - \frac{1}{r_2}\right).$$

This implies that the work W that the external agent must do is

$$W = W_e - (K_1 - K_2) = \frac{Qq}{8\pi\epsilon_0}\left(\frac{1}{r_1} - \frac{1}{r_2}\right) \quad \underline{\text{Ans.}}$$

26-38

(a) The initial potential energy V is

$$V = q^2/(4\pi\epsilon_0 d) = 0.225 \text{ J} \quad \underline{\text{Ans}},$$

since $q = 5 \times 10^{-6}$ C, d = 1 m.

(b) The force on each sphere after the string is cut is just the electrostatic force F. Since F = ma, the accelerations are

$$a_1 = \frac{1}{m_1}\cdot q^2/(4\pi\epsilon_0 d^2) = 45 \text{ m/s}^2 \quad \underline{\text{Ans}},$$

$$a_2 = 22.5 \text{ m/s}^2 \quad \underline{\text{Ans}},$$

since $m_1 = 0.005$ kg and $m_2 = 0.01$ kg $= 2m_1$.

(c) A long time after the string is cut the spheres are far enough apart so that the electrostatic potential energy is close to zero. Conservation of energy then implies that

$$0.225 = \tfrac{1}{2}m_1v_1^2 + \tfrac{1}{2}m_2v_2^2,$$

as the initial speeds, and therefore initial kinetic energy, were zero; the speeds in the above equation are those after a long time. Conservation of momentum gives a second relation between the speeds

$$m_1v_1 = m_2v_2.$$

With $m_2 = 2m_1$, this last equation requires that $v_1 = 2v_2$. Putting this into the energy equation yields

$$0.225 = \tfrac{1}{2}(\tfrac{1}{2}m_2)(2v_2)^2 + \tfrac{1}{2}m_2v_2^2 = \tfrac{3}{2}\cdot m_2v_2^2 = \tfrac{3}{2}(0.01)v_2^2,$$

$$v_2 = 3.873 \text{ m/s} \quad \underline{\text{Ans}},$$

$$v_1 = 2v_2 = 7.746 \text{ m/s} \quad \underline{\text{Ans}}.$$

26-41

(a) Let $d\ell$ be an element of length of the ring; due to this element, carrying a charge dq, the potential at P is, since $q/2\pi a$ is the charge per unit length on the ring,

$$dV = \frac{1}{4\pi\epsilon_0}\frac{dq}{r} = \frac{1}{4\pi\epsilon_0}\frac{(q/2\pi a)d\ell}{(a^2 + x^2)^{\frac{1}{2}}}.$$

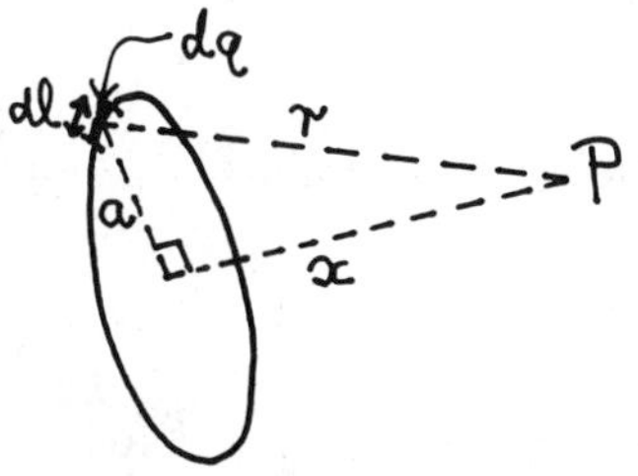

The total potential at P is

$$V = \int_{\text{ring}} dV = \frac{1}{4\pi\epsilon_0}\frac{(q/2\pi a)}{(a^2 + x^2)^{\frac{1}{2}}}\int d\ell,$$

noting that a and x are independent of the element's position on the ring. Now, $\int d\ell = 2\pi a$, the circumference of the ring, and hence

$$V = \frac{1}{4\pi\epsilon_0}\frac{q}{(a^2 + x^2)^{\frac{1}{2}}} \quad \underline{\text{Ans}}.$$

(b) By symmetry, the component of $\vec{E}$ at the axis is zero in any direction perpendicular to the axis. Thus,

$$E = E_x = -\frac{dV}{dx} = \frac{1}{4\pi\epsilon_0}\frac{xq}{(a^2 + x^2)^{3/2}} \quad \underline{\text{Ans}},$$

as found in the quoted example.

26-45

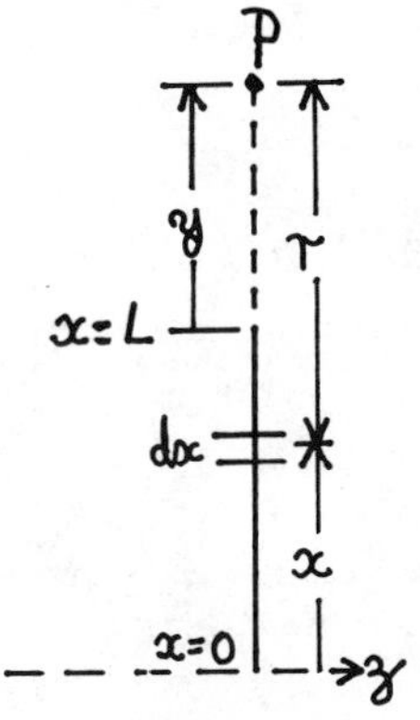

(a) With $\lambda dx = dq$ = charge on an element of length dx, the potential V at P is found from

$$4\pi\epsilon_0 V = \int\frac{\lambda dx}{r} = \int_0^L \frac{\lambda dx}{(L + y) - x},$$

$$4\pi\epsilon_0 V = -\lambda\cdot\ln(L + y - x)\Big|_0^L = -\lambda[\ln y - \ln(L + y)],$$

$$V = \frac{\lambda}{4\pi\epsilon_0}\ln\left(\frac{L + y}{y}\right) \quad \underline{\text{Ans}}.$$

(b) Now call the axis the y-axis and move the origin to the upper end of the charge segment, as shown. Then,

$$E_y = -\frac{dV}{dy} = \frac{\lambda}{4\pi\epsilon_0}\frac{L}{y(L + y)} \quad \underline{\text{Ans}}.$$

(c) The direction of the electric field at P due to any element of the segment of charge is directed along the y (or x)-axis; hence $E_z = 0$ Ans.

(This result cannot be derived from the potential obtained in (a) for that expression is valid only at one specific value of z, i.e. z = 0, and therefore dV/dz cannot be calculated from it.)

26-46

(a) The potential dV at P due to the element of length dx that carries a charge $dq = \lambda dx = kx dx$ will be

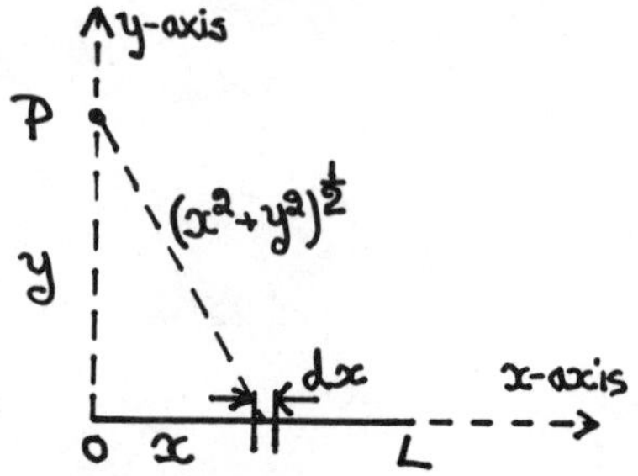

$$dV = \frac{1}{4\pi\epsilon_0}\frac{(kx)dx}{(x^2 + y^2)^{\frac{1}{2}}},$$

and therefore

$$V = \frac{k}{4\pi\epsilon_0}\int_0^L \frac{x\,dx}{(x^2 + y^2)^{\frac{1}{2}}} = \frac{k}{4\pi\epsilon_0}[(L^2 + y^2)^{\frac{1}{2}} - y] \quad \underline{\text{Ans}}.$$

(b) The y-component of the electric field at P is

$$E_y = -\frac{dV}{dy} = \frac{k}{4\pi\epsilon_0}\left[1 - \frac{y}{(L^2 + y^2)^{\frac{1}{2}}}\right] \quad \underline{\text{Ans}}.$$

(c) In order to calculate $E_x = -dV/dx$, V must be known as a function of x in a neighborhood about P, but in (a) V was computed for any point along a line, each point of which has the same single x coordinate, that is x = 0. Hence, the derivative cannot be calculated due to lack of "x information".

26-53

Since $r \ll a$, the disturbing effect on the assumed uniform charge distribution on each sphere due to the electric field from the other sphere can be disregarded; that is, $V = \pm 1500 = Q/4\pi\epsilon_0 r$, so that, with r = 0.15 m, $Q = \pm 2.5 \times 10^{-8}$ C Ans.

26-54

(a) Let r be the radius of each sphere and d their distance apart. Then, at the point midway ($\frac{1}{2}$d) from each,

$$V = \frac{q_1}{4\pi\epsilon_0(\frac{1}{2}d)} + \frac{q_2}{4\pi\epsilon_0(\frac{1}{2}d)} = -180 \text{ V} \quad \underline{\text{Ans}},$$

since $q_1 = 10^{-8}$ C and $q_2 = -3 \times 10^{-8}$ C.

(b) In evaluating the potential of each sphere, the potential due to the distant ($d \gg r$) sphere should be ignored (if it is not, the charge distribution on the spheres will not be uniform and the potential for spherically-symmetric charge distributions cannot be used). Hence,

$$V_1 = q_1/4\pi\epsilon_0 r = 3000 \text{ V}, \quad V_2 = -9000 \text{ V} \quad \underline{\text{Ans}}.$$

26-55

(a) The charge q_α of an α-particle is +2e and its mass $m_\alpha \approx 4m_p$, $e = 1.6 \times 10^{-19}$ C being the charge on the proton and m_p being the proton mass. Let V be the accelerating potential and K kinetic energy; then,

$$K_\alpha = q_\alpha V = (2e)V = 2(1.6 \times 10^{-19} \text{ C})(10^6 \text{ V}) = 3.2 \times 10^{-13} \text{ J} \quad \underline{\text{Ans}}.$$

(b) Similarly, $K_p = eV = 1.6 \times 10^{-13}$ J Ans.

(c) Evidently $K_\alpha = 2K_p$, and therefore

$$K_p = \tfrac{1}{2}m_p v_p^2 = \tfrac{1}{2}K_\alpha = \tfrac{1}{2}(\tfrac{1}{2}m_\alpha v_\alpha^2) = \tfrac{1}{2}[\tfrac{1}{2}(4m_p)v_\alpha^2],$$

$$v_p^2 = 2v_\alpha^2.$$

Thus the proton, due to its greater charge to mass ratio, has the greater final speed.

26-57

(a) The potential of the shell and the electric field near the shell's surface are

$$V = \frac{1}{4\pi\epsilon_0}\frac{q}{r}, \quad E = \frac{1}{4\pi\epsilon_0}\frac{q}{r^2}.$$

Hence,

$$E = V/r.$$

For $E < 10^8$ V/m, it is necessary that r be greater than $V/E = 9 \times 10^6 \text{ V}/10^8 \text{ V/m} = 0.09 \text{ m} = 9 \text{ cm}$.

(b) The work done in bringing up to the machine a charge Q is QV. Therefore, the power P supplied must be

$$P = \frac{dW}{dt} = V\frac{dQ}{dt} = (9 \times 10^6 \text{ V})(3 \times 10^{-4} \text{ C/s}) = 2700 \text{ W} \quad \underline{\text{Ans.}}$$

(c) If the surface charge density is σ and x denotes a length of the belt, then, since $Q = \sigma A = \sigma wx$,

$$\frac{dQ}{dt} = \sigma\frac{dA}{dt} = \sigma\frac{d(wx)}{dt} = \sigma wv,$$

$$\sigma = (dQ/dt)/wv = 2 \times 10^{-5} \text{ C/m}^2 \quad \underline{\text{Ans.}}$$

CHAPTER 27

27-6

The capacitance of a parallel flat-plate capacitor is $\epsilon_0 A/d$, where d is the plate separation. If the plates of the upper capacitor are a distance d apart, the plates of the lower capacitor must be separated by a - (b + d). Hence, the equivalent capacitance C_e is given by

$$\frac{1}{C_e} = \frac{1}{C_1} + \frac{1}{C_2},$$

$$\frac{1}{C_e} = \frac{1}{\epsilon_0 A/d} + \frac{1}{\epsilon_0 A/(a - b - d)} = \frac{1}{\epsilon_0 A}(d + a - b - d) = \frac{a - b}{\epsilon_0 A},$$

$$C_e = \frac{\epsilon_0 A}{a - b},$$

independent of d and therefore of the position of the center piece.

27-14

The initial charges are given to the left and final charges to the right of each capacitor in the figure. Clearly,

$$q = C_1 V_0.$$

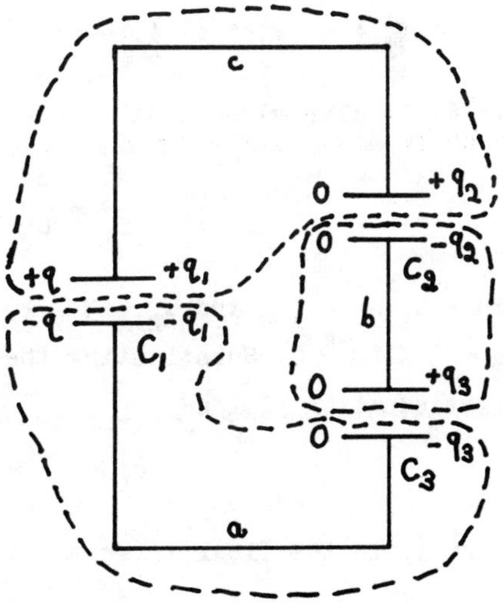

By conservation of charge applied to conductor a (if no charge jumps the gap):

$$-q = -q_1 - q_3;$$

from conductor b,

$$0 = -q_2 + q_3;$$

for conductor c,

$$+q = +q_1 + q_2.$$

Also, since electrostatic fields are conservative, the total drop in potential around the circuit must be zero. As the q's are taken to be positive, this indicates that

$$0 = \frac{q_2}{C_2} + \frac{q_3}{C_3} - \frac{q_1}{C_1}.$$

Solving the above equations simultaneously gives

$$q_1 = \frac{C_1C_2 + C_1C_3}{C_1C_2 + C_1C_3 + C_2C_3} C_1V_0 \quad \underline{\text{Ans}},$$

$$q_2 = q_3 = \frac{C_2C_3}{C_1C_2 + C_1C_3 + C_2C_3} C_1V_0 \quad \underline{\text{Ans}}.$$

<u>27-16</u>

Let primed quantities represent final values (switches closed). The final polarities must be the same since the capacitors are then connected in parallel. By the conservation of charge,

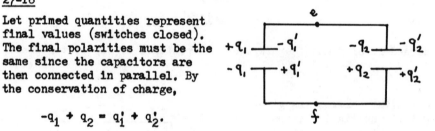

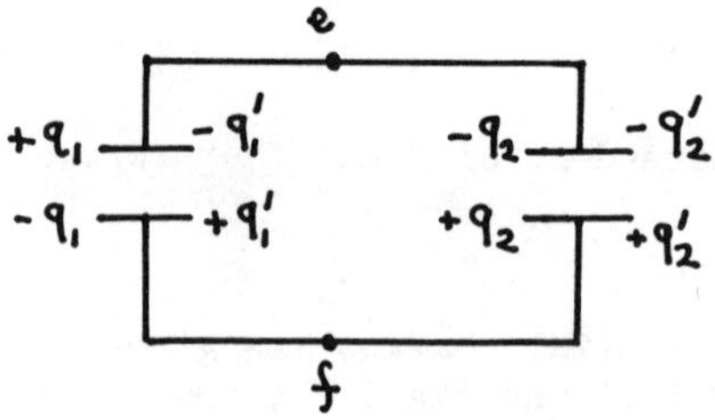

$$-q_1 + q_2 = q_1' + q_2'.$$

But it is also given that

$$\frac{q_1}{C_1} = \frac{q_2}{C_2} = 100 \text{ V}.$$

Since $C_1 = 10^{-6}$ F and $C_2 = 3 \times 10^{-6}$ F, the above gives $q_1 = 10^{-4}$ C, $q_2 = 3 \times 10^{-4}$ C. Substituting these numerical values into the very first equation yields

$$q_1' + q_2' = 2 \times 10^{-4}.$$

However, in the final arrangement also,

$$V' = \frac{q_1'}{C_1} = \frac{q_2'}{C_2}$$

so that

$$q_2' = 3q_1',$$

just as $q_2 = 3q_1$. Solving the two equations for q_1' and q_2' gives
(b)

$$q_1' = 0.5 \times 10^{-4}\ \text{C} \quad \underline{\text{Ans}},$$

(c)

$$q_2' = 1.5 \times 10^{-4}\ \text{C} \quad \underline{\text{Ans}}.$$

(a) Finally,

$$V_{ef} = V' = (1.5 \times 10^{-4}\ \text{C})/(3 \times 10^{-6}\ \text{F}) = 50\ \text{V} \quad \underline{\text{Ans}}.$$

27-18

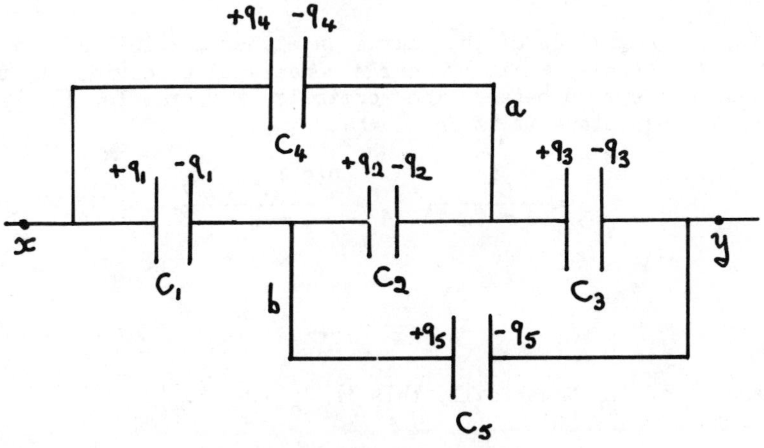

Apply a potential difference V between points x and y (say, by connecting a battery) thereby charging the capacitors. By the conservation of charge applied to conductors a and b:

$$-q_4 - q_2 + q_3 = 0,$$

$$-q_1 + q_2 + q_5 = 0,$$

the capacitors being considered originally uncharged. But the potential difference V between x and y is independent of the path taken to evaluate it (conservative field); hence, with $C_1 = C_3 = C_4 = C_5 = C$,

$$\frac{q_4}{C} + \frac{q_3}{C} = V,$$

$$\frac{q_1}{C} + \frac{q_2}{C_2} + \frac{q_3}{C} = V,$$

$$\frac{q_1}{C} + \frac{q_5}{C} = V.$$

From these five equations the charges q_1, q_2, q_3, q_4, q_5 can be found in terms of C, C_2, V. However, it is not necessary to solve for all of them since the equivalent capacitance C_e is

$$C_e = \frac{q_1 + q_4}{V} = \frac{q_3 + q_5}{V};$$

that is, the magnitude of the charge on either terminal x or y (they will be equal, again by charge conservation) divided by the potential difference between the terminals. For example, in solving the equations it is found that

$$q_3 = \frac{C^2}{C + C_2}\frac{V}{2}, \quad q_5 = \frac{C(C + 2C_2)}{C + C_2}\frac{V}{2}.$$

Thus,

$$C_e = \frac{CV}{V} = C,$$

independent of C_2. Numerically this is $C_e = 4\ \mu F$.

<u>27-21</u>

The new capacitance C' in terms of the new radius R' is $C' = 4\pi\epsilon_0 R'$.

But the new radius R' in terms of the old radius R is found from

$$\frac{4}{3}\pi R'^3 = 2\left(\frac{4}{3}\pi R^3\right),$$

$$R' = 2^{1/3}R$$

so that

$$C' = 2^{1/3}4\pi\epsilon_0 R = 5.04\ \pi\epsilon_0 R \quad \underline{\text{Ans}}.$$

27-22

(a) The capacitance is $C = q/V$ where q = magnitude of the charge on either facing surface and V = absolute value of the potential difference between inner and outer surfaces. But, from Chapter 26, the potential in the region between the shells is

$$V(r) = \frac{1}{4\pi\epsilon_0}\frac{q}{r} + \frac{1}{4\pi\epsilon_0}\frac{(-q)}{b},$$

assuming $+q$ charge on the outer surface of the inner shell and $-q$ on the inner surface of the outer shell. Hence,

$$V = V(a) - V(b) = \frac{q}{4\pi\epsilon_0}\left(\frac{1}{a} - \frac{1}{b}\right) = \frac{q}{4\pi\epsilon_0}\left(\frac{b - a}{ab}\right).$$

V may also be found by evaluating $-\int \vec{E}\cdot d\vec{r}$, $E = q/4\pi\epsilon_0 r^2$ from one shell to the other. The capacitance is

$$C = \frac{q}{V} = 4\pi\epsilon_0\frac{ab}{b - a}.$$

(b) If $a = R$ and b "=" ∞, the $a = R$ in the denominator can be deleted (since $R \ll \infty$) giving

$$C = 4\pi\epsilon_0 R,$$

as in Example 3 for the capacitance of an isolated sphere.

27-35

With the plates disconnected, no charge can be transferred from one plate to the other. If the plate separation remains small compared with the dimensions of the plates, the electric field will remain uniform and independent of the distance between the plates.

(a)

$$V_f = E_f d_f = E(2d) = 2(Ed) = 2V \quad \underline{\text{Ans.}}$$

(b) Initially,

$$U_i = \tfrac{1}{2}C_i V_i^2 = \tfrac{1}{2}(\epsilon_0 A/d)V^2 \quad \underline{\text{Ans.}}$$

Both the capacitance and potential difference change as the plates are pulled apart; hence,

$$U_f = \tfrac{1}{2}C_f V_f^2 = \tfrac{1}{2}(\epsilon_0 A/d_f)(2V)^2 = \tfrac{1}{2}(\epsilon_0 A/2d)(2V)^2 = \epsilon_0 AV^2/d \quad \underline{\text{Ans.}}$$

(c) The work W required to pull the plates apart is

$$W = U_f - U_i = U_i = \tfrac{1}{2}(\epsilon_0 A/d)V^2 \quad \underline{\text{Ans.}}$$

27-38

The energy stored in an electric field occupying a volume V is

$$U = \int u_E dV = \int \tfrac{1}{2}\epsilon_0 E^2 dV$$

where u_E is the electric field energy density. From Example 2,

$$E = \frac{q}{2\pi\epsilon_0 L}\frac{1}{r},$$

between the plates of a coaxial cylindrical capacitor of length L and charge q; r is measured from the axis. The volume element dV is taken to be the volume contained in a very thin cylindrical shell of radius r, thickness dr and length L, so that

$$dV = 2\pi rL\, dr.$$

Hence, the energy stored within a region between the inner shell of radius a and a cylindrical surface of radius s is

$$U_s = \int_a^s \tfrac{1}{2}\epsilon_0 \left(\frac{q}{2\pi\epsilon_0 Lr}\right)^2 2\pi rL\, dr = \frac{q^2}{4\pi\epsilon_0 L}\int_a^s \frac{dr}{r} = \frac{q^2}{4\pi\epsilon_0 L}\ln\left(\frac{s}{a}\right).$$

Using the expression for the capacitance C derived in Example 2, the energy stored between the plates is

$$U = \tfrac{1}{2}q^2/C = \frac{1}{2}\,\frac{q^2}{2\pi\epsilon_0 L}\ln\left(\frac{b}{a}\right).$$

Thus, the value of s such that $U_s = \tfrac{1}{2}U$ is found from

$$\frac{q^2}{4\pi\epsilon_0 L}\ln\left(\frac{s}{a}\right) = \frac{1}{2}\,\frac{q^2}{4\pi\epsilon_0 L}\ln\left(\frac{b}{a}\right),$$

$$s = (ab)^{\frac{1}{2}}.$$

27-40

Move one plate with zero acceleration holding the other plate in place. As the charge q on the plates does not change, neither does the electric field $E = q/A\epsilon_0$. But $V = Ex$ will increase; hence,

$$dW = Fdx = U_f - U_i = \tfrac{1}{2}q(V_f - V_i),$$

$$Fdx = \tfrac{1}{2}q[(x + dx)E - xE] = \tfrac{1}{2}qEdx,$$

$$F = \tfrac{1}{2}qE = \frac{1}{2}\,\frac{q^2}{A\epsilon_0} \quad \underline{\text{Ans}}.$$

27-42

The work done by the bubble in expanding is

$$W = \int p_b dV,$$

p_b being the pressure of the air in the bubble. If surface tension is ignored, then initially (uncharged bubble) this equals the atmospheric pressure p. Assuming this pressure to remain constant,

$$W = p\Delta V = p\,\frac{4\pi}{3}(R^3 - R_0^3).$$

The electric field set up by a charge q distributed uniformly over a spherical surface exists throughout all space external to the surface. If the sphere has radius s, the stored electric energy is

$$U = \int \tfrac{1}{2}\epsilon_0 E^2 dV = \int_s^\infty \tfrac{1}{2}\epsilon_0 \left(\frac{q}{4\pi\epsilon_0 r^2}\right)^2 (4\pi r^2 dr) = \frac{q^2}{8\pi\epsilon_0}\int_s^\infty r^{-2} dr = \frac{q^2}{8\pi\epsilon_0}\frac{1}{s}.$$

Thus, if the surface expands from radius R_0 to radius R with q held fixed, the stored energy decreases by

$$\Delta U = \frac{q^2}{8\pi\epsilon_0}\left(\frac{1}{R_0} - \frac{1}{R}\right).$$

Setting $W = \Delta U$ gives

$$\frac{q^2}{8\pi\epsilon_0}\frac{R - R_0}{RR_0} = \tfrac{4}{3}\pi p(R^3 - R_0^3),$$

$$q^2 = \frac{32}{3}\pi^2 \epsilon_0 pRR_0(R^2 + RR_0 + R_0^2).$$

This calculation is somewhat artificial since it is not likely that the pressure of the air inside the bubble remains constant; also, surface tension has been ignored. If it is assumed that the pressure in the bubble is inversly proportional to the volume of bubble, and surface tension still ignored, then the result obtained will be

$$q^2 = 32\pi^2\epsilon_0 pR(R^3 - R_0^3).$$

27-44

The capacitance of the capacitor is

$$C = 7 \times 10^{-8} = k\epsilon_0 A/d = 2.8\epsilon_0 A/d.$$

The dielectric strength is the maximum potential gradient possible without breakdown, so that

$$(V/d)_{max} = \frac{4000}{d} = 18 \times 10^6.$$

Hence, $d = 2.222 \times 10^{-4}$ m. Then, from the first equation,

$$A = (7 \times 10^{-8})(2.222 \times 10^{-4})/2.8\epsilon_0 = 0.628\ m^2 \quad \underline{\text{Ans.}}$$

27-46

Let q, -q be the charges on the plates separated by a distance d. If V is the potential difference across the plates, then,

$$C = \frac{q}{V}.$$

The uniform electric fields in the dielectrics are $\sigma/k\epsilon_0$ so that

$$E_1 = \frac{q/A}{k_1\epsilon_0},\quad E_2 = \frac{q/A}{k_2\epsilon_0}.$$

Therefore,

$$C = \frac{q}{E_1(\frac{1}{2}d) + E_2(\frac{1}{2}d)} = \frac{2q}{d}\left(\frac{q/A}{k_1\epsilon_0} + \frac{q/A}{k_2\epsilon_0}\right)^{-1} = \frac{2A\epsilon_0}{d}\,\frac{k_1k_2}{k_1 + k_2}.$$

For an air capacitor, $k_1 = k_2 = 1$, and the above reduces to $C = A\epsilon_0/d$, as expected. If $k_1 = k_2 = k$ then, as anticipated, the above becomes $C = k\epsilon_0 A/d$.

27-52

For the values before the slab is introduced, see Example 8.

(d) The capacitance with the slab in place does not depend on the conditions under which the slab is introduced, but only on the geometry of the capacitor and the dielectric between its plates. Hence, from Example 8, part (f), C = 16 pF <u>Ans.</u>

(a) Since the battery remains connected, V = 100 volts. Thus,

$$q = CV = (16 \times 10^{-12}\ F)(100\ V) = 1.6 \times 10^{-9}\ C \quad \underline{\text{Ans.}}$$

(b) The electric field in the gap is

$$E_0 = \frac{q}{A\epsilon_0} = \frac{1.6 \times 10^{-9}\ C}{(10^{-2}\ m^2)(8.85 \times 10^{-12}\ C^2/N\cdot m^2)} = 1.8 \times 10^4\ N/C \quad \underline{\text{Ans.}}$$

(c) The electric field in the dielectric is

$$E_d = \frac{1}{k}E_0 = \frac{1}{7}(1.8 \times 10^4 \text{ N/C}) = 2.6 \times 10^3 \text{ N/C} \quad \underline{\text{Ans}}.$$

27-54

Let the electric field in the gaps be E and the field in the dielectric be E_d. Now $E_d = E/k$ and therefore the potential difference across the plates is

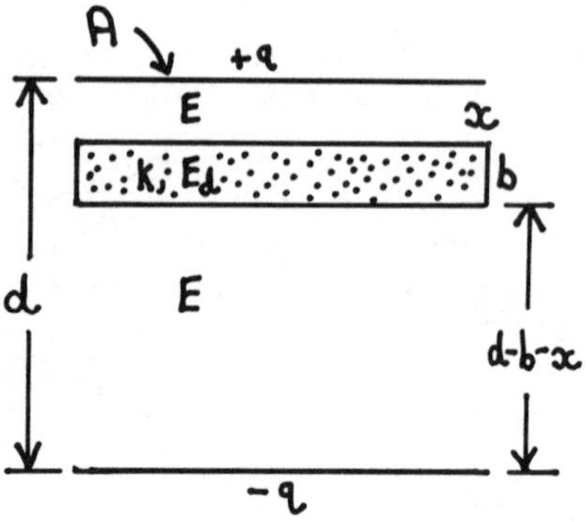

$$V = xE + bE_d + (d - b - x)E,$$

$$V = E(d - b) + E_d b,$$

$$V = E(d - b + \frac{b}{k}).$$

But, for a parallel-plate capacitor,

$$E = \sigma/\epsilon_0 = q/A\epsilon_0,$$

A the area of each plate. Thus,

$$V = \frac{q}{A\epsilon_0}[d + b(\frac{1}{k} - 1)] = \frac{q}{A\epsilon_0 k}[kd - b(k - 1)],$$

and hence the capacitance is

$$C = \frac{q}{V} = \frac{k\epsilon_0 A}{kd - b(k - 1)}.$$

The cases $b = 0$ and $k = 1$ each correspond to an air capacitor for which $C = \epsilon_0 A/d$ by the above formula, as expected. If $b = d$ the dielectric fills entirely the space between the plates giving $C = k\epsilon_0 A/d$, again as anticipated.

CHAPTER 28

28-3

(a) Since each alpha-particle carries a charge $2(1.6 \times 10^{-19}$ C) and 1 A = 1 C/s, the number of alpha-particles striking the surface each second is

$$\frac{0.25 \times 10^{-6} \text{ C/s}}{3.2 \times 10^{-19} \text{ C}} = 7.81 \times 10^{11}.$$

Hence, in three seconds, $3(7.81 \times 10^{11}) = 2.34 \times 10^{12}$ particles strike.

(b) The kinetic energy $K = (20 \text{ MeV})(1.6 \times 10^{-13} \text{ J/MeV}) = \frac{1}{2}mv^2$. The mass m of an alpha-particle is about four times a proton mass, so that $m = (4)(1.66 \times 10^{-27}$ kg). Hence,

$$3.2 \times 10^{-12} = \tfrac{1}{2}(6.64 \times 10^{-27})v^2,$$

$$v = 3.1046 \times 10^7 \text{ m/s}.$$

The time t required to travel 20 cm = 0.20 m is

$$t = \frac{0.20 \text{ m}}{3.1046 \times 10^7 \text{ m/s}} = 6.442 \times 10^{-9} \text{ s}.$$

From (a), 7.81×10^{11} particles pass any point each second; it follows that in time t, $(7.81 \times 10^{11})(6.442 \times 10^{-9}) = 5031$ particles pass, and therefore there are this many particles in a 20 cm length of the beam.

(c) If the accelerating potential is V, then K = qV, so that 20 MeV = (+2)V, giving $V = 10^7$ volts.

28-7

Let σ be the charge density sought. On a length L of the belt of width w there is present an amount of charge $Q = \sigma(wL)$. If the belt is moving, the current i due to the moving charges will be

$$i = \frac{dQ}{dt} = \frac{d(\sigma wL)}{dt} = \sigma w \frac{dL}{dt} = \sigma w v,$$

v the speed of the belt. Therefore,

$$10^{-4} = \sigma(0.5)(30),$$

$$\sigma = 6.67 \times 10^{-6}\ C/m^2 \quad \underline{\text{Ans}}.$$

28-9

The potential of a conducting sphere carrying a charge q is

$$V = \frac{1}{4\pi\epsilon_0}\frac{q}{R}.$$

If q is changing, V will also, at the rate

$$\frac{dV}{dt} = \frac{1/R}{4\pi\epsilon_0}\frac{dq}{dt} = \frac{1}{4\pi\epsilon_0 R}\,i = \frac{1}{4\pi\epsilon_0 R}(i_{in} - i_{out})$$

i being the net current. If i does not vary with time,

$$\frac{\Delta V}{\Delta t} = \frac{1}{4\pi\epsilon_0 R}\,i,$$

$$\frac{1000}{\Delta t} = (9 \times 10^9)(0.1)^{-1}(1.000002 - 1),$$

$$\Delta t = 5.56 \times 10^{-3}\ s \quad \underline{\text{Ans}}.$$

28-17

In order that the density of the material not be altered by the stretching, the volume V of the wire must be unchanged. If the length increases by a factor of 3, then since $V = AL$, the cross-sectional area must decrease by the factor of 3. Hence, the new resistance R is

$$R = \rho\frac{L}{A} = \rho\frac{3L_0}{A_0/3} = 9\left(\rho\frac{L_0}{A_0}\right) = 9(6) = 54\ \Omega \quad \underline{\text{Ans}}.$$

28-20

The length of the composite resistor is still ℓ, the length of any single resistor, but the area through which current can flow is now nine times the area A of a single wire. The new resistance is

$$R = \rho\frac{\ell}{A} = \frac{\rho\ell}{9\pi d^2/4}.$$

The area of the equivalent single wire is $\pi D^2/4$, but its length is still ℓ. For this wire to have the same resistance as the composite given above, the diameter D must be chosen so that

$$\frac{\rho\ell}{9\pi d^2/4} = \frac{\rho\ell}{\pi D^2/4},$$

$$D = 3d \quad \underline{\text{Ans}}.$$

28-21

Since the materials of the two conductors are the same, $\rho_A = \rho_B$. Also, it is given that $\ell_A = \ell_B$. Therefore, with $R = \rho\ell/A$ for each,

$$\frac{R_A}{R_B} = \frac{A_B}{A_A} = \frac{\pi(d_o^2 - d_i^2)/4}{\pi d^2/4} = \frac{d_o^2 - d_i^2}{d^2} = \frac{2^2 - 1^2}{1^2} = 3 \quad \underline{\text{Ans}}.$$

28-25

(a) Consider a thin slice, perpendicular to the axis of the cone, of thickness dx and at a distance x from the narrow end of the cone; the cross-sectional area of this slice is πr^2 and its resistance dR is

$$dR = \rho\frac{d\ell}{A} = \rho\frac{dx}{\pi r^2}.$$

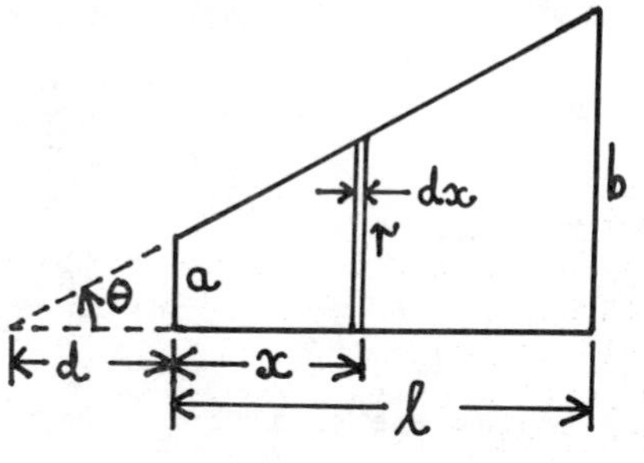

Let θ be the semi-angle of the cone from which the resistor was cut. Let the vertex of the cone be at a distance d from the narrow end of the resistor, as shown. Then,

$$\tan\theta = \frac{a}{d} = \frac{r}{x + d} = \frac{b}{\ell + d}.$$

From the first equation

$$\frac{a}{d} = \frac{r}{x + d},$$

it is found that

$$d = \frac{ax}{r - a}.$$

Substituting this into the second equation

$$\frac{r}{x + d} = \frac{b}{\ell + d},$$

gives

$$r = \frac{b - a}{\ell} x + a.$$

Thus, the resistance R of the object is

$$R = \int dR = \frac{\rho}{\pi}\int_0^{\ell}\left(\frac{b - a}{\ell} x + a\right)^{-2} dx = \rho\frac{\ell}{\pi ab} \quad \underline{\text{Ans}}.$$

(b) For zero taper a = b, the above gives $R = \rho\ell/\pi a^2 = \rho\ell/A$, as expected.

28-28

(a) With $R = \rho\ell/A$ and assuming small changes,

$$\Delta R = \frac{\partial R}{\partial \rho}(\Delta\rho) + \frac{\partial R}{\partial \ell}(\Delta\ell) + \frac{\partial R}{\partial A}(\Delta A),$$

and therefore

$$\frac{\Delta R}{R} = \frac{\Delta\rho}{\rho} + \frac{\Delta\ell}{\ell} - \frac{\Delta A}{A}.$$

Let β be the coefficient of thermal expansion, so that

$$\Delta\ell = \beta\ell(\Delta T), \quad \Delta A = 2\beta A(\Delta T),$$

where ΔT is the change in temperature. Therefore, the percent change in length is $\beta(\Delta T) = (1.7 \times 10^{-5})(1) = 0.0017\%$, and the percent change in area is twice that, or 0.0034%. The percent change in R is, from the above,

$$\frac{\Delta R}{R} = (\alpha + \beta - 2\beta)\Delta T = (\alpha - \beta)\Delta T = 0.3883\%,$$

since $\alpha = 3.9 \times 10^{-3}$ /C°.

(b) The relative change in resistivity is much larger than those in length and area.

28-29

The resistance of the conductor is $R = R_i + R_c$, where R_i, R_c are the resistances of the iron and carbon sections. If their cross-sectional areas are the same, then

$$R = \rho_i \ell_i/A + \rho_c \ell_c/A,$$

$$\Delta R = \frac{1}{A}(\ell_i \Delta\rho_i + \ell_c \Delta\rho_c) = 0,$$

the last by supposition. But

$$\Delta\rho_i = \alpha_i \rho_i \Delta T, \quad \Delta\rho_c = \alpha_c \rho_c \Delta T$$

so that the thicknesses ℓ_i, ℓ_c must be chosen to satisfy $\ell_i \alpha_i \rho_i + \ell_c \alpha_c \rho_c = 0$, giving

$$\ell_i/\ell_c = -\alpha_c \rho_c/\alpha_i \rho_i = 35 \quad \underline{\text{Ans}},$$

using the entries in Table 28-1.

28-33

(a) The electric field is $E = \rho j = (1.7 \times 10^{-8}\ \Omega\cdot\text{m})(5 \times 10^{6}\ \text{A/m}^2) = 0.085$ N/C Ans.

(b) The acceleration is given from $F = ma$ with $F = eE$. Hence, $a = \frac{e}{m}E = (1.76 \times 10^{11}\ \text{C/kg})(0.085\ \text{N/C}) = 1.496 \times 10^{10}\ \text{m/s}^2$ Ans.

(c) From rest, the time t required, since $v_d = at$, is $t = v_d/a = (3.7 \times 10^{-4}\ \text{m/s})/(1.496 \times 10^{10}\ \text{m/s}^2) = 2.47 \times 10^{-14}$ s Ans, which is essentially the same as found in Example 4.

28-43

(a) Since the power P dissipated is $P = iV$, the current i must be

$$i = \frac{1250\ W}{115\ V} = 10.87\ A \quad \underline{Ans.}$$

(b) By Ohm's law $V = iR$ so that

$$R = \frac{V}{i} = \frac{115\ V}{10.87\ A} = 10.58\ \Omega \quad \underline{Ans.}$$

(c) Since $P = 1250$ J/s, in one hour the heater generates

$$(1250\ J/s)(3600\ s/h\ ) = 4.5 \times 10^6\ J \quad \underline{Ans.}$$

28-46

The rate of Joule (thermal) heating is $P = V^2/R$. As the temperature drops, R changes but V does not; hence,

$$\Delta P \approx \frac{\partial P}{\partial R}(\Delta R) = -\frac{V^2}{R^2}(\Delta R) = -\frac{P}{R}(\Delta R).$$

But $R = \rho L/A$ so that $\Delta R = (\Delta\rho)L/A$; dividing these equations gives

$$\frac{\Delta R}{R} = \frac{\Delta\rho}{\rho} = \alpha(\Delta T).$$

Therefore,

$$\Delta P = -P\alpha(\Delta T) = -(500\ W)(4 \times 10^{-4}\ /C^\circ)(-600\ C^\circ) = +120\ W,$$

and it follows that at 200°C, $P = 500 + 120 = 620$ W $\underline{Ans.}$

28-47

(a) The charge q accelerated during each pulse is

$$q = iT = (0.5\ A)(0.1 \times 10^{-6}\ s) = 5 \times 10^{-8}\ C.$$

Each electron carries 1.6×10^{-19} C, so the number n of electrons accelerated is

$$n = (5 \times 10^{-8})/(1.6 \times 10^{-19}) = 3.125 \times 10^{11} \quad \underline{Ans.}$$

(b) In one second a total charge

$$Q = (500 \text{ pulses})(5 \times 10^{-8} \text{ C/pulse}) = 25 \times 10^{-6} \text{ C}$$

is accelerated. The average current is

$$\bar{I} = \frac{Q}{t} = 25 \times 10^{-6} \text{ A} \quad \underline{\text{Ans.}}$$

(c) The accelerating voltage V must be

$$V = K/e = (50 \text{ MeV})/(1) = 5 \times 10^{7} \text{ volts},$$

since the charge on the electron is one quantum unit. The power output is $P = iV$; per pulse this is

$$P = (0.5 \text{ A})(5 \times 10^{7} \text{ V}) = 2.5 \times 10^{7} \text{ W} \quad \underline{\text{Ans}};$$

over one second the average power is

$$\bar{P} = \bar{i}V = (25 \times 10^{-6} \text{ A})(5 \times 10^{7} \text{ V}) = 1250 \text{ W} \quad \underline{\text{Ans.}}$$

CHAPTER 29

29-7

(a) The current in the circuit is

$$i = \frac{\mathcal{E}}{r + R},$$

and therefore the rate of thermal (Joule) heating is

$$P = i^2R = \frac{\mathcal{E}^2 R}{(r + R)^2}.$$

To find the value of R that maximizes P, set dP/dR equal to zero and solve for R:

$$\frac{dP}{dR} = 0 = \mathcal{E}^2 \frac{(r - R)}{(r + R)^3},$$

which gives R = r Ans.

(b) The maximum power dissipation is

$$P(R = r) = i^2 r = \mathcal{E}^2 r/(r + r)^2 = \mathcal{E}^2/4r \quad \text{Ans.}$$

29-10

Let there be n^2 resistors. They cannot be connected either all in series or all in parallel since the equivalent resistance in the former case is n^2R and in the latter R/n^2 (R the resistance of each resistor), and these equal R only for n = 1. But a single resistor cannot tolerate a current greater than $(P/R)^{\frac{1}{2}} = (1\ W/10\ \Omega)^{\frac{1}{2}} = 10^{-\frac{1}{2}}$ A, and a current equal to $(5\ W/10\ \Omega)^{\frac{1}{2}} = 2^{-\frac{1}{2}}$ A is required. Suppose the n^2 resistors are arranged into n sets, each set being n resistors connected in parallel, the sets being connected in series. The equivalent resistance $R_e = n(R/n) = R = 10\ \Omega$. The current i entering the combination must be $2^{-\frac{1}{2}}$ A in order that $i^2R_e = 5$ W as demanded. The current passing through each resistor is $i/n = 1/2^{\frac{1}{2}}n$, so that each resistor dissipates

$$P = (1/2^{\frac{1}{2}}n\ A)^2(10\ \Omega) = 5/n^2\ W.$$

This must be 1 W at the most: hence, $5/n^2 \leq 1$, so that the minimum n = 3. Thus, at least nine resistors are needed, connected as shown below.

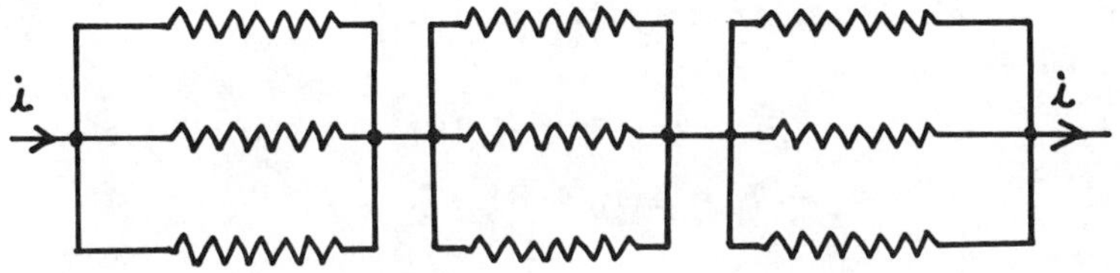

29-13

(a) Let the potential difference be $V = V_A - V_B > 0$ since energy is removed from the circuit. As $P = iV$, 50 W = (1 A)V, giving V = 50 volts Ans.

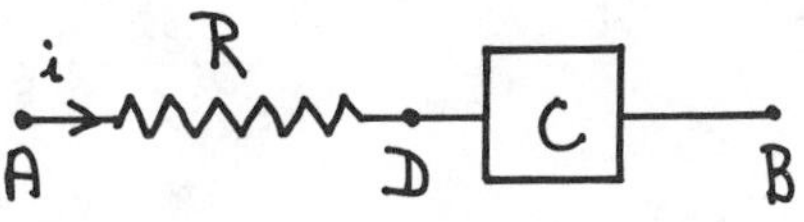

(b) Consider a point D between the resistor and element C; now

$$V_A - V_D = iR = 2 \text{ volts},$$

so that

$$V_A - V_B = (V_A - V_D) + (V_D - V_B),$$

$$50 = 2 + (V_D - V_B),$$

$$V_D - V_B = 48 \text{ volts} \quad \underline{\text{Ans.}}$$

(c) Since $V_D > V_B$, the negative terminal must be at B.

29-14

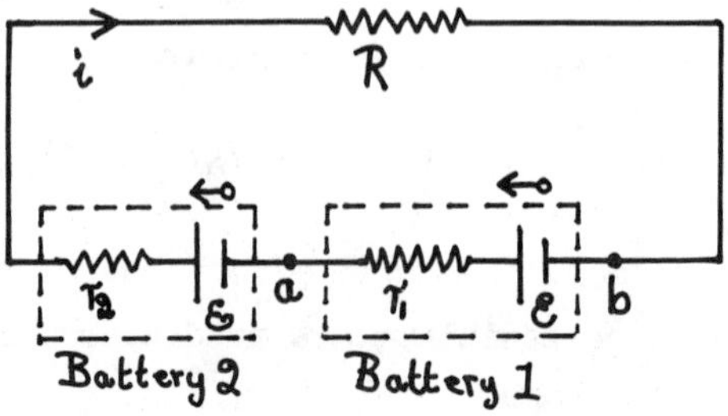

Let the points a and b represent the terminals of battery 1. The current i flows clockwise in the diagram and

$$V_a + \mathcal{E} - ir_2 - iR = V_b,$$

$$V_a - V_b = i(r_2 + R) - \mathcal{E}.$$

If $V_a - V_b = 0$, then the last equation gives

$$i(r_2 + R) = \mathcal{E}.$$

But, by the loop theorem, $i = 2\mathcal{E}/(r_1 + r_2 + R)$ so that the condition becomes

$$\frac{2\mathcal{E}}{r_1 + r_2 + R}(r_2 + R) = \mathcal{E},$$

$$R = r_1 - r_2 \quad \text{Ans.}$$

29-16

The brighter bulb indicates a higher value of i^2R.

(a) In parallel, $i_2 = \mathcal{E}/R_2$ so that $i_2^2R_2 = \mathcal{E}^2/R_2$. Similarly, $i_1^2R_1 = \mathcal{E}^2/R_1$. Since $R_2 < R_1$, $\mathcal{E}^2/R_2 > \mathcal{E}^2/R_1$, so the R_2 bulb is brighter.

(b) When connected in series the currents through the bulbs are equal; therefore $i^2R_1 > i^2R_2$ since $R_1 > R_2$, and R_1 is brighter.

29-23

By the loop theorem, applied to the left and right hand branches respectively, $\mathcal{E}_1 + \mathcal{E}_2 + I_2 R_2 = I_3 R_3$

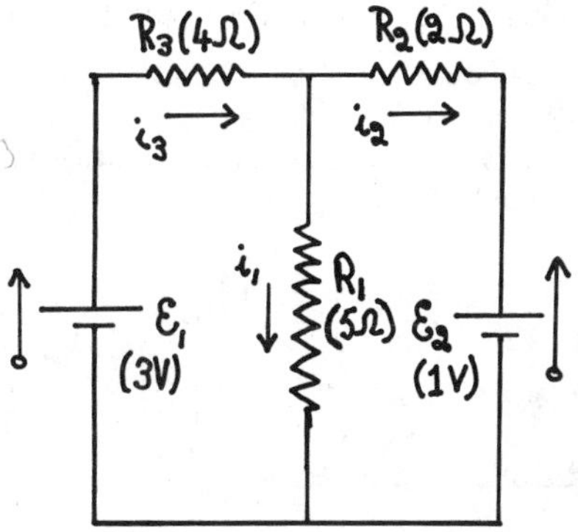

$$\mathcal{E}_1 - i_3 R_3 - i_1 R_1 = 0,$$

$$\mathcal{E}_2 + i_2 R_2 - i_1 R_1 = 0,$$

assuming the currents are directed as shown. The junction equation further provides that

$$i_3 = i_1 + i_2.$$

These three equations may be solved for the currents to yield (writing $r^2 = R_1 R_2 + R_1 R_3 + R_2 R_3$),

$$r^2 i_1 = \mathcal{E}_1 R_2 + \mathcal{E}_2 R_3,$$

$$r^2 i_2 = \mathcal{E}_1 R_1 - \mathcal{E}_2 (R_1 + R_3),$$

$$r^2 i_3 = \mathcal{E}_1 (R_1 + R_2) - \mathcal{E}_2 R_1.$$

Numerically, these give

$$i_1 = \frac{5}{19} \text{ A}, \quad i_2 = \frac{3}{19} \text{ A}, \quad i_3 = \frac{8}{19} \text{ A}.$$

(a) The rates at which thermal energy appears in the resistors are

$$P_1 = i_1^2 R_1 = 0.346 \text{ W}, \quad P_2 = i_2^2 R_2 = 0.050 \text{ W}, \quad P_3 = i_3^2 R_3 = 0.709 \text{ W}.$$

(b) The powers supplied by the batteries are

$$P_{b1} = i_3 \mathcal{E}_1 = 1.263 \text{ W}, \quad P_{b2} = -i_2 \mathcal{E}_2 = -0.158 \text{ W}$$

negative in the second case as the current i_2 flows through battery 2 in the direction opposite to the battery's emf.

(c) Energy is supplied to the circuit by battery 1 and is dissipated as heat in the three resistors and stored in battery 2: as expected then, $1.263 \approx 0.346 + 0.050 + 0.709 + 0.158$, in W.

29-28

Let the currents be directed as shown, with the currents already labelled to satisfy the junction equation. Since $V_a = V_b$,

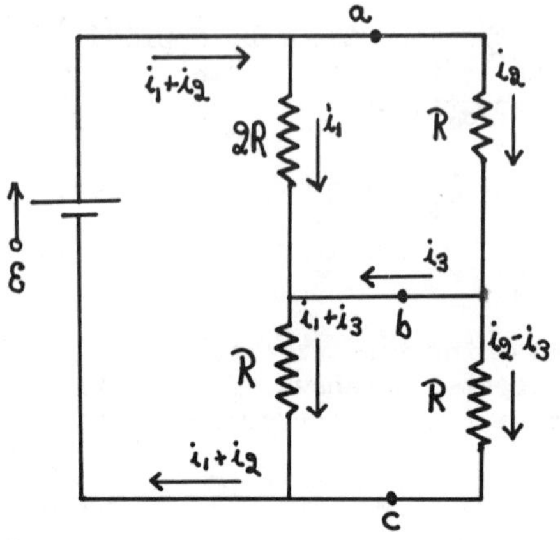

$$i_1(2R) = i_2(R),$$

$$i_2 = 2i_1.$$

Similarly, with $V_b = V_c$,

$$(i_1 + i_3)R = (i_2 - i_3)R,$$

$$i_2 - i_1 = 2i_3.$$

Eliminating i_1 between these two current equations above gives $i_2 = 4i_3$. But, applying the loop theorem to the outside loop,

$$\mathcal{E} - i_2R - (i_2 - i_3)R = 0.$$

Substituting $i_2 = 4i_3$ found above yields $i_3 = \mathcal{E}/7R$ Ans.

29-33

The "brute force" method is illustrated by solving part (b) for the equivalent resistance of a face diagonal. Connect a battery between points B and C, between which the equivalent resistance is desired. This equivalent resistance R_e is defined by

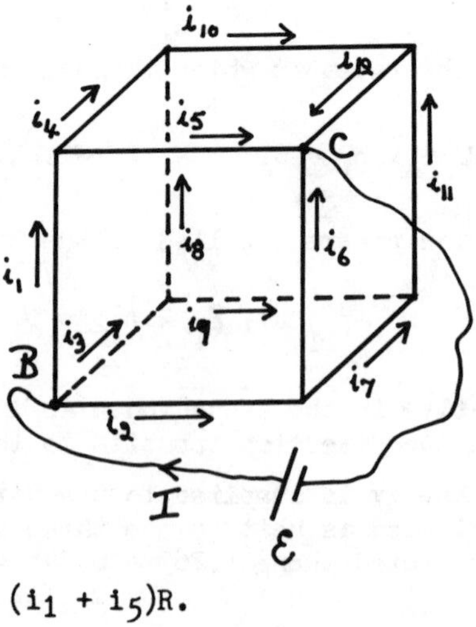

$$\mathcal{E} = IR_e.$$

But $I = i_1 + i_2 + i_3$ by the junction equation, and

$$\mathcal{E} - i_1R - i_5R = 0$$

by the loop equation, so that $IR_e = (i_1 + i_5)R$.

Now write down all the loop equations; since the resistances are identical, the R in each equation will drop out, leaving

$$i_1 + i_5 = i_2 + i_6, \quad i_7 + i_{11} = i_6 - i_{12},$$
$$i_3 + i_9 = i_7 + i_2, \quad i_8 + i_{10} = i_9 + i_{11},$$
$$i_1 + i_4 = i_8 + i_3, \quad i_4 + i_{10} = i_5 - i_{12}.$$

The junction equations are

$$i_1 = i_4 + i_5, \quad i_{10} + i_{11} = i_{12},$$
$$i_{10} = i_4 + i_8, \quad i_2 = i_6 + i_7,$$
$$i_3 = i_8 + i_9, \quad i_{11} = i_7 + i_9.$$

Solving these 12 equations for the 12 currents gives

$$i_1 = i_2 = i_5 = i_6 = i,$$
$$i_3 = i_{12} = \frac{2}{3}i,$$
$$i_4 = i_7 = 0,$$
$$i_8 = i_9 = i_{10} = i_{11} = \frac{1}{3}i.$$

Thus $I = i_1 + i_2 + i_3 = \frac{8}{3}i$, and $i_1 + i_5 = 2i$. Therefore

$$2iR = \frac{8}{3}i\,R_e,$$
$$R_e = \frac{3}{4}R \quad \underline{\text{Ans}}.$$

<u>29-34</u>

With $R_V = \infty$, $i_V = 0$ and

$$i = i_1 = \frac{\mathcal{E}}{r + R_1 + R_2} = 0.04545\ \text{A};$$

hence, the voltmeter reading is $iR_1 = 2.2727$ V. When R_V is finite, $i_1R_1 = i_VR_V$ = voltmeter reading. But in this case also,

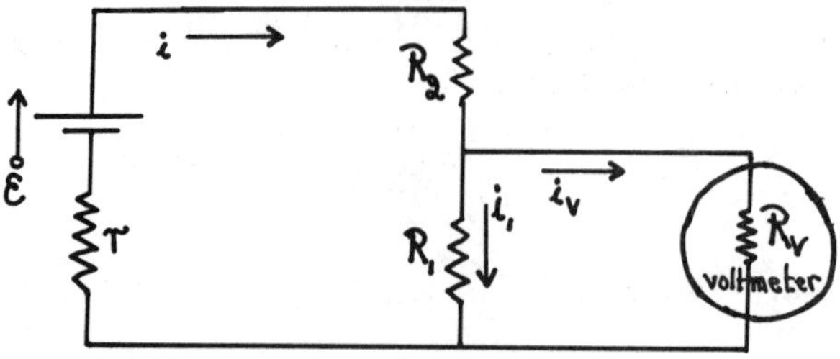

$$\mathcal{E} - (r + R_2)i - i_1R_1 = 0, \quad i = i_1 + i_V.$$

The three equations yield

$$i_1R_1 = \frac{R_1R_V\mathcal{E}}{rR_V + R_2R_V + R_1R_V + R_1R_2 + R_1r} = 2.2124 \text{ V}.$$

Thus, the error is

$$\frac{2.2727 - 2.2124}{2.2727} = 2.7 \% \quad \underline{\text{Ans.}}$$

29-36

Ignore the battery and R_0 and focus on the rhombus. By the loop theorem,

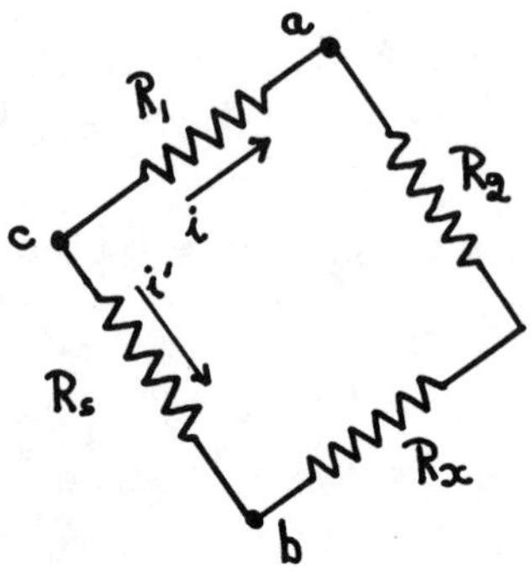

$$i(R_1 + R_2) = i'(R_x + R_s).$$

Also, $V_{ac} = V_{bc}$, so that

$$iR_1 = i'R_s.$$

Eliminate i, say, between these two equations to get

$$i'R_s(R_1 + R_2) = i'R_1(R_x + R_s),$$

$$R_x = R_s(R_2/R_1).$$

29-44

The charge on the capacitor is $q = q_0 e^{-t/\tau}$ and therefore the potential difference across it is

$$V = q/C = (q_0/C)e^{-t/\tau} = V_0 e^{-t/\tau}.$$

When $t = 0$, $V = 100$ volts so that, by the above equation, $V_0 = 100$ volts also; hence,

$$V = 100e^{-t/\tau}.$$

Furthermore, at $t = 10$ s, $V = 1$ volt; substituting these into the above gives

$$1 = 100e^{-10/\tau}.$$

(a) At $t = 20$ s,

$$V = 100e^{-20/\tau} = 100(e^{-10/\tau})^2 = 100(\tfrac{1}{100})^2 = 0.01 \text{ V} \quad \underline{\text{Ans}}.$$

(b) As shown above, $1 = 100e^{-10/\tau}$, and therefore

$$\ln(100) = 10/\tau,$$

$$\tau = 10/\ln(100) = 2.17 \text{ s} \quad \underline{\text{Ans}}.$$

29-46

(a) The charge on the capacitor and the energy stored within it are

$$q = q_0 e^{-t/\tau}, \quad U_C = \tfrac{1}{2}q^2/C.$$

At $t = 0$, $U_C = 0.5$ J, so that $0.5 = \frac{1}{2}q_0^2/(10^{-6})$, giving $q_0 = 10^{-3}$ C.

(b) The current i is found by differentiating the charge q with respect to time t:

$$i = \frac{dq}{dt} = -(q_0/\tau)e^{-t/\tau}.$$

Therefore, disregarding the sign,

$$i(0) = q_0/\tau = q_0/RC = \frac{10^{-3}}{(10^6)(10^{-6})} = 10^{-3} \text{ A} \quad \underline{\text{Ans}}.$$

(c) The voltage across the capacitor is

$$V_C = q/C = (q_0/C)e^{-t/\tau}.$$

But, from (b), the time constant is evidently $\tau = RC = 1$ s, and $q_0/C = (10^{-3}\ C)/(10^{-6}\ F) = 10^3$ V. All this gives

$$V_C = 1000e^{-t}\ V \quad \underline{\text{Ans}}.$$

From the loop theorem, $V_C + V_R = 0$, and thus $V_R = -1000e^{-t}$ volts.
(d) The rate of generation of thermal energy in the resistor is

$$P = i^2R = [(-q_0/\tau)e^{-t/\tau}]^2R = (q_0^2R/\tau^2)e^{-2t/\tau} = e^{-2t}\ W \quad \underline{\text{Ans}},$$

using the numerical results from the other parts of the problem.

<u>29-47</u>
(a) The time constant is $\tau = RC = (3 \times 10^6\ \Omega)(10^{-6}\ F) = 3$ s. Also

$$q = \mathcal{E}C(1 - e^{-t/\tau}),\ i = (\mathcal{E}/R)e^{-t/\tau}.$$

Hence, for t = 1 s,

$$\frac{dq}{dt} = i = (4\ V)(e^{-1/3})/(3 \times 10^6\ \Omega) = 9.5538 \times 10^{-7}\ A \quad \underline{\text{Ans}}.$$

(b) The rate at which energy is being stored in the capacitor is

$$dU_C/dt = d(q^2/2C)/dt = \frac{q}{C}\frac{dq}{dt} = iq/C.$$

Now $q(1) = (10^{-6}\ F)(4\ V)(1 - e^{-1/3}) = 1.1339 \times 10^{-6}$ C; $i(1)$ was found in (a) and therefore

$$dU_C/dt = (1.1339 \times 10^{-6}\ C)(9.5538 \times 10^{-7}\ A)/(10^{-6}\ F),$$

$$dU_C/dt = 1.0833 \times 10^{-6}\ W \quad \underline{\text{Ans}}.$$

(c) For the resistor,

$$dU_R/dt = i^2R = (9.5538 \times 10^{-7}\ A)^2(3 \times 10^6\ \Omega) = 2.7383 \times 10^{-6}\ W.$$

(d) Finally, the battery delivers energy at the rate

$$dU_B/dt = i\mathcal{E} = (9.5538 \times 10^{-7}\ \text{A})(4\ \text{V}) = 3.8215 \times 10^{-6}\ \text{W} \quad \underline{\text{Ans.}}$$

Notice that, within round-off error,

$$\frac{dU_B}{dt} = \frac{dU_C}{dt} + \frac{dU_R}{dt},$$

as required by the loop theorem.

29-49

(a) At t = 0 the capacitor exerts no influence and therefore

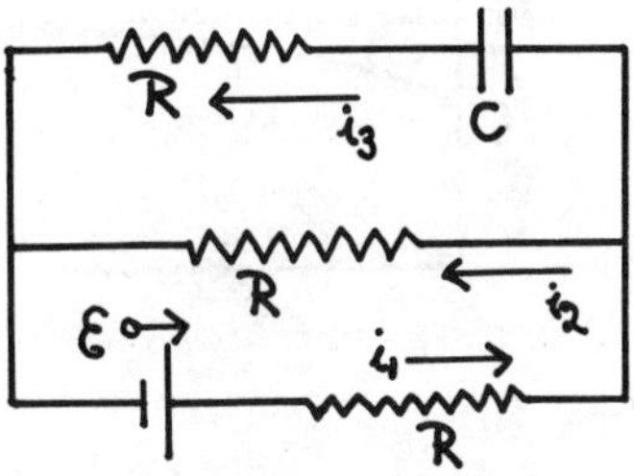

$$\mathcal{E} - i_1R - i_2R = 0,$$

$$-i_3R + i_2R = 0,$$

$$i_1 = i_2 + i_3.$$

Solving for the currents gives

$$i_2 = i_3 = \mathcal{E}/3R = 5.48 \times 10^{-4}\ \text{A} \quad \underline{\text{Ans,}}$$

$$i_1 = 2\mathcal{E}/3R = 1.10 \times 10^{-3}\ \text{A} \quad \underline{\text{Ans.}}$$

At $t = \infty$ the capacitor prevents the flow of current in its branch of the circuit: $i_3 = 0$ Ans. In the other branch, now just a series circuit,

$$i_1 = i_2 = i = \mathcal{E}/2R = 8.22 \times 10^{-4}\ \text{A} \quad \underline{\text{Ans.}}$$

(b) For other values of t,

$$i_1 = i_2 + i_3,$$

$$\mathcal{E} - i_1R - i_2R = 0,$$

$$-\frac{q}{C} - i_3R + i_2R = 0.$$

Using the first two equations to eliminate i_2 in the third gives

$$\frac{3}{2}R\frac{dq}{dt} + \frac{q}{C} = \tfrac{1}{2}\mathcal{E},$$

since $i_3 = dq/dt$. The solution of this equation, for $q(0) = 0$, is

$$q = \tfrac{1}{3}\mathcal{E}C(1 - e^{-2t/3RC}).$$

Therefore $i_3 = (\mathcal{E}/3R)e^{-2t/3RC}$. Using the third equation in (b) above, the potential drop across the resistor is found to be

$$V_R = i_2R = \frac{\mathcal{E}}{6}(3 - e^{-2t/3RC}).$$

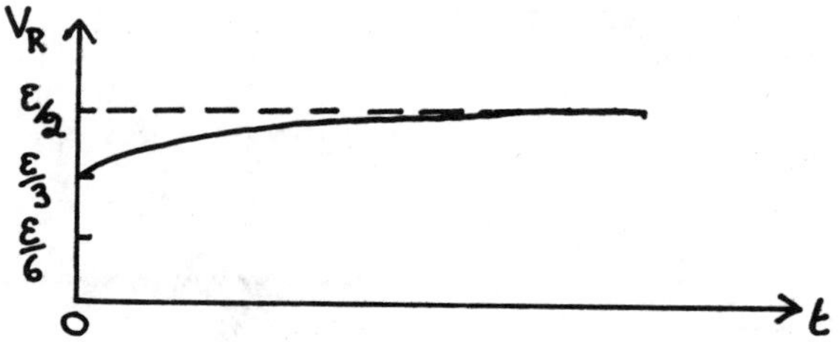

(c) From (b)

$$V_R(0) = \mathcal{E}/3 = 400\text{ V},\quad V_R(\infty) = \mathcal{E}/2 = 600\text{ V} \quad \underline{\text{Ans.}}$$

(d) The time constant is

$$\tau = \frac{3}{2}RC = \frac{3}{2}(7.3 \times 10^5\ \Omega)(6.5 \times 10^{-6}\text{ F}) = 7.1\text{ s}.$$

After many time constants i_3 is close to zero and at $t = \infty$ falls below measurable values.

CHAPTER 30

<u>30-3</u>

(a) The charge on the electron is $q = -e$, with $e = 1.6 \times 10^{-19}$ C. Therefore, $\vec{F} = q\vec{v} \times \vec{B} = -e\vec{v} \times \vec{B}$ $= e\vec{B} \times \vec{v}$, and this points to the east <u>Ans</u>.

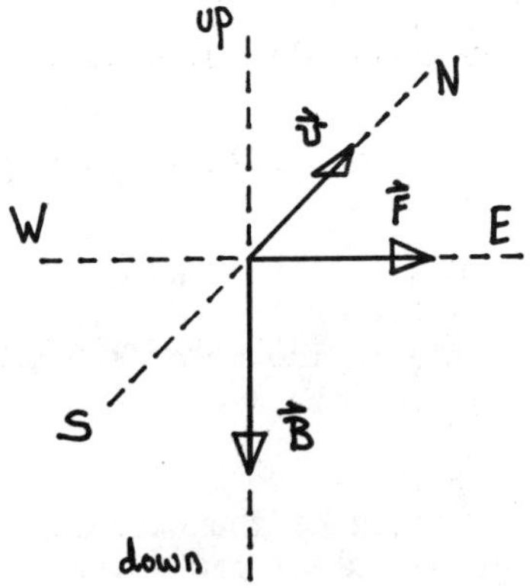

(b) In magnitude, since $\measuredangle(\vec{v},\vec{B}) = 90^\circ$,

$$F = evB = ma,$$

$$a = \left(\frac{e}{m}\right)vB\,.$$

But $v = (2K/m)^{\frac{1}{2}}$ and the kinetic energy $K = (12000)(1.6 \times 10^{-19}\text{ J})$ which, with $m = 9.11 \times 10^{-31}$ kg, gives $v = 6.492 \times 10^{7}$ m/s. Hence,

$$a = (1.76 \times 10^{11}\text{ C/kg})(6.492 \times 10^{7}\text{ m/s})(5.5 \times 10^{-5}\text{ T}),$$

$$a = 6.28 \times 10^{14}\text{ m/s}^2 \quad \underline{\text{Ans}}.$$

(c) The electrons follow a circular path of radius R at a constant speed v; for uniform circular motion,

$$mv^2/R = evB,$$

$$R = \frac{mv}{eB} = 6.7066\text{ m}.$$

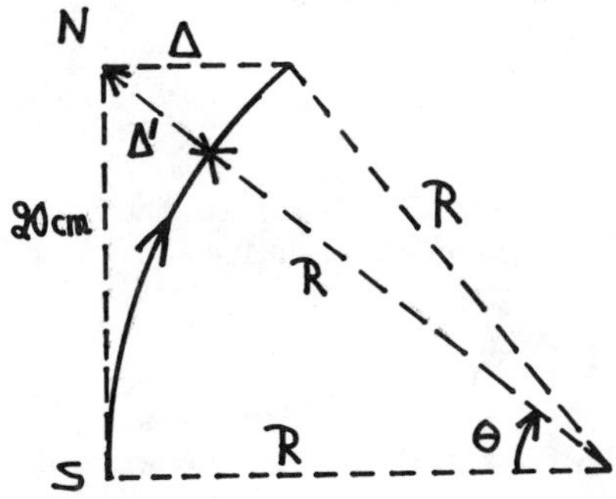

The deflection sought is Δ. From the figure,

$$\tan\theta = \frac{0.20\text{ m}}{6.7066\text{ m}} = 0.02982,$$

so that $\theta = 1°43'$. Also

$$R^2 = (R + \Delta')^2 + \Delta^2 - 2\Delta(R + \Delta')\cos\theta.$$

Now $\Delta' = R(\sec\theta - 1)$, and therefore this equation becomes

$$\Delta^2 - 2R\Delta + R^2\tan^2\theta = 0,$$

or, numerically,

$$\Delta^2 - 13.4132\,\Delta + 0.039996 = 0,$$

$$\Delta = 0.002983,\ 2R - 0.002983,$$

in meters. Thus the required deflection is about 3.0 mm. The second solution is the deflection from the same N-S line to the other side of the upper semicircular path.

30-6

If the radius of the circular path of the electrons after they emerge from the tube is less than d, the electrons will not reach the plate. For radius d,

$$evB = mv^2/d.$$

Since $K = \frac{1}{2}mv^2$, this equation in terms of K rather than v is

$$e\left(\frac{2K}{m}\right)^{\frac{1}{2}}B = \frac{2K}{d};$$

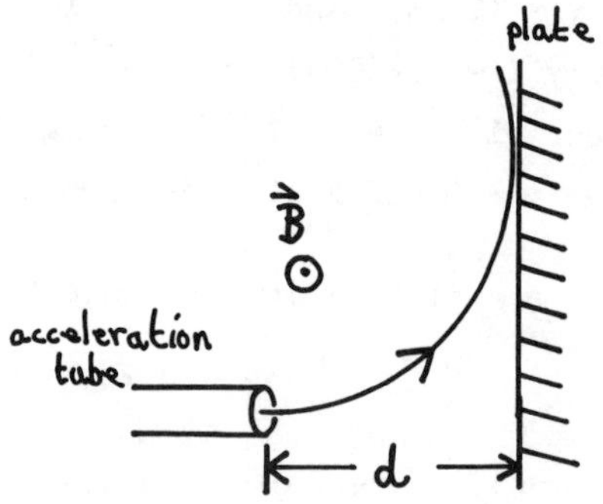

hence, choose

$$B > (2Km/e^2d^2)^{\frac{1}{2}},$$

and arrange the field so that it is perpendicular to the velocity of the emergent electrons.

30-11

Clearly, to remove the tension in the leads the force $\vec{F}$ exerted by the magnetic field on the wire of length L must equal $-m\vec{g}$. Since

$\vec{F} = i\vec{L} \times \vec{B}$, the current must pass from left to right. Then,

$$F = iLB\sin 90^\circ = iLB = mg,$$

$$i = \frac{mg}{LB} = \frac{(10^{-2}\ \text{kg})(9.8\ \text{m/s}^2)}{(0.6\ \text{m})(0.4\ \text{T})} = 0.41\ \text{A} \quad \underline{\text{Ans}}.$$

<u>30-13</u>

Let $\vec{B}$ make an angle θ with the vertical. The magnetic force $F = iLB$ since $\vec{B}$ is perpendicular to the horizontal rod. For the rod to be on the verge of moving,

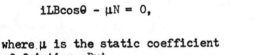

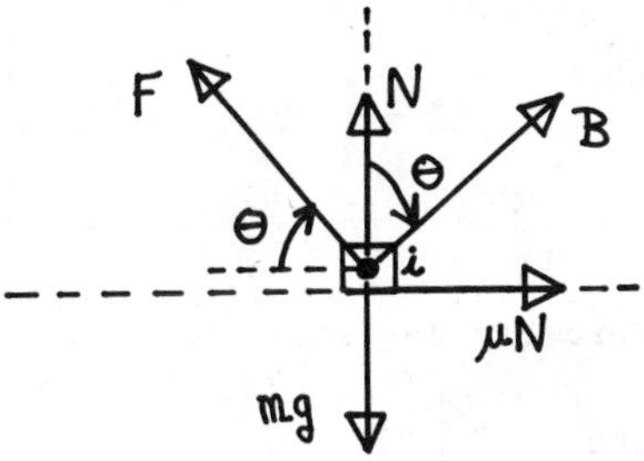

$$iLB\cos\theta - \mu N = 0,$$

where μ is the static coefficient of friction. But

$$N + iLB\sin\theta - mg = 0$$

so that $N = mg - iLB\sin\theta$. Hence,

$$iLB\cos\theta - \mu(mg - iLB\sin\theta) = 0,$$

$$B = \frac{\mu mg/iL}{\cos\theta + \mu\sin\theta}.$$

To find the minimum field, set $\frac{dB}{d\theta} = 0$; this gives

$$\theta = \tan^{-1}\mu = \tan^{-1}(0.6) = 31^\circ \quad \underline{\text{Ans}}.$$

Numerically, $\mu mg/iL = (0.6)(1)(9.8)/(50)(1) = 0.1176$ T. Using the equation for B above, with the value of θ just determined, gives $B_{min} = 0.101$ T <u>Ans</u>.

<u>30-15</u>

Look at the ring edge on. The forces $d\vec{F}$ on each of two small elements of the ring, each element of length $d\ell$, situated opposite each other are shown. Evidently, the horizontal components cancel, leaving only the vertical components. Hence,

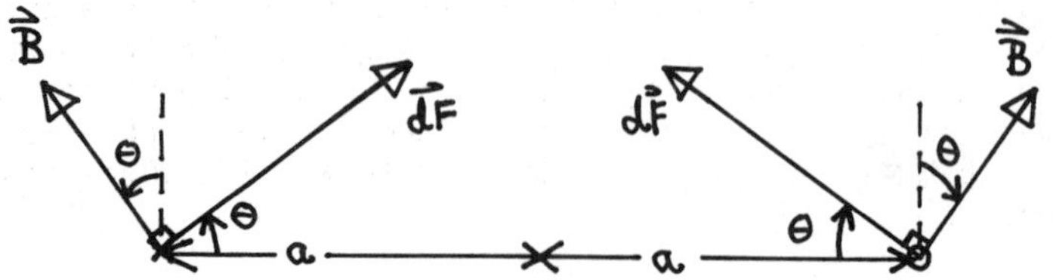

$$F = \int dF\sin\theta = \int i(d\ell)B\sin 90^\circ \sin\theta = iB\sin\theta\int d\ell,$$

$$F = iB\ell\sin\theta = iB(2\pi a)\sin\theta \quad \underline{\text{Ans}},$$

vertically upward.

30-19

(a) The deflection of a galvanometer is proportional to the current passing through it. If the device obeys Ohm's law, this current is proportional to the voltage across the instrument. Hence, the deflection is proportional to the voltage also. Thus, if the total resistance of the galvanometer plus auxiliary resistor is R,

$$1V = (0.00162\ A)R,$$

$$R = 617.3\ \Omega.$$

Thus, connect the auxiliary resistor r in series with the galvanometer, choosing $r = 617.3 - 75.3 = 542\ \Omega$ Ans.

(b) In this case attach the auxiliary resistor in parallel with the galvanometer so that a current $i = 0.00162$ A flows through the galvanometer when it is attached to a circuit branch in which the current is $I = 0.050$ A, as shown. For a parallel connection,

$$i_a r_a = i r_G;$$

furthermore,

$$I = i + i_a.$$

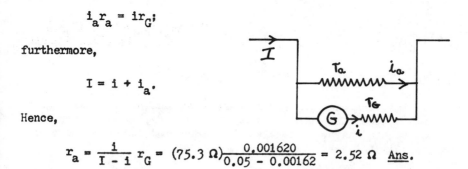

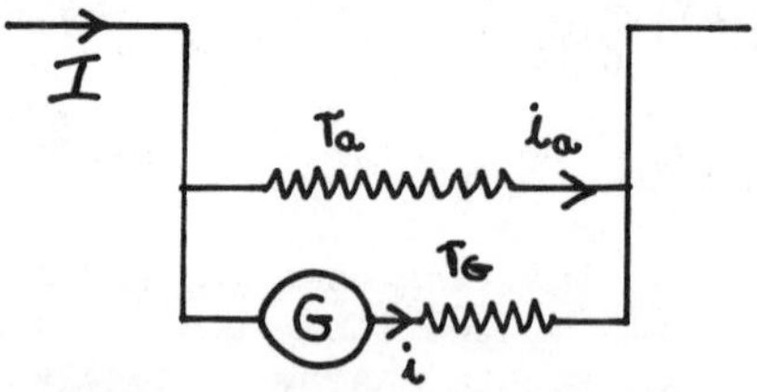

Hence,

$$r_a = \frac{i}{I - i} r_G = (75.3\ \Omega)\frac{0.001620}{0.05 - 0.00162} = 2.52\ \Omega \quad \underline{\text{Ans}}.$$

30-21

Let the hinge lie along the z-axis. It is evident that the only force that can exert a torque along the hinge is the one exerted on the side of the rectangle opposite the hinge. (The forces on the other-5 cm-sides are in the z-direction and, by the cross-product rule, their torques will be perpendicular to z.) The magnitude F of this force is

$$F = ibB\sin 90^\circ = ibB.$$

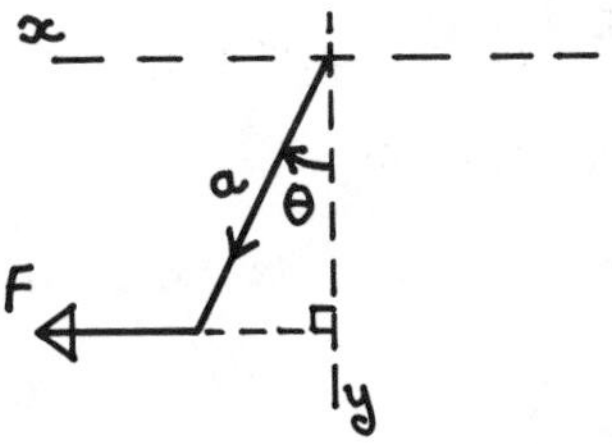

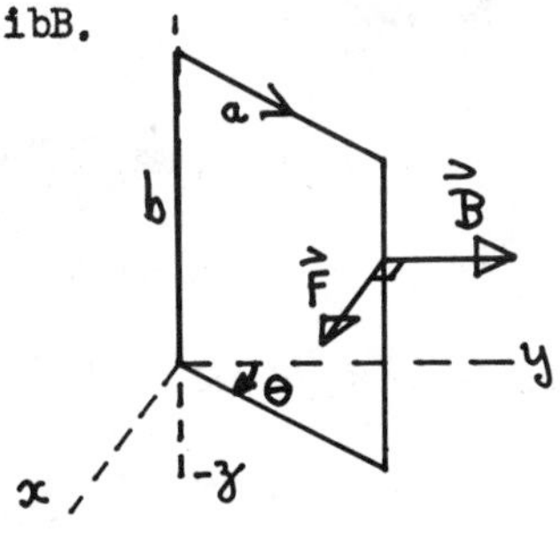

From the sketch the appropriate moment arm is seen to be $a\cos\theta$ so that the torque will be, for N loops,

$$\tau = N(ibB)(a\cos\theta) = 4.33 \times 10^{-3}\ \text{N}\cdot\text{m} \quad \underline{\text{Ans}}.$$

From the right-hand rule, it is seen that this is directed in the -z direction, i.e., down.

30-23

If N closed loops are formed from the wire of length ℓ, the circumference of each loop is ℓ/N, the radius of each is $\ell/2\pi N$ and therefore the area A of each is

$$A = \pi(\ell/2\pi N)^2 = \ell^2/4\pi N^2.$$

For maximum torque orient the plane of the loops parallel to the field lines so that $\sin\sphericalangle(\vec{\mu},\vec{B}) = \sin 90^\circ = 1$. Since there are N loops

$$\tau = NiAB = Ni(\ell^2/4\pi N^2)B = i\ell^2 B/4\pi N.$$

Clearly, then, the maximum torque is $i\ell^2 B/4\pi$ for $N = 1$.

30-26

The current density is $j = nev_d$ and therefore, since $j = \sigma E = E/\rho$,

$$E = \rho j = nev_d\rho$$

where E is the electric field responsible for the current. If E_H is the Hall field then, for the charge carriers to move along straight lines, on the average,

$$\vec{F} = e\vec{E}_H + e\vec{v}_d \times \vec{B} = 0,$$

$$E_H = v_d B$$

for a magnetic field perpendicular to the wire. Hence,

$$\frac{E_H}{E} = \frac{v_d B}{nev_d\rho} = \frac{B}{ne\rho}.$$

30-32

(a) At the equator, the velocity will be perpendicular to the field B; hence $F = qvB = mv^2/R$ so that $v = qBR/m$; numerically,

$$v = \frac{(1.6 \times 10^{-19}\ C)(41 \times 10^{-6}\ T)(6.4 \times 10^{6}\ m)}{(1.67 \times 10^{-27}\ kg)} = 2.5 \times 10^{10}\ m/s.$$

This is, however, greater than the speed of light $c = 3 \times 10^8$ m/s and the situation thereby calls for a relativistic calculation in which the proton rest mass m in the above equations is replaced by its relativistic mass m_r given by

$$m_r = \frac{m}{(1 - \beta^2)^{\frac{1}{2}}},$$

where $\beta = v/c$. Substituting this for m in the equation for R gives

$$\beta = \frac{A}{(1 + A^2)^{\frac{1}{2}}}$$

with $A = RqB/mc = 83.8004$. This

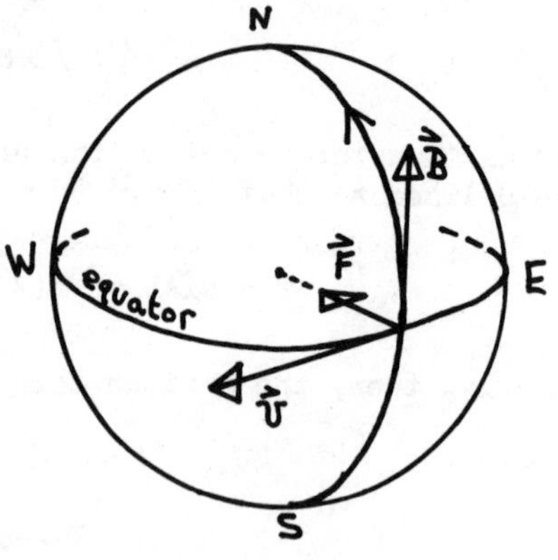

gives $v = \beta c = 2.9998 \times 10^8$ m/s <u>Ans</u>.

(b) The arrangement of the vectors is shown on the sketch on the previous page. Note that the magnetic SOUTH pole (where the field lines enter the earth) is in the geographic NORTHERN hemisphere.

<u>30-36</u>

The ion enters the spectrometer with a speed v related to the accelerating potential V by

$$\tfrac{1}{2}mv^2 = qV.$$

Inside the instrument the ion undergoes uniform circular motion, the speed v remaining unchanged; thus

$$mv^2/r = qvB.$$

But $v^2 = 2qV/m$ from the first equation and $r = \tfrac{1}{2}x$; substituting these gives

$$\frac{m(2qV/m)^{\frac{1}{2}}}{\frac{1}{2}x} = qB,$$

$$m = B^2qx^2/8V.$$

<u>30-38</u>

(a) The orientation of the vectors $\vec{v}$, $\vec{B}$, and $\vec{F} = q\vec{v} \times \vec{B}$ $(q > 0)$ is shown in the sketch. Let $\vec{v} = v_y\vec{j} + v_z\vec{k}$; then $\vec{F} = v_yB\vec{i}$, independent of v_z. If $v_z = 0$, uniform circular motion results with $v_y = v\sin\theta$ the governing speed. If $v_z \neq 0$, the additional uniform rectilinear motion along $\vec{B}$ streches the circle into a helix.

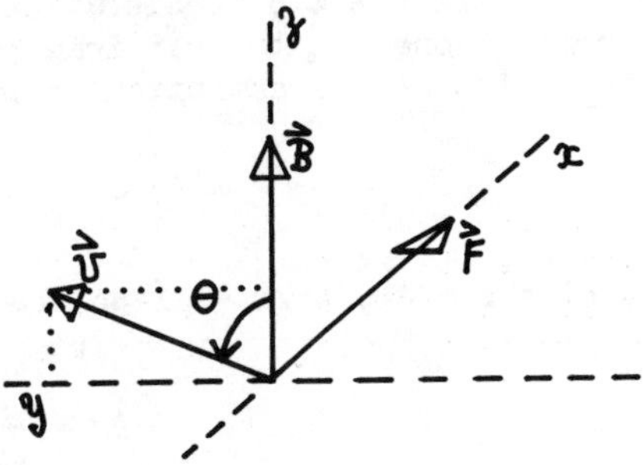

(b) The equation of motion for the circular motion in the x,y-plane is

$$\frac{m(v\sin\theta)^2}{r} = q(v\sin\theta)B.$$

The period $T = 2\pi r/(v\sin\theta)$; eliminating r from these two equations gives

$$T = \frac{2\pi}{B(q/m)} = 2\pi/(0.1)(1.76 \times 10^{11}) = 3.57 \times 10^{-10}\ \text{s} \quad \underline{\text{Ans}}.$$

(c) The pitch p is the distance travelled parallel to the magnetic field in one period T: i.e.,

$$p = (v\cos\theta)T = (2K/m)^{\frac{1}{2}}T\cos\theta,$$

K the kinetic energy, $K = 3.2 \times 10^{-16}$ J. Putting in the other numbers gives $p = 0.17$ mm <u>Ans</u>.

(d) From (a),

$$r = \frac{mv\sin\theta}{qB} = 1.5\ \text{mm} \quad \underline{\text{Ans}}.$$

<u>30-45</u>

From Example 6, the final deuteron energy is 17 MeV and the dee radius is 0.53 m. Thus, the number of revolutions made by a deuteron is

$$\frac{17\ \text{MeV}}{2(80\ \text{keV})} = 106$$

since there are two accelerations per revolution. Over one of these accelerations, as the particle passes from one dee to the other, the radius of its semicircular path jumps from r to r' where

$$\tfrac{1}{2}m(r\omega)^2 + qV = 2\pi^2 m\nu^2 r^2 + qV = 2\pi^2 m\nu^2 r'^2.$$

If $r' = r + \Delta r$, $\Delta r/r \ll 1$ so that $r'^2 \approx r^2(1 + 2\Delta r/r)$, then

$$\Delta r = \frac{qV}{4\pi^2 m\nu^2}\,\frac{1}{r}.$$

Thus the deuteron spends more of the 106 revolutions at radii greater than (0.53 m)/2 than at smaller radii: hence, as an average radius, choose $(0.53\ \text{m})(2^{\frac{1}{2}}/2)$ say, for which the circumference of the corresponding circle is $2\pi(0.37\ \text{m}) = 2.3$ m. Since the deuteron makes 106 revolutions, the path travelled must be about $(106)(2.3\ \text{m}) = 240$ m <u>Ans</u>.

30-49

If the electron travels in a straight line, the net force on it must be zero (assuming constant speed), and therefore

$$evB = eE,$$

$$B = \frac{E}{v} = \frac{\mathcal{E}}{dv}.$$

But $\frac{1}{2}mv^2 = eV$ so that

$$B = \frac{\mathcal{E}}{d}\left(\frac{m}{2eV}\right)^{\frac{1}{2}}.$$

Putting $\mathcal{E}$ = 100 volts, V = 1000 volts and d = 0.02 m gives B = 2.67 X 10^{-4} T Ans.

31-11

By the Biot-Savart law, the contribution to B from a short section $d\ell$ of the wire is

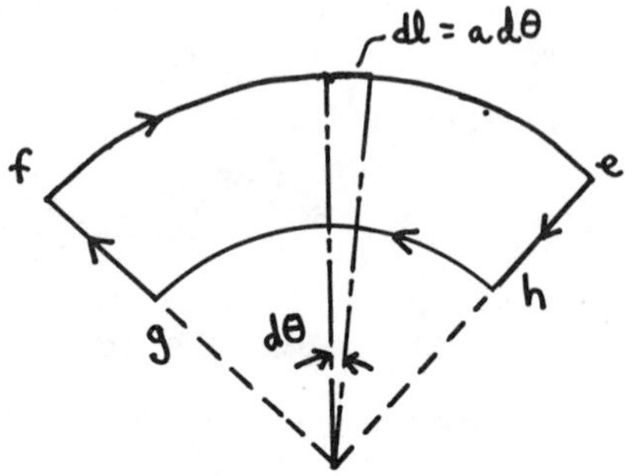

$$\overrightarrow{dB} = \frac{\mu_0 i}{4\pi}\frac{\overrightarrow{d\ell} \times \vec{r}}{r^3}.$$

Now sections he and fg give nothing since $\overrightarrow{d\ell} \times \vec{r} = 0$. Along section ef,

$$B_2 = \frac{\mu_0 i}{4\pi}\int\frac{r\,d\ell\,\sin(\vec{r},\overrightarrow{d\ell})}{r^3},$$

$$B_2 = \frac{\mu_0 i}{4\pi}\int\frac{a(ad\theta)\sin(\pi/2)}{a^3} = \frac{\mu_0 i}{4\pi a}\int_0^\theta d\theta = \frac{\mu_0 i\theta}{4\pi a}.$$

Clearly, the contribution due to gh is $B_1 = (\mu_0 i\theta/4\pi b)$. Also, $B_1 > B_2$ since $b < a$, and $\vec{B}_1$ is out of, $\vec{B}_2$ into, the page. Thus,

$$B = \frac{\mu_0 i\theta}{4\pi}\left(\frac{1}{b} - \frac{1}{a}\right),$$

out of the page.

31-13

Let the wire rest along the x-axis, its midpoint at the origin. Since the wire is of finite length, the Biot-Savart law is called for. All elements of the wire give rise to a field directed into the paper, the magnitude of the total field being

$$B = (\mu_0 i/4\pi)\int_{x=-\frac{1}{2}\ell}^{x=\frac{1}{2}\ell}\frac{(d\ell)(r)\sin\theta}{r^3}.$$

But

$$\sin\theta = \frac{R}{r}, \; r^2 = x^2 + R^2,$$

and therefore

$$B = (\mu_0 iR/4\pi)\int_{-\frac{1}{2}\ell}^{\frac{1}{2}\ell}(x^2 + R^2)^{-3/2}dx = \frac{\mu_0 i}{2\pi R}\frac{\ell}{(\ell^2 + 4R^2)^{\frac{1}{2}}}$$

If $R \ll \ell$ neglect R^2 in the denominator to obtain $B = \mu_0 i/2\pi R$, the field due to a very long wire: from points extremely close to the wire, the wire appears to be very long.

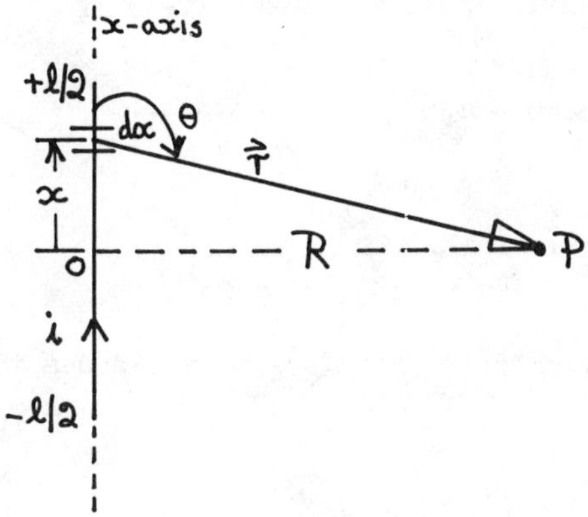

31-16

The field due to the square is the vector sum of the fields due the four sides of the square. Consider, then, one side. The point at which the field is desired lies along the perpendicular bisector to that side, at a distance R given by

$$R = \tfrac{1}{2}(4x^2 + a^2)^{\frac{1}{2}}.$$

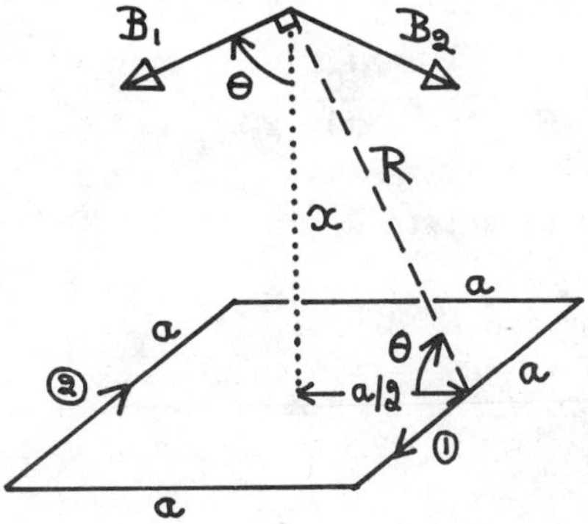

Hence, using the result of Problem 31-13, the field from one side is

$$B = (\mu_0 i/2\pi R)\,\frac{a}{(a^2 + 4R^2)^{\frac{1}{2}}},$$

putting a for ℓ as the length of each wire segment. Substituting for R from the first equation gives

$$B = \frac{\mu_0 i}{\pi}\,\frac{a}{(4x^2 + a^2)^{\frac{1}{2}}(4x^2 + 2a^2)^{\frac{1}{2}}}.$$

The direction of this field is perpendicular to the plane that contains the side considered and the perpendicular bisector of that side. The component perpendicular to the normal to the square will be cancelled by the analogous component of the field due to the opposite side of the square. Thus, the total field is

$$B_T = 4B\cos\theta = 4B\,\frac{a/2}{R} = \frac{4aB}{(4x^2 + a^2)^{\frac{1}{2}}},$$

$$B_T = \frac{4\mu_0 i}{\pi}\,\frac{a^2}{(4x^2 + a^2)(4x^2 + 2a^2)^{\frac{1}{2}}}.$$

As expected, for $x = 0$ (center of square), this reduces to the result of Problem 31-14.

31-18

(a) The field due to the n sides is n times the field due to one of them; hence, by Problem 31-13,

$$B_T = nB = \frac{n\mu_0 i}{2\pi R}\,\frac{\ell}{(\ell^2 + 4R^2)^{\frac{1}{2}}}.$$

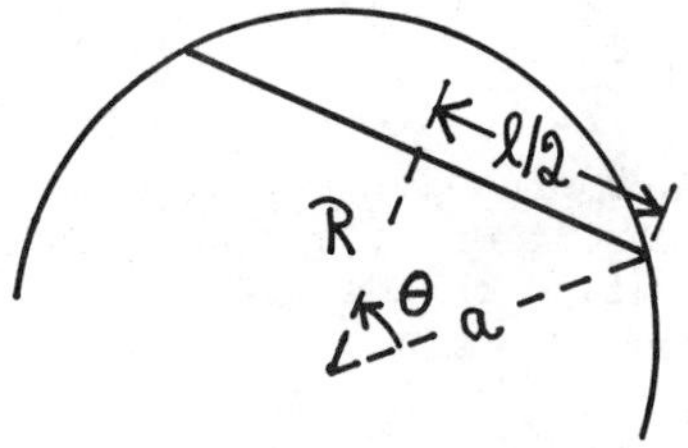

To calculate R, note that

$$n(2\theta) = 2\pi,$$

$$n\cdot\tan^{-1}\left(\frac{\ell/2}{R}\right) = \pi.$$

Also,

$$R^2 + (\tfrac{1}{2}\ell)^2 = a^2,$$

$$a = \tfrac{1}{2}(4R^2 + \ell^2)^{\frac{1}{2}}.$$

Substituting for ℓ/R and $(\ell^2 + 4R^2)^{\frac{1}{2}}$ in the equation for B_T gives

$$B_T = (n\mu_0 i/2\pi a)\tan(\pi/n).$$

(b) The field can be rewritten as

$$B_T = (\mu_0 i/2a)\frac{\tan(\pi/n)}{(\pi/n)}.$$

Let $\theta = \pi/n$. As n approaches infinity, θ goes to zero. But, Lim $\tan\theta/\theta = 1$ as θ goes to zero. Therefore, the field due to the polygon approaches $\mu_0 i/2a$ as n approaches infinity. See Eq. 31-8 for $x = 0$.

31-20

$B_Q = 0$ since, in using the Biot-Savart law, $\vec{d\ell}$ is parallel to $\vec{r}$ so that $\vec{d\ell} \times \vec{r} = 0$. As for B_P, it is directed into the page and has a magnitude

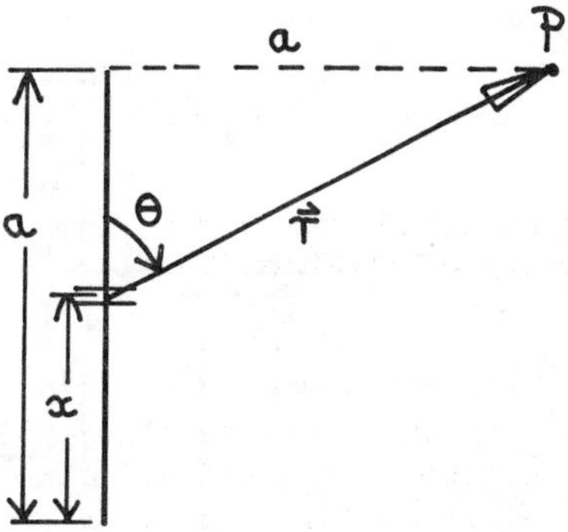

$$B_P = \frac{\mu_0 i}{4\pi}\int\frac{r\,d\ell\,\sin\theta}{r^3} = \frac{\mu_0 ia}{4\pi}\int\frac{d\ell}{r^3},$$

since $a = r\sin\theta$. Expressing the variables in terms of x gives

$$B_P = (\mu_0 ia/4\pi)\int_0^a \frac{dx}{[a^2 + (a - x)^2]^{3/2}} = \frac{\mu_0 i 2^{\frac{1}{2}}}{8\pi a}.$$

31-27

Let the current in the wire be i_w and the current in the rectangle be i_r. The force $\vec{F}$ on the rectangle is the vector sum of the forces on the four sides of the rectangle, labelled 1, 2, 3, 4. Then,

$$\vec{F}_1 = i_r\vec{\ell} \times \vec{B},$$

$\vec{\ell}$ pointing in the direction of the current in side 1. From Ampere's law, the field B due to the wire is

$$B = \mu_0 i_w/2\pi a,$$

directed into the paper; thus,

$$F_1 = \mu_0 i_r i_w \ell/2\pi a, \text{ to the left.}$$

Similarly,

$$F_3 = \mu_0 i_r i_w \ell/2\pi(a + b), \text{ to the right,}$$

since this side of the rectangle is at a distance $(a + b)$ from the wire and its current is directed oppositely from the current in side 1. F_2 is more difficult to compute since different parts of

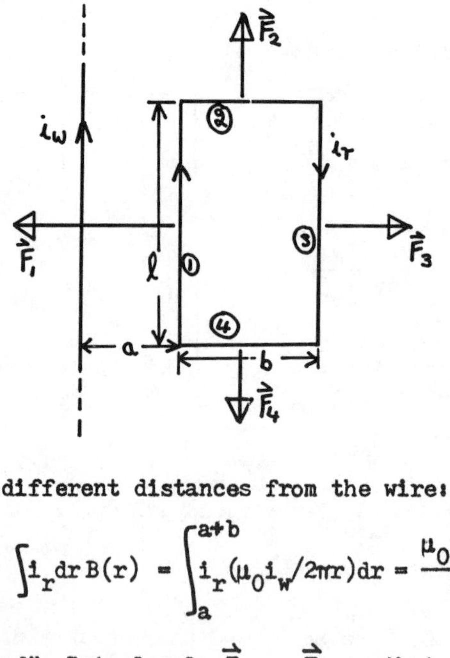

side 2 are at different distances from the wire:

$$F_2 = \int dF_2 = \int i_r dr\, B(r) = \int_a^{a+b} i_r(\mu_0 i_w/2\pi r)dr = \frac{\mu_0 i_r i_w}{2\pi}\ln(1 + \frac{b}{a}),$$

pointing "upward". But clearly $\vec{F}_4 = -\vec{F}_2$ so that, with $\vec{F} = \vec{F}_1 + \vec{F}_2 + \vec{F}_3 + \vec{F}_4$,

$$F = i_r i_w \mu_0 \ell/2\pi a - i_r i_w \mu_0 \ell/2\pi(a + b) = \frac{\mu_0 i_r i_w \ell b}{2\pi a(a + b)},$$

$$F = \frac{(20\ \text{A})(30\ \text{A})(4\pi \times 10^{-7}\ \text{T}\cdot\text{m/A})(0.3\ \text{m})(0.08\ \text{m})}{2\pi(0.01\ \text{m})(0.09\ \text{m})},$$

$$F = 3.2 \times 10^{-3}\ \text{N, to the left} \quad \underline{\text{Ans}}.$$

31-34

Let P represent a point in the hole. By superposition, the field $\vec{B}$ at P = (field $\vec{B}_1$ at P due to a wire of radius a, current I, containing no hole.) - (field $\vec{B}_2$ at P due to the hole *if* it carried a uniform current of current density $I/\pi R^2$ in the same direction as i). That is,

$$\vec{B} = \vec{B}_1 - \vec{B}_2.$$

Now, with equal current densities

$$\frac{i}{\pi a^2 - \pi b^2} = \frac{I}{\pi a^2}.$$

Also,

$$B_1 = \frac{\mu_0 I}{2\pi a^2} r, \quad B_2 = \frac{\mu_0 I}{2\pi a^2} R.$$

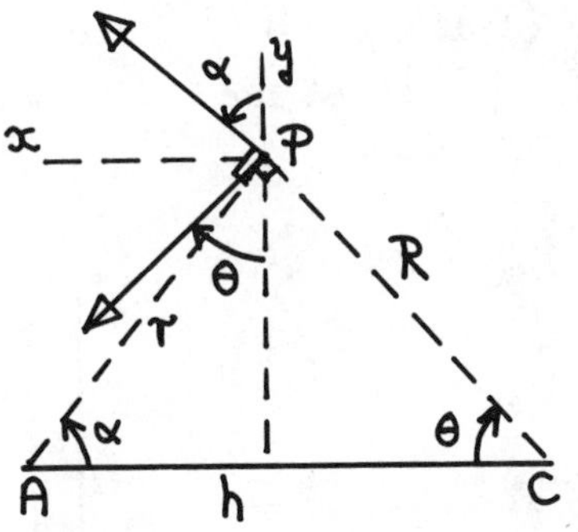

Let A be the center of the wire and C the center of the hole; then,

$$B_y = B_1\cos\alpha + B_2\cos\theta = \frac{\mu_0 I}{2\pi a^2}(r\cos\alpha + R\cos\theta) = \frac{\mu_0 I}{2\pi a^2} h;$$

$$B_x = B_1\sin\alpha - B_2\sin\theta = \frac{\mu_0 I}{2\pi a^2}(r\sin\alpha - R\sin\theta) = 0.$$

But

$$I = \frac{a^2}{a^2 - b^2} i,$$

and therefore

$$B = \frac{\mu_0 i}{2\pi} \frac{h}{a^2 - b^2},$$

perpendicular to AC, independent of the location of P in the hole.

31-36

Apply Ampere's law by integrating around the dotted rectangle shown. Since there is no current cutting the area bounded by the path of integration, it is expected that

$$\oint \vec{B}\cdot\vec{d\ell} = 0.$$

On section 2, $\int \vec{B}\cdot\vec{d\ell} = 0$ since $B = 0$ there. On sections 1 and 3, the integral is zero because either $B = 0$ or $\vec{B}$ is perpendicular to $\vec{d\ell}$. On section 4, however,

$$\int \vec{B}\cdot\vec{d\ell} = B_4 b.$$

Thus $\oint \vec{B}\cdot\vec{d\ell} = B_4 b \neq 0$, contradicting Ampere's law. Hence, the geometry assumed for the magnetic field must be in error.

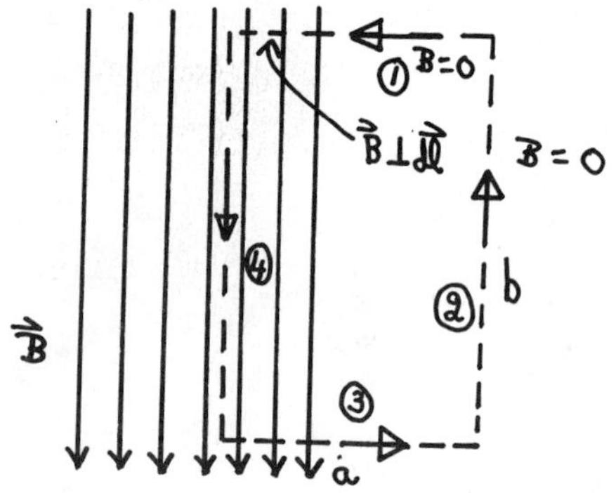

31-38

(a)

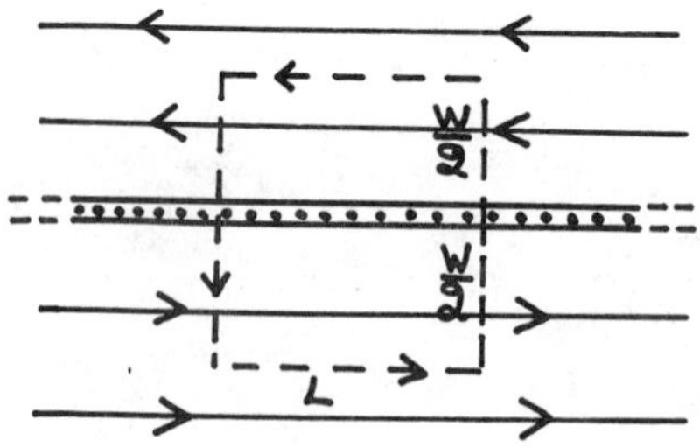

Consider a large number of long wires carrying identical currents and arranged in a plane, as in the sketch on the right (previous page). By the right-hand rule, the fields above the plane all are directed to the left and below the plane all are directed to the right. Between the wires the fields due to neighboring wires are oppositely pointing. It seems reasonable that as the wires are brought closer together the already opposing fields between them will vanish (cancel), yielding the pattern shown in the left-hand sketch.

(b) Selecting as a path of integration a rectangle of length L and width W aligned as shown, perpendicular to the current sheet, it is clear that $\vec{B}$ is at right angles to $\vec{d\ell}$ along the sides W of the rectangle, so the line integral in Ampere's law gives nil on these sides, while along the top side L,

$$\int \vec{B}\cdot\vec{d\ell} = \int B\,d\ell \cos 0^\circ = B\int d\ell = BL.$$

Since both of these sides (length L, parallel to the sheet) are at the same distance from the sheet, B is the same at each of them, so that Ampere's law gives

$$\oint \vec{B}\cdot\vec{d\ell} = 2BL = \mu_0 i_{enc} = \mu_0(\lambda L),$$

$$B = \tfrac{1}{2}\mu_0\lambda.$$

31-39

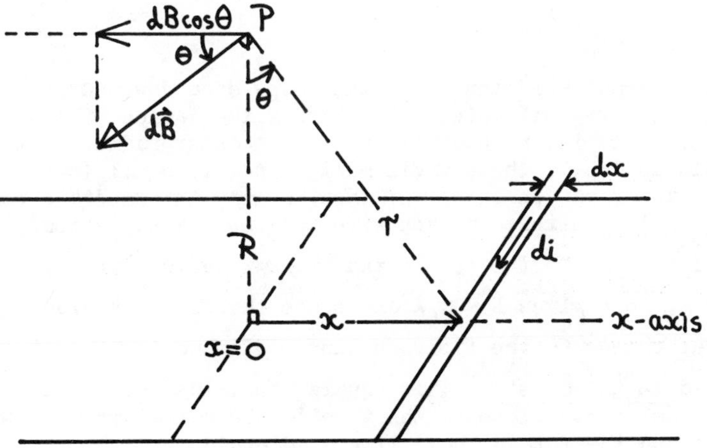

Let the point P at which the field is to be evaluated be a distance R above the plane, directly above $x = 0$, the x-axis running at a right angle to the current. Subdivide the plane into long infinitesimal filaments of width dx, each of which can be considered a long wire carrying a current $di = \lambda dx$. The field contribution dB at P due to the filament shown is

$$dB = (\mu_0 di)/2\pi r = (\mu_0 \lambda/2\pi r)dx = (\mu_0 \lambda/2\pi R\sec\theta)dx.$$

Note that $\vec{dB}$ is perpendicular to $\vec{r}$. Only the component of $\vec{dB}$ that is parallel to the plane, $dB\cos\theta$, is effective, however; the perpendicular component is canceled by the contribution associated with a symmetrically located filament on the other side of $x = 0$. Hence,

$$B = \int dB\cos\theta = \int (\mu_0 \lambda/2\pi R)dx/\sec^2\theta.$$

But

$$x = R\tan\theta,$$
$$dx = R\sec^2\theta d\theta.$$

Since the plane "extends to infinity", the limits on θ are $\pm\pi/2$. Therefore

$$B = (\mu_0 \lambda/2\pi)\int_{-\pi/2}^{+\pi/2} d\theta = \tfrac{1}{2}\mu_0 \lambda.$$

31-43

A toroid is a solenoid bent into the shape of a doughnut. The length of the original solenoid is $2\pi r$, r the radius of the toroid. Of course, there is an inner radius and an outer radius, but if the toroid is ideal, these radii must be nearly equal (skinny doughnut) and r can be either of them, or any intermediate radius. Since i_0 is the current in each wire and there are N wires, Ni_0 is the total current in the toroid and $Ni_0/2\pi r$ is the current per unit length $= \lambda$. This gives $B = \mu_0 \lambda$ and since the field outside is zero, $\mu_0 \lambda$ is the change in the field encountered in moving from inside the toroid to the outside. This equals the value for the solenoid because from points close to the toroid, the curvature is not perceived and the toroid looks like a solenoid.

31-48

The field $\vec{B}$ at C = field $\vec{B}_w$ due to an infinite straight wire + field $\vec{B}_c$ due to a circle of wire; these are

$$B_w = \mu_0 i/2\pi R, \quad B_c = \mu_0 i/2R,$$

the former from Eq. 31-3 and the latter from Eq. 31-8 with x = 0.

(a) Here $\vec{B}_w$ and $\vec{B}_c$ both are directed normal to and out of the page; thus,

$$B = B_c + B_w = (\mu_0 i/2R)(1 + \frac{1}{\pi}), \text{ out of page } \underline{\text{Ans.}}$$

(b) Now $\vec{B}_c$ and $\vec{B}_w$ are at right angles and therefore

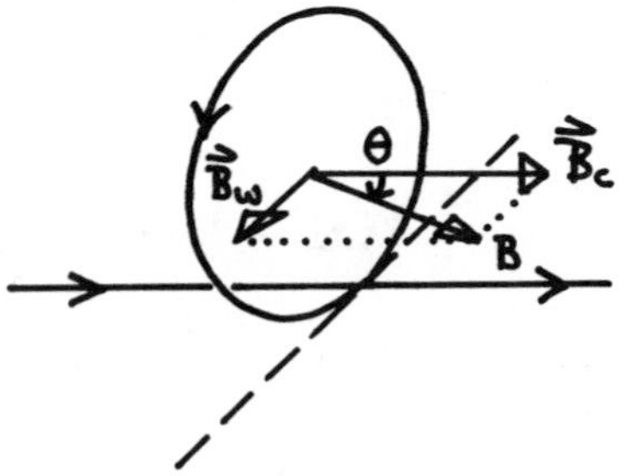

$$B = (B_w^2 + B_c^2)^{\frac{1}{2}},$$

$$B = (\mu_0 i/2R)[1^2 + (\frac{1}{\pi})^2]^{\frac{1}{2}},$$

$$B = (\mu_0 i/2\pi R)(1 + \pi^2)^{\frac{1}{2}} \quad \underline{\text{Ans.}}$$

The direction of $\vec{B}$ is at an angle θ out of the page, where

$$\theta = \tan^{-1}(\frac{B_w}{B_c}) = \tan^{-1}(\frac{1}{\pi}) = 18° \quad \underline{\text{Ans.}}$$

31-50

(a) Consider a narrow ring of radius r and width dr; it carries an amount of charge

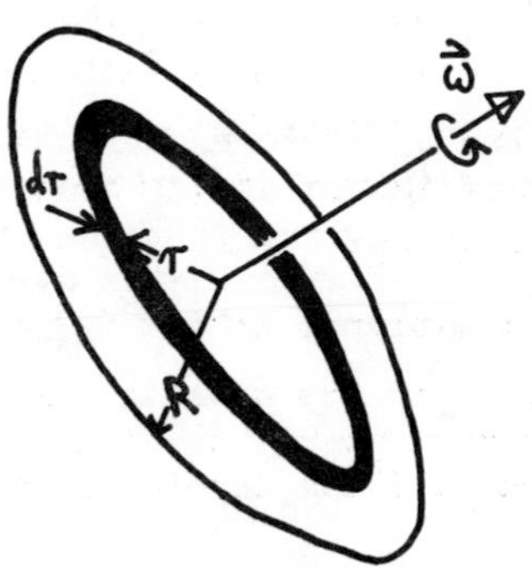

$$dq = \frac{q}{\pi R^2}(2\pi r\,dr),$$

i.e., the charge per unit area of the disc times the area of the ring. In time $2\pi/\omega$ all this charge passes any fixed point near the ring, so that the equivalent current di is

$$di = \frac{dq}{\tau} = \frac{2\pi qr\ dr/\pi R^2}{2\pi/\omega} = q\omega r\ dr/\pi R^2.$$

By Eq. 31-8 ($x = 0$) this ring sets up a field dB at the center of the disc given by

$$dB = \frac{\mu_0 di}{2r} = \frac{\mu_0}{2r}(q\omega r\ dr/\pi R^2).$$

Thus the total field is

$$B = \int dB = \frac{\mu_0 q\omega}{2\pi R^2}\int_0^R dr = \mu_0 q\omega/2\pi R.$$

(b) The dipole moment is

$$\mu = \int A\,di = \int_0^R (\pi r^2)\ \frac{q\omega r\ dr}{\pi R^2} = \frac{q\omega}{R^2}\int_0^R r^3 dr = \tfrac{1}{4}q\omega R^2.$$

31-51

To find the field at P, compute the contributions B_R and B_L due to the segments of the solenoid lying to the right and to the left of P; since the current in the turns flows in the same direction, the fields are parallel and the net field is $B = B_R + B_L$. Due to the strip of width dx shown in the sketch, the field is

$$dB_L = \frac{\mu_0 R^2}{2(R^2 + x^2)^{3/2}}\ ni\ dx,$$

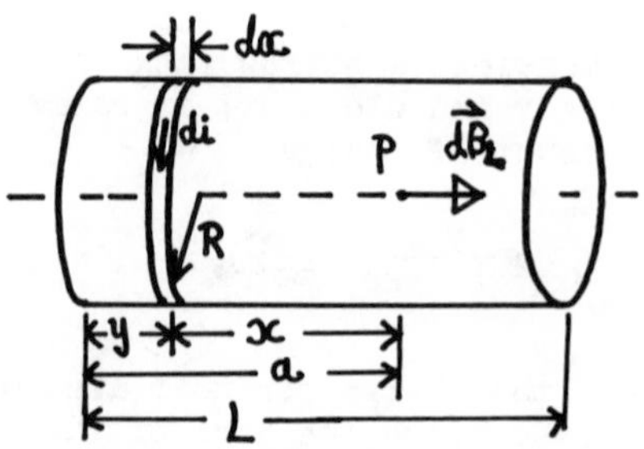

n the number of turns per unit length of the solenoid, each turn carrying a current i; there are n dx turns of wire in the strip being considered. Since

$$y + x = a,$$

then dy = dx, the minus sign being of no importance. Then,

$$dB_L = \frac{\mu_0 n i R^2 dy}{2(R^2 + a^2 - 2ay + y^2)^{3/2}},$$

and the total field at P from that part of the solenoid to the left of P is

$$B_L = \tfrac{1}{2}\mu_0 n i R^2 \int_0^a \frac{dy}{(R^2 + a^2 - 2ay + y^2)^{3/2}} = \tfrac{1}{2}\mu_0 n i \frac{a}{(R^2 + a^2)^{\frac{1}{2}}} .$$

It follows directly from this that

$$B_R = \tfrac{1}{2}\mu_0 n i \frac{L - a}{[R^2 + (L - a)^2]^{\frac{1}{2}}},$$

the segment on the right having a length L - a. But for an ideal solenoid $L \gg R$ and also $a \gg R$: ignoring the terms in R and adding gives

$$B = B_R + B_L = \tfrac{1}{2}\mu_0 n i + \tfrac{1}{2}\mu_0 n i = \mu_0 n i.$$

32-3

Choose $\vec{dS}$, always perpendicular to the surface over which the flux is to be computed, in the same direction as $\vec{B}$ as it cuts the surface. From Example 3, Chapter 31,

$$B = \mu_0 i_0 r/2\pi R_0^2.$$

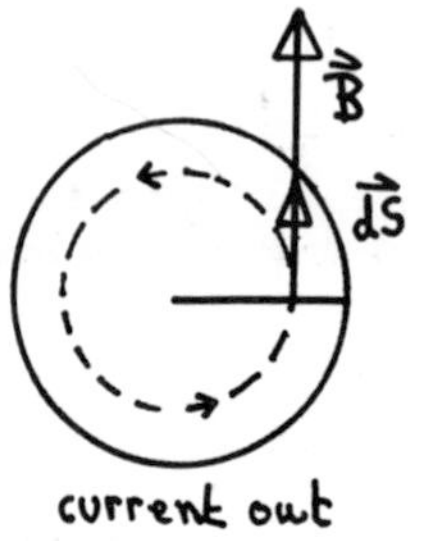

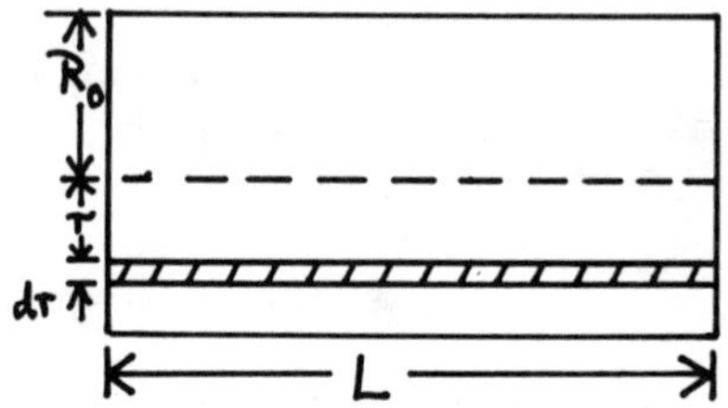

Thus,

$$\Phi_B = \int \vec{B}\cdot\vec{dS} = \int B(dS)\cos 0° = \int B\, dS.$$

Now B varies across the surface (r direction) but not over the narrow strip, since all parts of the strip are at the same distance r from the axis of the wire. Hence,

$$\Phi_B = \int_0^{R_0} (\mu_0 i_0 r/2\pi R_0^2)L\, dr = \mu_0 i_0 L/4\pi,$$

and therefore the flux per unit length of the wire is just $\Phi_B/L = \mu_0 i_0/4\pi$.

32-5

(a) Call the radius of each wire b and let a be the distance

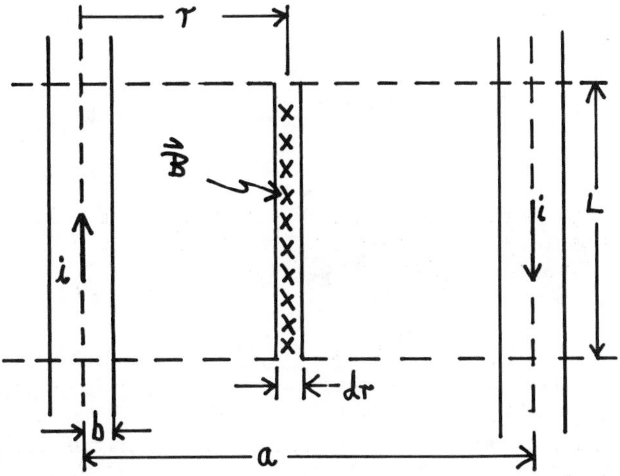

between their axes. Consider a single wire. The flux inside the wire is, by Problem 32-3,

$$\Phi_i = \mu_0 iL/4\pi.$$

The flux outside, out to a distance r = a from the axis is

$$\Phi_o = \int B\, dS = \int_b^a (\mu_0 i/2\pi r)L\, dr = (\mu_0 iL/2\pi)\ln(\frac{a}{b}),$$

choosing $\vec{dS}$ into the paper, in the same direction as $\vec{B}$ from the left-hand wire. The wires and currents being identical, then, by the right-hand rule, the fields due to the two wires are in the same direction since their currents are opposite. Hence, the total flux per unit length is

$$\Phi = (2\mu_0 i/4\pi)[1 + \ln(\frac{a}{b})^2].$$

Numerically, $a/b = 20/1.25 = 16$, $i = 10$ A and $\mu_0 = 4\pi \times 10^{-7}$ Wb/A·m giving $\Phi = 13.09 \times 10^{-6}$ Wb/m Ans.

(b) Consider the left-hand wire, say. Part of its flux for r > b lies inside the other wire: this flux is

$$\Phi' = \int B\, dS = (\mu_0 iL/2\pi)\ln(\frac{a}{a - b}).$$

A similar flux lies inside the left-hand wire due to the right-hand wire. Hence, the total flux inside the wires is

$$2\Phi_i + 2\Phi' = (2\mu_0 iL/4\pi)[1 + \ln(\frac{a}{a - b})^2].$$

Numerically, per unit length this is 2.258 X 10^{-6} Wb/m. Therefore, the fraction desired is

$$\frac{2.258}{13.09} = 17.25\% \quad \underline{\text{Ans}}.$$

(c) If the currents are in the same direction, the net magnetic field between the wires is antisymmetrical about a line parallel to the axes of the wires and midway between them. That is, at equal distances to the left and right of this line the magnetic fields are equal in strength but oppositely directed. The net flux, then, is zero.

32-10

(a) By Faraday's law $\mathcal{E} = -N\cdot d\Phi_B/dt$ and therefore the flux through the rectangular loop at any time t must be computed first. This is

$$\Phi_B = \int \vec{B}\cdot d\vec{S} = \int B(dS)\cos\theta = B\cos\theta\int dS = abB\cos\theta,$$

since B is uniform and θ, the angle between $\vec{B}$ and $d\vec{S}$, has at any instant the same value at all points "on" the rectangle. If the loop rotates at the constant rate ν, then $\theta = 2\pi\nu t$ and

$$d\Phi_B/dt = d(abB\cos 2\pi\nu t)/dt = -2\pi\nu abB\sin 2\pi\nu t$$

and therefore $\mathcal{E} = 2\pi\nu NabB\sin 2\pi\nu t = \mathcal{E}_0\sin 2\pi\nu t$.

(b) If $\mathcal{E}_0 = 150$ V, $\nu = 60$ Hz and B = 0.50 T, then

$$2\pi Nab = \mathcal{E}_0/\nu B = 5\ m^2$$

and any loop built according to the specification $Nab = 5/2\pi\ m^2$ will produce the desired effect.

32-11

(a) By Faraday's law the current i induced in the coil is

$$i = \frac{\mathcal{E}}{R} = \frac{N}{R}\, d\Phi_B/dt$$

in which N = 100, R = 5 Ω. Now $\Phi_B = \int \vec{B}\cdot d\vec{S}$, the integral taken over the cross-section of the coil, or any area bounded by the coil, and B is the field due to the solenoid. Since $B = \mu_0 n i_s$ inside the solenoid (i_s = current in the solenoid) and B = 0 outside the solenoid, the integral for the flux over the area between the coil and solenoid will be zero and

$$\Phi_B = \int_{\substack{\text{area of}\\ \text{solenoid}}} (\mu_0 n i_s)dS + 0 = \mu_0 n i_s A_s,$$

where A_s is the cross-sectional area of the solenoid. Thus

$$i = \frac{N}{R}\, d(\mu_0 n i_s A_s)/dt = \frac{N}{R}\, \mu_0 n A_s d(i_s)/dt = \frac{N}{R}\, \mu_0 n A_s 2 i_0/t.$$

Numerically, N = 100, R = 5 Ω, $\mu_0 = 4\pi \times 10^{-7}$ Wb/A·m, n = 20,000 m^{-1}, $A_s = \pi(0.015\text{ m})^2$, i_0 = 1.5 A, t = 0.05 s; these give $i = 2.1 \times 10^{-2}$ A Ans.

(b) The magnetic flux is entirely confined to the internal volume of the solenoid only if the current in the solenoid is constant; if the current increases, say, additional lines of flux must snake into the solenoid, to reflect the increased field strength, cutting the coil as they do so.

32-21

(a) By Faraday's law $\mathcal{E} = -d\Phi_B/dt$ so that $i = \mathcal{E}/R = (-1/R)d\Phi_B/dt$. But $i = dq/dt$ and therefore

$$\frac{dq}{dt} = -\frac{1}{R}\, d\Phi_B/dt,$$

$$\int_0^q \frac{dq}{dt}\, dt = -\frac{1}{R}\int_{\Phi_B(0)}^{\Phi_B(t)} (d\Phi_B/dt)dt,$$

$$q = -\frac{1}{R}[\Phi_B(t) - \Phi_B(0)],$$

where the limits on the integrals are the values of charge and flux at times t and 0.

(b) If the flux at time t and the flux at t = 0 both are zero, the equation derived in (a) implies that the net charge through the circuit is zero; if the flux changed during the time interval, this indicates that equal amounts of charge flowed clockwise through the circuit for part of the time, say, and counterclockwise for the remaining part of the time interval, giving a net charge flow of zero. This means that induced currents of varying directions could have existed during the entire time interval.

32-23

(a) The emf induced is $\mathcal{E} = d\Phi/dt = d(BS)/dt = d(B\ell x)/dt = B\ell dx/dt = B\ell v$, where x measures the distance of the rod from the closed (right) end of the rails. Thus, $\mathcal{E} = (1.2\ \mathrm{T})(0.1\ \mathrm{m})(5\ \mathrm{m/s}) = 0.60\ \mathrm{V}$.

(b) The induced current $i = \mathcal{E}/R = (0.6\ \mathrm{V})/(0.4\ \Omega) = 1.5\ \mathrm{A}$. As the rod moves to the left, an increasing, upward directed flux becomes contained within the rod-rail circuit. The induced current will flow so that the magnetic field it sets up will point downward in this same region bounded by the circuit. This requires a clockwise induced current.

(c) The rate is $P = i^2R = (1.5\ \mathrm{A})^2(0.4\ \Omega) = 0.90\ \mathrm{W}$.

(d) The force $\vec{F}$ exerted by the external agent must overcome the $i\vec{\ell} \times \vec{B}$ force exerted by the external magnetic field B on the induced current i in the moving rod. This latter force points to the right; for zero acceleration set $F = i\ell B$ exactly (zero net force); then $F = (1.5\ \mathrm{A})(0.1\ \mathrm{m})(1.2\ \mathrm{T}) = 0.18\ \mathrm{N}$.

(e) The rate of doing work is $d(Fx)/dt = F(dx/dt) = Fv = (0.18\ \mathrm{N})\cdot(5\ \mathrm{m/s}) = 0.90\ \mathrm{W}$, equal to (c) by energy conservation.

32-27

(a) The forces acting on the wire are its weight mg, the normal force N, and the force F exerted by the magnetic field on the current induced in the wire by virtue of its motion induced by gravity. The equation of motion of the wire is therefore

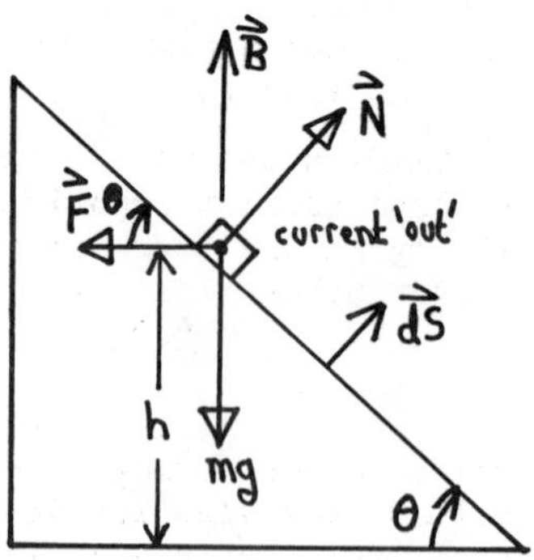

$$mg\sin\theta - F\cos\theta = ma.$$

But $F = i\ell B = (\mathcal{E}/R)\ell B$, ℓ the

length of the wire rod. Since the angle between $\vec{dS}$ (the normal to the inclined plane) and $\vec{B}$ is θ, the angle of the incline, $\mathcal{E} = B\ell v \cdot \cos\theta$ and therefore

$$mg\sin\theta - (B^2\ell^2v/R)\cos^2\theta = ma.$$

Initially $v = 0$; as the wire rod accelerates down the incline v will increase until it reaches a value v_t given by

$$v_t = \frac{mgR\sin\theta}{B^2\ell^2\cos^2\theta},$$

at which point $ma = 0$ and the wire stops accelerating, as the net force on it is now zero. Subsequently, then, the rod slides down at the constant speed v_t.

(b) The kinetic and gravitational potential energies are $\frac{1}{2}mv^2 = K$ and $mgh = U$. Let x be the distance of the rod from the bottom of the incline measured along the incline. Then $h = x\sin\theta$. By the conservation of energy it is expected that

$$\frac{dU}{dt} = \frac{dK}{dt} + P,$$

the last term being the rate of generation of thermal energy. As $v = dx/dt$ and $a = dv/dt$, this equation becomes

$$mgv\sin\theta = mv\frac{dv}{dt} + i^2R,$$

$$mg\sin\theta = ma + (B^2\ell^2v/R)\cos^2\theta,$$

since $i = \mathcal{E}/R = B\ell v\cos\theta/R$. But this last equation agrees with (a); hence the result in (a) is consistent with energy conservation. After the rod has reached its terminal speed, $dK/dt = 0$ so that $dU/dt = P$, as asserted.

(c) If $\vec{B}$ is directed down instead of up the induced current will flow in the direction opposite from that in (a), but $\vec{F} = (-i)\vec{\ell} \times (-\vec{B})$ will be in the same direction as in (a), so the motion of the rod will be unchanged.

32-29

(a) Consider a conduction electron in the rod. As a result of the rod's motion, the electron too is moving to the left with the same velocity $\vec{v}$. Hence, the magnetic field exerts a force $\vec{F} = q\vec{v} \times \vec{B} =$

$(-e)\vec{v} \times \vec{B}$ on the electron. This is directed away from the long wire as shown. From Ampere's law,

$$B = \mu_0 i/2\pi r$$

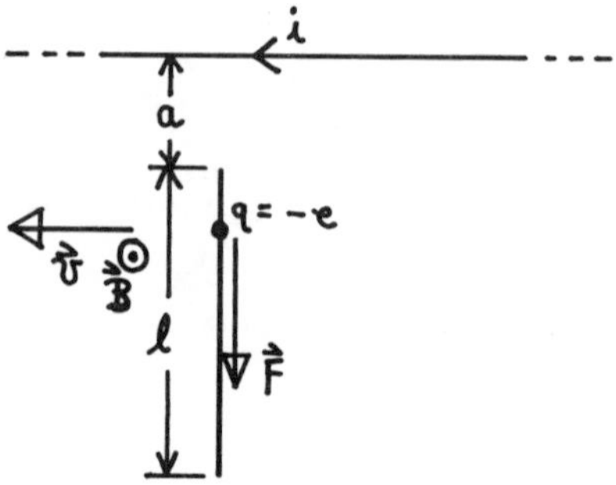

so that

$$F = e\mu_0 iv/2\pi r.$$

If the electron started from the end of the rod closest to the wire, the total work W done by the magnetic field in moving the electron to the other end is

$$W = \int F dr = (e\mu_0 iv/2\pi)\int_a^{a+\ell} \frac{dr}{r} = (e\mu_0 iv/2\pi)\ln\left(\frac{a+\ell}{a}\right).$$

But $W = e\mathcal{E}$, so that

$$\mathcal{E} = \frac{\mu_0 iv}{2\pi}\ln\left(\frac{a+\ell}{a}\right) = 2.398 \times 10^{-4}\ \text{V} \quad \underline{\text{Ans.}}$$

(b) No steady current flows in the rod in this situation since there is no closed circuit. A separation of charge does occur, the end of the rod closer to the wire becoming positive, the other end negative. Motion of the rod must be maintained to generate an emf so that there is always present a $q\vec{v} \times \vec{B}$ force to oppose the electrostatic attraction between the oppositely charged ends of the rod. The work done by the external agent must always equal (except in sign) the work continually done by the electrostatic forces in trying to pull the electrons from the end of the rod on which they have collected. If the rod is put on rails, completing a circuit, a current will flow and the external agent must do work against the forces that resist the current.

<u>32-32</u>

(a) Far away from the larger loop the field may be taken, across the area of the small loop, as approximately uniform and equal to the value B on the axis. By Example 6, Chapter 31, this is

$$B = \mu_0 iR^2/2x^3,$$

so that the flux through the small loop is

$$\Phi = B(\pi r^2) = \mu_0 i R^2 \pi r^2 / 2x^3 \quad \underline{\text{Ans}}.$$

(b) For a single loop the emf is

$$\mathcal{E} = d\Phi/dt = (\mu_0 i R^2 \pi r^2/2)\,\frac{dx^{-3}}{dt}.$$

But $dx^{-3}/dt = -3x^{-4}dx/dt = -3v/x^4$. Hence, except for sign,

$$\mathcal{E} = \frac{3}{2}\mu_0 i \pi R^2 r^2 v/x^4 \quad \underline{\text{Ans}}.$$

(c) The field due to i in the larger loop, near its axis, points away from the large loop toward the small loop. Therefore, as the small loop moves away from the larger, the small loop sees a steadily decreasing upward directed flux cutting across it. To oppose this, the induced emf will seek to set up an induced current the magnetic field of which will be directed upward also, in the area bounded by the small loop. By the right-hand rule, this will require a counterclockwise current as seen looking down from above the small loop (i.e., in the same direction as i in the large loop).

CHAPTER 33

33-8

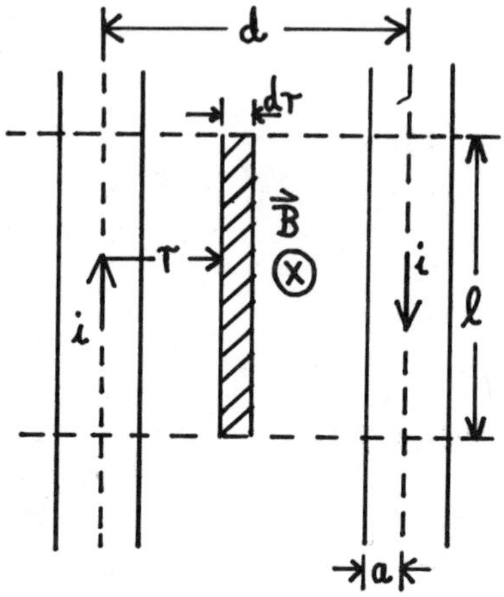

The inductance L is defined by

$$-L\,\frac{di}{dt} = -d\Phi_B/dt,$$

where the flux is

$$\Phi_B = \int \vec{B}\cdot d\vec{S}.$$

The area of integration for the flux is the area of the loop formed by imagining two short additional wires connecting those given, to form a closed circuit. The lengths of these new wires being very small compared to the original wires, their fluxes may be neglected. Then the field $\vec{B}$ is the sum of the fields set up by the two given wires: both of these are into the paper and therefore, by Ampere's law,

$$B = \frac{\mu_0 i}{2\pi r} + \frac{\mu_0 i}{2\pi(d - r)}.$$

B does not vary in a direction parallel to the wires and therefore for $d\vec{S}$ take a skinny rectangle of length ℓ and width dr; choose the direction of $d\vec{S}$ as into the paper (same as $\vec{B}$). Then,

$$\Phi_B = \int B(dS)\cos 0° = \int_a^{d-a}\left[\frac{\mu_0 i}{2\pi r} + \frac{\mu_0 i}{2\pi(d - r)}\right]\ell\,dr = \frac{\mu_0 i\ell}{\pi}\ln\left(\frac{d - a}{a}\right).$$

Hence,

$$d\Phi_B/dt = \frac{\mu_0\ell}{\pi}\,\frac{di}{dt}\,\ln\left(\frac{d - a}{a}\right) = L\,\frac{di}{dt},$$

and therefore, with the flux within the wires themselves ignored,

$$L = \frac{\mu_0 \ell}{\pi}\ln(\frac{d - a}{a}).$$

33-11

The current is, by Eq.33-13

$$i = \frac{\mathcal{E}}{R}(1 - e^{-t/\tau})$$

where $\tau = L/R$ is the inductive time constant. The equilibrium value of the current is $\mathcal{E}/R$ and therefore, if t is the time being sought,

$$0.999\frac{\mathcal{E}}{R} = \frac{\mathcal{E}}{R}(1 - e^{-t/\tau}),$$

$$0.999 = 1 - e^{-t/\tau},$$

$$e^{-t/\tau} = 1 - 0.999 = 0.001,$$

$$\ln(e^{-t/\tau}) = \ln(0.001),$$

$$-t/\tau = -6.908,$$

$$t = 6.908\tau,$$

or that 6.9 time constants, approximately, must elapse.

33-16

The current in an LR circuit drops exponentially:

$$i = i_0 e^{-t/\tau},$$

where i_0 is the current at $t = 0$ and $\tau = L/R$. At $t = 0$ the current is 1 A so that $i_0 = 1$ A. At any subsequent time, then,

$$i = e^{-t/\tau}.$$

When $t = 1$ s, the current is 10 mA = 0.01 A. Hence,

$$0.010 = e^{-1/\tau},$$

$$\ln(0.01) = -1/\tau,$$

$$-\ln(100) = -R/L,$$

$$R = L\cdot\ln(100) = 46\ \Omega \quad \underline{\text{Ans}}.$$

33-18

(a) The inductor "breaks" the right-hand branch: $i_1 = i_2 = i$. But

$$i = \frac{\mathcal{E}}{R_1 + R_2} = \frac{100}{10 + 20} = 3.33\ \text{A} \quad \underline{\text{Ans}}.$$

(b) Now the inductor has no effect and therefore

$$\mathcal{E} - i_1R_1 - i_1R_e = 0,$$

where

$$\frac{1}{R_e} = \frac{1}{R_2} + \frac{1}{R_3}$$

giving $R_e = 12\ \Omega$. Thus $i_1 = (100\ \text{V})/(22\ \Omega) = 4.55$ A <u>Ans</u>.
Furthermore,

$$\mathcal{E} - i_1R_1 - i_2R_2 = 0,$$

$$100 = (4.545)(10) + i_2(20),$$

$$i_2 = 2.73\ \text{A} \quad \underline{\text{Ans}}.$$

(c) The left-hand branch is now broken so that $i_1 = 0$. The current through R_2 equals the current through R_3 since the remaining elements form a series circuit. The initial value of this current equals the current through R_3 a long time after S was originally closed. From (b) this is $i_2 = 4.545 - 2.727 = 1.82$ A.

(d) There are now no sources of emf in the circuit and therefore all currents are zero.

33-19

(I) With the switch S just closed, the inductor effectively breaks the right-hand branch: $i_2 = 0$.

(a) $i_1 = \mathcal{E}/R_1 = (10\ V)/(5\ \Omega) = 2.0\ A$ Ans.

(b) $i_2 = 0$ Ans.

(c) $i = i_1 + i_2 = 2 + 0 = 2.0\ A$ Ans.

(d) $V_2 = i_2R_2 = (0)(10\ \Omega) = 0$ Ans.

(e) By the loop theorem, $V_L + V_2 + V_1 = 0$ giving $V_L = -i_1R_1 + i_2R_2$ (going clockwise around the circuit). Hence $V_L = -(2\ A)(5\ \Omega) + 0$ $= -10\ V$ Ans.

(f) In absolute value, $V_L = L\ di_2/dt$, so that $10\ V = (5\ H)di_2/dt$, or $di_2/dt = 2.0\ A/s$ Ans.

(II) After a long time the inductor has no effect and the circuit reduces to two resistors connected in parallel across a battery, and all currents are independent of the time.

(a) $i_1 = \mathcal{E}/R_1 = 2.0\ A$ Ans.

(b) $i_2 = \mathcal{E}/R_2 = 1.0\ A$ Ans.

(c) $i = i_1 + i_2 = 3.0\ A$ Ans.

(d) $V_2 = i_2R_2 = 10\ V$ Ans.

(e) $V_L = -L\ di_2/dt = 0$ Ans.

(f) $di_2/dt = 0$ Ans.

33-22

Suppose the switch to have been in position a for a time T before being thrown to b. The energy stored in the inductor the instant the switch is thrown is

$$U_B(T) = \tfrac{1}{2}Li_T^2$$

where $i_T = \frac{\mathcal{E}}{R}(1 - e^{-T/\tau})$ and $\tau = L/R$. With the switch in position b the current in the circuit becomes

$$i = i_Te^{-(t - T)/\tau},$$

time still being measured from the instant the switch was closed on a. The energy dissipated in the resistor after the switch is thrown to b over all subsequent time is

$$E = \int_T^{\infty} i^2 R\, dt = i_T^2 R \int_T^{\infty} e^{-2(t - T)/\tau} dt = i_T^2 R\tau/2,$$

$$E = i_T^2 RL/2R = \tfrac{1}{2} L i_T^2,$$

proving the assertion.

33-23

(a) The current is

$$i = \frac{\mathcal{E}}{R}(1 - e^{-t/\tau}) = \frac{50}{10,000}(1 - e^{-t/\tau}),$$

$$i = 0.005(1 - e^{-t/\tau}).$$

At $t = 5$ ms, $i = 2$ mA; hence,

$$0.002 = 0.005(1 - e^{-0.005/\tau}),$$

$$0.4 = 1 - e^{-0.005/\tau},$$

$$e^{-0.005/\tau} = 0.6,$$

$$\tau = -\frac{0.005}{\ln(0.6)} = 9.79 \times 10^{-3} = L/R,$$

$$L = (9.79 \times 10^{-3})(10^4) = 97.9 \text{ H} \quad \underline{\text{Ans.}}$$

(b) The energy stored is $U = \tfrac{1}{2} L i^2 = \tfrac{1}{2}(97.9)(0.002)^2 = 1.96 \times 10^{-4}$ J.

33-29

(a) The energy density is given by

$$u = \tfrac{1}{2} B^2/\mu_0 = (\mu_0 i N/2\pi r)^2/2\mu_0 = \mu_0 i^2 N^2/8\pi^2 r^2 \quad \underline{\text{Ans.}}$$

(b) Since u depends on r, take as a volume element the volume between two coaxial circular cylinders, radii r and r + dr, the axes of which coincide with the axis of the toroid. That is $dV = 2\pi rh\, dr$; thus, the stored energy is

$$U = \int u\,dV = \int_a^b (\mu_0 i^2N^2/8\pi^2r^2)2\pi rh\ dr = (\mu_0 i^2N^2h/4\pi)\int_a^b \frac{dr}{r},$$

$$U = \frac{\mu_0 i^2N^2h}{4\pi}\ln(\frac{b}{a}) = 1.73 \times 10^{-4}\ \mathrm{J}\quad \underline{\text{Ans}}.$$

(c) The energy U is also

$$U = \tfrac{1}{2}[\frac{\mu_0 N^2h}{2\pi}\ln(\frac{b}{a})]i^2 = \tfrac{1}{2}Li^2$$

and this is seen to agree with (b).

33-35

By Equation 33-23, $\mathcal{E}_2 = -M(di_1/dt)$. Compute $\mathcal{E}_2$ by Faraday's law. The flux linking the inner with the outer solenoid is

$$\Phi_2 = \int \vec{B}_1 \cdot d\vec{S},$$

where B_1 is the field set up by a current i_1 in the inner solenoid and the integral is over the cross-sectional area of the outer solenoid. But $B_1 = \mu_0 n_1 i_1$ inside solenoid 1 and zero outside. Thus there is no contribution to the integral from the area between the solenoids, so that

$$\Phi_2 = B_1(\pi R_1^2) = \mu_0 n_1 i_1 \pi R_1^2.$$

Since there are $n_2\ell$ turns of solenoid 2 in a length ℓ, Faraday's law gives

$$\mathcal{E}_2 = -(n_2\ell)(d\Phi_2/dt) = -\mu_0 n_1 n_2 \ell\pi R_1^2(di_1/dt) = -M(di_1/dt),$$

$$M = \mu_0 n_1 n_2 \pi R_1^2 \ell.$$

33-36

(a) Connect the coils to a battery. The resistors in the circuit sketch represent the resistance of the coils. By the loop theorem,

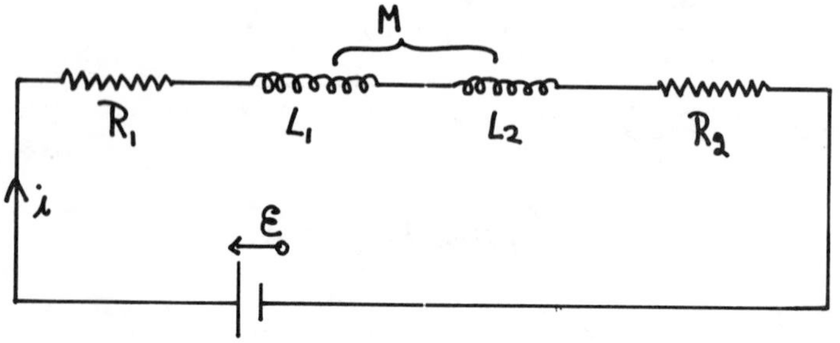

$$\mathcal{E} - iR_1 - L_1\frac{di}{dt} \pm M\frac{di}{dt} - L_2\frac{di}{dt} \pm M\frac{di}{dt} - iR_2 = 0,$$

$$\mathcal{E} = i(R_1 + R_2) + (L_1 + L_2 \pm 2M)\frac{di}{dt},$$

and therefore the second term in parenthesis is the equivalent inductance, for the parallel equation for a circuit containing only one inductor would be

$$\mathcal{E} = iR_e + L_e\frac{di}{dt}.$$

(b) If the coils are wound so that their fluxes are in the same direction, use the + sign, whereas if the windings of one are reversed so that the fluxes tend to cancel, use the - sign.

34-6

(a) In a time $2\pi/\omega$, one period of rotation, all of the charge q on the ring passes any fixed point near the ring; thus the equivalent current i is

$$i = \frac{q}{2\pi/\omega} = \frac{\omega q}{2\pi}.$$

Therefore,

$$\mu = NiA = (1)\left(\frac{\omega q}{2\pi}\right)(\pi r^2) = \tfrac{1}{2}\omega q r^2 \quad \underline{\text{Ans}}.$$

(b) By the right-hand rule, the magnetic moment vector is parallel to the angular velocity $\vec{\omega}$.

34-10

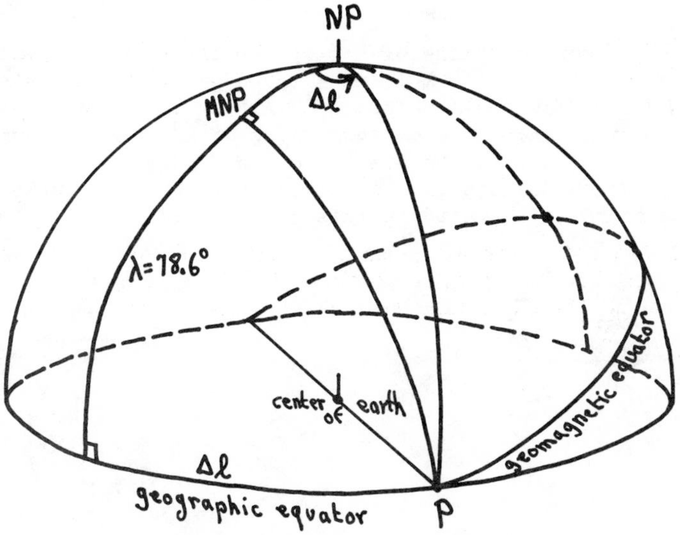

Every point on the geographic equator is 90° from the geographic north pole NP; similarly, every point on the geomagnetic equator is 90° from the geomagnetic north pole MNP. Hence, the point P where these equators intersect is 90° from both NP and MNP. In the sketch λ = latitude of MNP, so that the angular distance between the two poles is $90° - \lambda$. If $\Delta\ell$ is the difference in longitude between P and MNP, then the law of cosines applied to the triangle formed by P and the two poles gives

$$\cos 90° = \cos(90° - \lambda)\cos 90° + \sin(90° - \lambda)\sin 90° \cos(\Delta\ell),$$

$$0 = \cos\lambda \cos(\Delta\ell),$$

$$\Delta\ell = 90°.$$

This result can be anticipated by noticing that P bears the same relation to the great circle through NP and MNP as NP bears to the geographic equator, and MNP bears to the geomagnetic equator: to wit, lines of "longitude" intersect the appropriate equator at right angles.
Since $\Delta\ell = 90°$, the longitude of P is 70°W + 90° = 160°W. The other intersection point, on the other side of the earth, has a longitude 160° + 180° = 340°W = 360° - 340° = 20°E.

34-14

Examine the maximum value of B/T in the range of observation. This will be (0.5 tesla)/(10 kelvin) = (5000 gauss)/(10 kelvin) = 500 gauss/kelvin. Now the abscissa of Fig. 34-7 is in units of 1000 gauss/kelvin. Hence, the maximum value of B/T in these units is 0.5. This is close to the origin in the Figure, well within the range in which Curie's law is obeyed and therefore Curie's law will be found to be valid in this test.

34-16

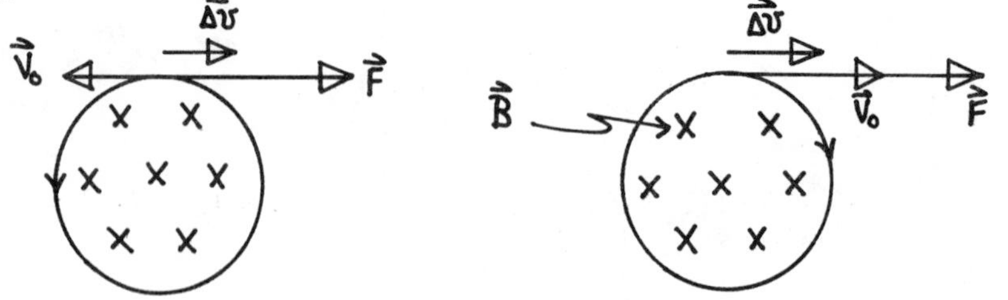

As the magnetic field is introduced, B increases and a counterclockwise electric field is induced; by Example 32-3 this field is $E = \frac{r}{2}\frac{dB}{dt}$ and the electrons feel an electrical force, directed as shown. Suppose the magnetic field increases by an amount B in time T. Then, for each electron,

$$\Delta v = aT = \frac{F}{m}T = \frac{eE}{m}T = \frac{e}{m}\left(\frac{r}{2}\frac{B}{T}\right)T = \frac{erB}{2m}.$$

The new velocities become

$$v = v_0 \pm \frac{erB}{2m},$$

+ for the clockwise circulating electron and - for the other. Dividing by r, assumed to remain constant, gives

$$\omega = \omega_0 \pm \frac{eB}{2m};$$

the changed angular velocity leads to an increase or decrease in the orbital magnetic moment. Also, the existence of the diamagnetic effect in a constant magnetic field can be "explained" by noting that the circulating electron still "cuts" lines of magnetic flux.

34-17

(a) Let the radius of the core be r; the core volume is $V = 4\pi r^3/3$ and the mass is $m = V\rho$, where ρ = density in the core = 14000 kg/m^3. Since the atomic mass of iron is 56 the number of atoms in the core is m/56u, where u is the atomic mass unit. If μ is the magnetic dipole moment of one iron atom, then,

$$\frac{(4\pi r^3/3)(\rho)}{56u}\mu = 8 \times 10^{22},$$

$$\frac{(4\pi r^3/3)(14000)}{(56)(1.66 \times 10^{-27})}(2.1 \times 10^{-23}) = 8 \times 10^{22},$$

$$r = 1.82 \times 10^5 \text{ m} = 182 \text{ km} \quad \underline{\text{Ans.}}$$

(b) The fraction f of the volume of the earth, radius R, occupied by this core is

$$f = \left(\frac{r}{R}\right)^3 = \left(\frac{182}{6370}\right)^3 = 2.33 \times 10^{-5} \quad \underline{\text{Ans.}}$$

34-19

(a) The field in a toroid is $B = (\mu_0 iN)/2\pi r$, where N is the total number of turns. This is a nonuniform field, but consider the field to be approximately uniform and equal to the actual value at the center of the toroid tube. It is desired that this field B_0 equal 2×10^{-4} T:

$$2 \times 10^{-4} = (4\pi \times 10^{-7})(i)(400)/2\pi(0.055),$$

$$i = 0.14 \text{ A} \quad \underline{\text{Ans.}}$$

(b) With the iron present inside the toroid the field is $B_M + B_0 = 801\ B_0$. If A is the cross-sectional area of the toroid then, by Problem 32-21, the charge induced in a coil of N_c turns and resistance R_c is

$$q = N_c[\Phi_B(\text{final}) - \Phi_B(\text{initial})]/R_c = N_c(B_0 + B_M)A/R_c,$$

$$q = (50)(2 \times 10^{-4})\pi(0.005)^2(801)/(8),$$

$$q = 7.86 \times 10^{-5} \text{ C} \quad \underline{\text{Ans.}}$$

CHAPTER 35

35-5

(a) If $U_E = \frac{1}{2}U_B$, then since $U_E + U_B = \frac{1}{2}q_m^2/C$,

$$U_E = \frac{1}{3}(\tfrac{1}{2}q_m^2/C),$$

$$\tfrac{1}{2}q^2/C = \frac{1}{3}(\tfrac{1}{2}q_m^2/C),$$

$$q = q_m/3^{\frac{1}{2}} \quad \underline{\text{Ans}}.$$

(b) The charge as a function of time, with $q = q_m$ at $t = 0$, is

$$q = q_m \cos\omega t.$$

Substituting the expression for q from (a) yields

$$3^{-\frac{1}{2}} = \cos\omega t,$$

$$t = \frac{1}{\omega}\cos^{-1}(3^{-\frac{1}{2}}) = \frac{1}{\omega}(0.9553 \text{ rad}).$$

But $\omega = 2\pi/T$, T the period. Hence

$$t = \frac{0.9553}{2\pi}T = 0.152T \quad \underline{\text{Ans}}.$$

35-11

(a) The frequencies that can be tuned are $\nu = 1/2\pi(LC)^{\frac{1}{2}}$; thus,

$$\frac{\nu_{max}}{\nu_{min}} = \frac{(LC_{min})^{-\frac{1}{2}}}{(LC_{max})^{-\frac{1}{2}}} = (C_{max}/C_{min})^{\frac{1}{2}} = (365/10)^{\frac{1}{2}} = 6.04 \quad \underline{\text{Ans}}.$$

(b) Let C be the added capacitance. For capacitors in parallel the equivalent capacitance C' is the sum of the individual capacitances and therefore C' ranges between (C + 10) and (C + 365) pF. Hence,

$$\frac{1.60}{0.54} = \left(\frac{365 + C}{10 + C}\right)^{\frac{1}{2}},$$

$$C = 35.63 \text{ pF} \quad \underline{\text{Ans}}.$$

The inductance must be given by

$$\nu = \frac{\omega}{2\pi} = \frac{1}{2\pi}[(L)(C + 10)]^{-\frac{1}{2}} = 1.6 \times 10^6,$$

$$L = 2.2 \times 10^{-4} \text{ H} \quad \underline{\text{Ans}},$$

using the value of C (expressed in farads) found in the above.

35-14

The energy needed to raise the 100 μF capacitor to 300 V is $\frac{1}{2}C_1V^2 = 4.5$ J (since 1 μF = 10^{-6} F). Initially, the 900 μF capacitor has an energy $\frac{1}{2}C_9V^2 = 4.5$ J also, since V = 100 volts. First, transfer all of this energy to the inductor by leaving S_1 open, closing S_2 and waiting a time $\frac{1}{4}T_9$, where T_9 is the period of the LC_9 system. Then, before the energy starts to flow back to C_9, open S_2 and close S_1 simultaneously. Wait a time $T_1/4$ for the energy to flow from the inductor to C_1. Finally, open S_1 to prevent energy from flowing back to the inductor. The periods T_1 and T_9 are

$$T_1 = 2\pi/\omega_1 = 2\pi(LC_1)^{\frac{1}{2}} = 0.199 \text{ s},$$

$$T_9 = 2\pi(LC_9)^{\frac{1}{2}} = 0.596 \text{ s},$$

so the "manipulating" might be a little tricky in practice.

35-15

(a) The charge q as a function of time is

$$q = q_m \sin\omega t,$$

choosing sine rather than cosine so that q(0) = 0 as required. The current i, then, is

$$i = \frac{dq}{dt} = \omega q_m \cos\omega t.$$

Thus the maximum current $i_m = \omega q_m$ and therefore

$$q_m = i_m/\omega = i_m(LC)^{\frac{1}{2}} = (2)[(3 \times 10^{-3})(2.7 \times 10^{-6})]^{\frac{1}{2}},$$

$$q_m = 1.8 \times 10^{-4}\ \text{C} \quad \underline{\text{Ans}}.$$

(b) The energy stored in the capacitor at any time t is

$$U = \tfrac{1}{2}q^2/C = (\tfrac{1}{2}q_m^2/C)\sin^2\omega t.$$

The rate of increase is

$$\frac{dU}{dt} = (\tfrac{1}{2}q_m^2/C)(2\omega)\sin\omega t\cdot\cos\omega t = (\tfrac{1}{2}\omega q_m^2/C)\sin 2\omega t.$$

This reaches its greatest value when $\sin 2\omega t = 1$, or

$$2\omega t = \frac{\pi}{2},$$

$$2(\frac{2\pi}{T})t = \frac{\pi}{2},$$

$$t = T/8 \quad \underline{\text{Ans}}.$$

(c) The greatest rate of increase is, from (b),

$$(\frac{dU}{dt})_{max} = \tfrac{1}{2}\omega q_m^2/C.$$

Numerically, $\omega = (LC)^{-\frac{1}{2}} = (9 \times 10^{-5})^{-1}$, and $q_m = 1.8 \times 10^{-4}$ C, from (a). Also, $C = 2.7 \times 10^{-6}$ F. These give $(dU/dt)_{max} = 66.7$ W.

35-17

The charge q on the capacitor as a function of time t in a damped LC, or LCR, circuit is, by Eq.35-12,

$$q = q_m e^{-Rt/2L}\cos\omega' t = Q(t)\cos\omega' t$$

where q_m is the charge q at $t = 0$. The energy present over one oscillation, assuming $2\pi/\omega' \ll 2L/R$ (small damping) is

$$U = \tfrac{1}{2}Q^2/C = (\tfrac{1}{2}q_m^2/C)e^{-Rt/L} = [U(0)]e^{-Rt/L}.$$

The required time is found by setting $U = \frac{1}{2}U(0)$:

$$\tfrac{1}{2}U(0) = [U(0)]e^{-Rt/L},$$

$$\ln 2 = Rt/L,$$

$$t = \frac{L}{R}\ln 2 \quad \underline{\text{Ans}}.$$

36-7

(a) The capacitative reactance is $X_C = 1/\omega C$; hence

$$X_C = \frac{1}{\omega(10 \times 10^{-6})} = \frac{10^5}{\omega}.$$

For the inductor,

$$X_L = \omega L = \omega(6 \times 10^{-3}).$$

Setting these equal gives

$$\frac{10^5}{\omega} = \omega(6 \times 10^{-3}),$$

$$\omega^2 = \frac{10^5}{6 \times 10^{-3}} = \frac{1}{6} \times 10^8,$$

$$\nu = \frac{\omega}{2\pi} = 650 \text{ Hz} \quad \underline{\text{Ans}}.$$

(b) The reactance, calculated for the inductor say, is $X_L = (2\pi)\cdot(650)(0.006) = 24.5\ \Omega$ Ans.

(c) Using the formulae for X_C and X_L in (a) and setting $X_L = X_C$ yields

$$\omega L = \frac{1}{\omega C},$$

$$\omega = (LC)^{-\frac{1}{2}}$$

which is the frequency of an LC circuit.

36-11

(a) The current amplitude is

$$i_m = \frac{\mathcal{E}_m}{[R^2 + (\omega L - 1/\omega C)^2]^{\frac{1}{2}}}.$$

To find the maximum value of i_m, set $di_m/d\omega = 0$;

$$\frac{1}{\mathcal{E}_m}\frac{di_m}{d\omega} = \tfrac{1}{2}[R^2 + (\omega L - 1/\omega C)^2]^{-3/2}(2)(\omega L - \frac{1}{\omega C})(L + \frac{1}{\omega^2 C}) = 0,$$

$$(\omega L - \frac{1}{\omega C}) = 0,$$

$$\omega = (LC)^{-\frac{1}{2}} = \omega_0,$$

$$\omega_0 = [(1)(20 \times 10^{-6})]^{-\frac{1}{2}} = 223.61 \text{ rad/s} \quad \underline{\text{Ans}}.$$

(b) The maximum value of the current is $\mathcal{E}_m/R$, from (a) with $\omega = (LC)^{-\frac{1}{2}}$.

(c) To find ω_1 and ω_2, put $i_m = \frac{1}{2}\mathcal{E}_m/R$:

$$\tfrac{1}{2}\mathcal{E}_m/R = \frac{\mathcal{E}_m}{[R^2 + (\omega L - 1/\omega C)^2]^{\frac{1}{2}}},$$

$$[R^2 + (\omega L - 1/\omega C)^2] = 4R^2,$$

$$\omega^2 L^2 - 2L/C + 1/\omega^2 C^2 = 3R^2,$$

$$\omega^4 L^2 C^2 - 2LC\omega^2 + 1 = 3R^2 C^2 \omega^2,$$

$$(L^2C^2)\omega^4 - (3R^2C^2 + 2LC)\omega^2 + 1 = 0.$$

Substituting the values of L, C, R gives

$$(4 \times 10^{-10})\omega^4 - (3 \times 10^{-8} + 4 \times 10^{-5})\omega^2 + 1 = 0.$$

Multiply through by 10^5 to obtain

$$(4 \times 10^{-5})\omega^4 - (4.003)\omega^2 + 10^5 = 0.$$

Treat this as a quadratic equation with the unknown $x = \omega^2$. The quadratic formula gives

$$x = 5.197435 \times 10^4;\quad 4.810065 \times 10^4.$$

Taking the positive square roots ($\omega > 0$) yields

$$\omega_1 = 227.98 \text{ rad/s} \quad \underline{\text{Ans}};$$

$$\omega_2 = 219.32 \text{ rad/s} \quad \underline{\text{Ans}}.$$

(d) Taking ω_0 from (a), it is found that

$$(\omega_1 - \omega_2)/\omega_0 = 0.0387 \quad \underline{\text{Ans}}.$$

36-14

As noted in Problem 36-11(b), the maximum value of i_m is $\mathcal{E}_m/R$. To find the half-width, then, set $i_m = \frac{1}{2}\mathcal{E}_m/R$:

$$\tfrac{1}{2}\mathcal{E}_m/R = \frac{\mathcal{E}_m}{[R^2 + (\omega L - 1/\omega C)^2]^{\frac{1}{2}}}.$$

Squaring to solve for the term in parenthesis gives

$$\omega L - \frac{1}{\omega C} = 3^{\frac{1}{2}}R.$$

Now set $\omega = \omega_0 + \delta\omega = \omega_0(1 + \delta\omega/\omega_0)$; this should yield an accurate result if the half-width is small compared to the resonant frequency (sharp peak). To this same degree of approximation,

$$\frac{1}{\omega} \approx \frac{1}{\omega_0}(1 - \frac{\delta\omega}{\omega_0}).$$

The previous equation now becomes

$$\omega_0(1 + \frac{\delta\omega}{\omega_0})L - \frac{1}{\omega_0 C}(1 - \frac{\delta\omega}{\omega_0}) = 3^{\frac{1}{2}}R,$$

$$(\omega_0 L - \frac{1}{\omega_0 C}) + \frac{\delta\omega}{\omega_0}(\omega_0 L + \frac{1}{\omega_0 C}) = 3^{\frac{1}{2}}R.$$

But $\omega_0^2 = 1/LC$; hence the first term vanishes and

$$\omega_0 L + \frac{1}{\omega_0 C} = \frac{\omega_0^2 LC + 1}{\omega_0 C} = \frac{2}{\omega_0 C} = 2\left(\frac{L}{C}\right)^{\frac{1}{2}}.$$

Thus

$$\frac{\delta\omega}{\omega_0} = \tfrac{1}{2}3^{\frac{1}{2}}\left(\frac{C}{L}\right)^{\frac{1}{2}}R.$$

The half-width $\Delta\omega = 2\delta\omega$, so that

$$\frac{\Delta\omega}{\omega_0} = \left(\frac{3C}{L}\right)^{\frac{1}{2}}R.$$

For the values of L,C,R from Problem 36-11, this gives $\Delta\omega/\omega_0$ = 0.0387, the same as in that problem.

36-16

The average power dissipated is

$$P_{av} = \mathcal{E}_{rms} i_{rms} \cos\phi = \tfrac{1}{2}\mathcal{E}_m i_m \cos\phi = \tfrac{1}{2}\mathcal{E}_m\left(\frac{\mathcal{E}_m}{Z}\right)\left(\frac{R}{Z}\right) = \tfrac{1}{2}\,\frac{\mathcal{E}_m^2 R}{Z^2},$$

$$P_{av} = \tfrac{1}{2}\mathcal{E}_m^2 R\,\frac{1}{R^2 + (\omega L - 1/\omega C)^2}.$$

To find the desired values of C set $dP_{av}/dC = 0$: this gives

$$-\frac{\mathcal{E}_m^2 R}{Z^4}\left(\omega L - \frac{1}{\omega C}\right)(1/\omega C^2) = 0.$$

Two values of C satisfy this equation and they are

(a) $C = 1/L\omega^2$ = 117 μF Ans; (b) $C = \infty$ Ans.

The latter implies that the capacitor has been shorted-out, yielding an LR circuit. (C = 0 is an additional solution, but this means a broken circuit with P = 0 because nothing is happening.)

(c) Substituting the values of C from (a) and (b) into P_{av} gives

$$P_{av} = \tfrac{1}{2}\mathcal{E}_m^2/R = 90\ \text{W} \quad \underline{\text{Ans}};$$

$$P_{av} = \tfrac{1}{2}\mathcal{E}_m^2 R \frac{1}{R^2 + \omega^2 L^2} = 4.19\ \text{W} \quad \underline{\text{Ans}}.$$

(d) For $C = 1/L\omega^2$ the impedance $Z = R$ and therefore $\cos\phi = R/Z = 1$ so that $\phi = 0$ Ans.

On the other hand, as C approaches infinity, Z approaches $(R^2 + \omega^2 L^2)^{\frac{1}{2}}$ and thus in this case $\cos\phi = R/(R^2 + \omega^2 L^2)^{\frac{1}{2}} = 0.22$, giving $\phi = 78°$ Ans.

(e) The power factors are $\cos\phi$, so that from (d) these are 1 and 0.22 Ans.

36-17

(a) Writing $i = i_m \sin(\omega t + \phi)$, the power factor is $\cos\phi = \cos 42° = 0.743$.

(b) Since $\phi > 0$, $\omega t + \phi > \omega t$, so that the current leads the emf.

(c) With $\phi = 42°$, $\tan\phi = 0.900$. But

$$\tan\phi = (X_C - X_L)/R = 0.9 > 0,$$

so that $X_C > X_L$, and the circuit is capacitative.

(d) At resonance $X_C = X_L$, giving $\tan\phi = 0$, $\phi = 0$. Since $\phi = 42° \neq 0$ the circuit is not in resonance.

(e) If the box contained a resistor only, $\mathcal{E}$ and i would be in phase, or $\phi = 0$. A capacitor only or an inductor only would lead to currents depending exponentially on the time. If there is no resistor, but only a capacitor and an inductor (LC circuit), the current would be out of phase by 90° (set R = 0 in the equation for $\tan\phi$). The possibilities, then, are LR, CR, LCR. But the circuit was found to be capacitative; hence, there must be a capacitor. But an inductor is not necessary, for a pair of values for C, R can be found to yield the given current amplitude and phase angle with L = 0. The circuit, then, must contain a capacitor and a resistor, but may or may not possess an inductor.

(f) The average power is

$$P_{av} = \tfrac{1}{2}\mathcal{E}_m i_m \cos\phi = \tfrac{1}{2}(75)(1.2)(0.743) = 33.4\ \text{W} \quad \underline{\text{Ans}}.$$

(g) Although the power factor $\cos\phi$ depends on the values of L, C, R, and ω through

$$\tan\phi = (X_C - X_L)/R = (\frac{1}{\omega C} - \omega L)/R,$$

the value of ϕ has been given and does not have to be computed: the value of ω is therefore not needed. The power is an average over many cycles and $\langle \sin\omega t \cdot \sin(\omega t + \phi)\rangle = \tfrac{1}{2}\cos\phi$, independent of ω when the average is so taken.

36-23

The power output at the primary is

$$P_{av} = V_1 i_1 \cos\phi.$$

However, if Φ_1 and Φ_2 are the fluxes in the core due to the primary and secondary windings, then

$$\cos\phi = (\Phi_2/\Phi_1) = (N_2 i_2)/(N_1 i_1).$$

But $V_1/V_2 = N_1/N_2$ and therefore

$$P_{av} = V_1 i_1 (N_2/N_1)(i_2/i_1) = V_1 i_1 (V_2/V_1)(i_2/i_1) = V_2 i_2;$$

hence,

$$V_1 i_1 \cos\phi = V_2 i_2,$$

$$i_1 = \frac{N_2}{N_1} i_2/\cos\phi.$$

With $i_2 = V_2/R = (N_2 V_1/N_1)/R$ this becomes

$$i_1 = (N_2^2 V_1/N_1^2)/(R\cos\phi).$$

If $\phi = 0$ this can be written

$$i_1 = \frac{V_1}{(N_1/N_2)^2 R}$$

proving the assertion.

36-25

The resistance added to the amplifier by virtue of the connected transformer is, by Problem 36-23,

$$r_t = (N_1/N_2)^2 R$$

and thus the total resistance of the amplifier is $r + r_t$ in effect. The average power delivered to R is

$$P_{av} = i_{2,rms}^2 R = (\frac{N_1}{N_2} i_{1,rms})^2 R = (\frac{N_1}{N_2})^2 \frac{\mathcal{E}_m^2 \sin^2 \omega t}{[r + (N_1/N_2)^2 R]^2} R.$$

If $x = (N_1/N_2)^2$ this is, as far as dependence on x is concerned,

$$P_{av} = \frac{x}{(r + xR)^2}.$$

To find the value of x leading to maximum P_{av}, calculate dP_{av}/dx:

$$dP_{av}/dx = \frac{r - xR}{(r + xR)^3} = 0,$$

$$x = \frac{r}{R} = (N_1/N_2)^2 = \frac{1000\ \Omega}{10\ \Omega},$$

$$N_1/N_2 = 10 \quad \underline{\text{Ans}},$$

that is, sketch a coil with a turn ratio of 10 (amplifier) to 1 (speaker).

CHAPTER 37

37-4

The displacement current i_d is by definition

$$i_d = \epsilon_0 \frac{d\Phi_E}{dt} = \epsilon_0 \frac{d(EA)}{dt} = \epsilon_0 A \frac{dE}{dt}.$$

Now let x be the separation between the plates; thus $E = V/x$, V the potential difference across the plates. Then, if the plates are fixed in position, so that x does not change,

$$i_d = \epsilon_0 A \frac{d(V/x)}{dt} = \frac{\epsilon_0 A}{x} \frac{dV}{dt}.$$

But for a parallel-plate capacitor (k = 1), $C = \epsilon_0 A/x$ giving

$$i_d = C \frac{dV}{dt}.$$

37-6

(a) By Problem 37-4,

$$i_d = C \frac{dV}{dt} = C \frac{d}{dt}(V_m \sin\omega t) = \omega C V_m \cos\omega t.$$

The maximum displacement current is, therefore,

$$i_{d,max} = \omega C V_m = 2\pi\nu C V_m = (2\pi)(50)(10^{-10})(174{,}000),$$

$$i_{d,max} = 0.00547 \text{ A} \quad \underline{\text{Ans}}.$$

(b) From (a) it is apparent that i_d is proportional to V_m, and therefore a large value of the latter yields a more easily measurable displacement current.

37-8

The displacement current is defined by

$$i_d = \epsilon_0 \frac{d\Phi_E}{dt} = \epsilon_0 \frac{d}{dt}\int \vec{E}\cdot d\vec{S}.$$

If the area over which the integral is taken is aligned so that $\vec{E}$ and $d\vec{S}$ are parallel,

$$i_d = \epsilon_0 \frac{d}{dt}\int E\, dS = \int\left(\epsilon_0 \frac{dE}{dt}\right)dS,$$

and the current density j_d becomes

$$j_d = \frac{di_d}{dS} = \epsilon_0 \frac{dE}{dt}.$$

37-10

(a) The displacement current i_d in the gap between the plates is equal to the conduction current i in the wires; hence,

$$i_{max} = i_{d,max} = 8.9 \times 10^{-6}\ \text{A} \quad \underline{\text{Ans.}}$$

(b) By definition,

$$i_d = \epsilon_0 (d\Phi_E/dt),$$

so that

$$(d\Phi_E/dt)_{max} = i_{d,max}/\epsilon_0 = 8.9 \times 10^{-6}/8.85 \times 10^{-12},$$

$$(d\Phi_E/dt)_{max} = 1.0056 \times 10^6\ \text{V}\cdot\text{m/s} \quad \underline{\text{Ans.}}$$

(c) By Problem 37-4,

$$i_d = \frac{\epsilon_0 A}{d}\frac{dV}{dt} = \frac{\epsilon_0 A}{d}\frac{d\mathcal{E}}{dt} = \frac{\epsilon_0 A}{d}\mathcal{E}_m \omega \cos\omega t,$$

the potential across the capacitor, except for sign, being equal to the battery emf, by the loop theorem. The plate separation, then, must be

$$d = \epsilon_0 A \mathcal{E}_m \omega / i_{d,max} = \frac{(8.85 \times 10^{-12})(0.1)(200)(100)}{8.9 \times 10^{-6}},$$

$$d = 1.99 \times 10^{-3} \text{ m} = 1.99 \text{ mm} \quad \underline{\text{Ans.}}$$

(d) In the gap between the plates the conduction current $i = 0$. Ampere's law reduces to

$$\oint \vec{B} \cdot \vec{d\ell} = \mu_0 I_d$$

where I_d is the displacement current flowing across the area bounded by the circular path of integration of radius R. It is easy to verify that R is less than the radius of the plates. Then, if the current density j_d is uniform,

$$I_d = j_d(\pi R^2) = \frac{i_d}{A}(\pi R^2).$$

Returning to Ampere's law,

$$B(2\pi R) = \mu_0 I_d,$$

$$B_{max}(2\pi R) = \mu_0 I_{d,max} = \mu_0 (i_{d,max} \pi R^2 / A),$$

$$B_{max} = \tfrac{1}{2}\mu_0 R i_{d,max}/A = \tfrac{1}{2}(4\pi \times 10^{-7})(0.1)(8.9 \times 10^{-6})/(0.1),$$

$$B_{max} = 5.59 \times 10^{-12} \text{ T} \quad \underline{\text{Ans.}}$$

37-13

(a) Since $i = dq/dt$, the charge $q(t)$ on the faces will be

$$q = \int i\,dt = \int \alpha t\,dt = \tfrac{1}{2}\alpha t^2 \quad \underline{\text{Ans;}}$$

(with $q(0) = 0$ the integration constant is zero).

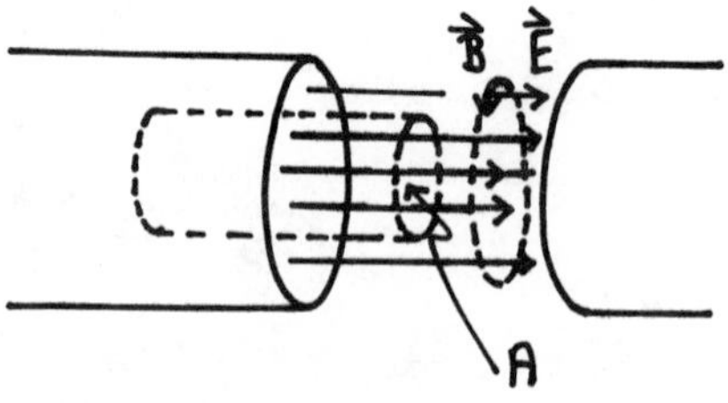

(b) Use the pill-box Gaussian surface shown. The charges on the rod faces do not contribute to the electric field in the rod, so set $E = 0$ there. $\vec{E}$ is parallel to the x-axis in the gap and Gauss's law gives under these circumstances

$$\epsilon_0 EA = qA/\pi R^2,$$

$$E = q/\pi\epsilon_0 R^2 = \tfrac{1}{2}\alpha t^2/\pi\epsilon_0 R^2 \quad \underline{\text{Ans}},$$

using the result from (a).

(c) From an examination of Fig. 37-1, the lines of B are concentric circles, centered on and perpendicular to the x-axis.

(d) Choosing as path of integration a circle of radius r coincident with a line of B,

$$\oint \vec{B}\cdot d\vec{\ell} = B(2\pi r).$$

In the gap $i = 0$; hence

$$2\pi r B = \mu_0\epsilon_0 \frac{d\Phi_E}{dt} = \mu_0\epsilon_0 \frac{d}{dt}(ES) = \mu_0\epsilon_0(\pi r^2)\frac{dE}{dt} = \mu_0\epsilon_0 \pi r^2\left(\frac{\alpha t}{\pi\epsilon_0 R^2}\right),$$

$$B = \frac{\tfrac{1}{2}\mu_0 r\alpha t}{\pi R^2} \quad \underline{\text{Ans}}.$$

(e) The displacement current in the gap equals the conduction current in the rod; thus the magnetic field in the gap and rod are the same by Ampere's law.

CHAPTER 38

38-5

It is necessary to calculate the speed v of the electrons. The relativistic kinetic energy K is

$$K = mc^2 - m_0c^2.$$

For the electron the rest energy $m_0c^2 = 0.511$ MeV. Since K = 18 GeV = 18×10^9 eV = 18,000 MeV, this implies that $m_0c^2 \ll mc^2$. To a good degree of approximation, then, drop the rest energy and set

$$K \approx mc^2 = \frac{m_0c^2}{(1 - v^2/c^2)^{\frac{1}{2}}},$$

$$18000 = \frac{0.511}{(1 - v^2/c^2)^{\frac{1}{2}}},$$

$$v^2/c^2 = 1 - 8.059 \times 10^{-10},$$

$$v/c \approx 1 - 4.02950 \times 10^{-10},$$

$$v = (1 - 4.0295 \times 10^{-10})c.$$

Let D be the length of the course. The time required for the electron is D/v and for light D/c. The light beam wins, of course, since c > v; the time Δt by which the light wins is

$$\Delta t = \frac{D}{v} - \frac{D}{c} = \frac{D}{c}\left[\frac{1}{1 - 4.0295 \times 10^{-10}} - 1\right],$$

$$\Delta t \approx \frac{D}{c}[(1 + 4.0295 \times 10^{-10}) - 1] = \frac{1}{186,000}(4.0295 \times 10^{-10}),$$

$$\Delta t = 2.17 \times 10^{-15} \text{ s} \quad \underline{\text{Ans}}.$$

38-9

The energy density due to an electric field is $u_E = \frac{1}{2}\epsilon_0 E^2$ if the field resides in a vacuum. Using the equation for a traveling wave this becomes

$$u_E = \tfrac{1}{2}\epsilon_0 E_m^2 \sin^2(kx - \omega t).$$

But $E_m/B_m = c$ and therefore

$$u_E = \tfrac{1}{2}\epsilon_0 c^2 B_m^2 \sin^2(kx - \omega t) = \tfrac{1}{2}\epsilon_0 c^2 B^2.$$

Finally, invoke the relation $c^2 = 1/\epsilon_0\mu_0$ to obtain

$$u_E = \tfrac{1}{2}\epsilon_0 \frac{1}{\epsilon_0\mu_0} B^2 = \tfrac{1}{2}B^2/\mu_0 = u_B.$$

In a traveling electromagnetic wave the electric and magnetic fields are in phase; hence there does not seem to be any need to resort to a time-average.

38-20

(a) The received intensity is the Poynting vector $\vec{S}$. Since $\vec{E}$ and $\vec{B}$ are perpendicular and $E = cB$ (the airplane receiving a wave that is practically plane),

$$\overline{S} = E_m B_m/2\mu_0 = E_m^2/2c\mu_0.$$

Hence,

$$E_m = (2c\mu_0\overline{S})^{\frac{1}{2}} = 0.0868 \text{ V/m} \quad \underline{\text{Ans}},$$

since $\overline{S} = 10^{-5}$ W/m^2.

(b) As $E = cB$, $E_m = cB_m$ (the fields being in phase). Therefore

$$B_m = E_m/c = 2.89 \times 10^{-10} \text{ T} \quad \underline{\text{Ans}}.$$

(c) With the transmitter radiating spherical waves,

$$\overline{P} = \overline{S}\cdot 4\pi r^2 = (10^{-5})(4\pi)(10^4)^2 = 12.6 \text{ kW} \quad \underline{\text{Ans}}.$$

38-22

(a) As $\vec{S} = (\vec{E} \times \vec{B})/\mu_0$ $\vec{S}$ is directed in the negative z-direction. Thus, the only faces across which $\vec{S}$ cuts are the two that are parallel to the x,y-plane. Energy at the rate EBa^2/μ_0 "enters" the top face and "leaves" through the bottom face (a^2 the area of each face) according to the Poynting vector picture. If the fields E and B are static, however, this model does not really make sense.

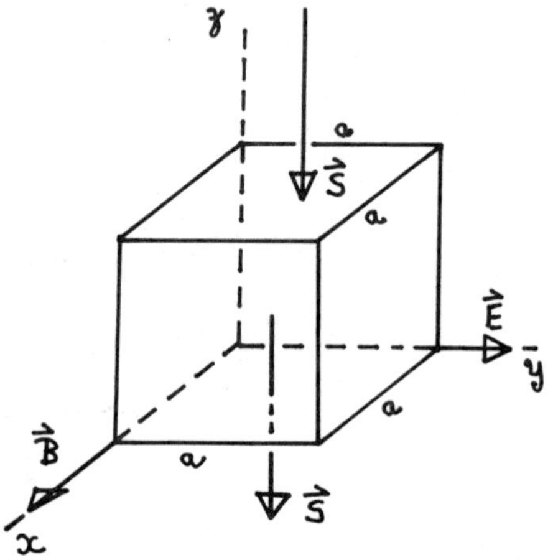

(b) Since energy appears to enter and leave at the same rate, the energy in the cube does not change, even in the Poynting point of view. Hence, the Poynting vector picture makes sense when the energy flow through a closed surface is examined, but sometimes not if only part of such a surface is considered.

38-24

(a) Sighting along the resistor in the direction of the current flow, $\vec{E}$ points directly away from the observer while $\vec{B}$ is directed transverse and at right angles to the resistor axis in the clockwise sense. Then, by the right-hand rule, $\vec{E} \times \vec{B}$ and therefore $\vec{S}$ are pointed radially inward.

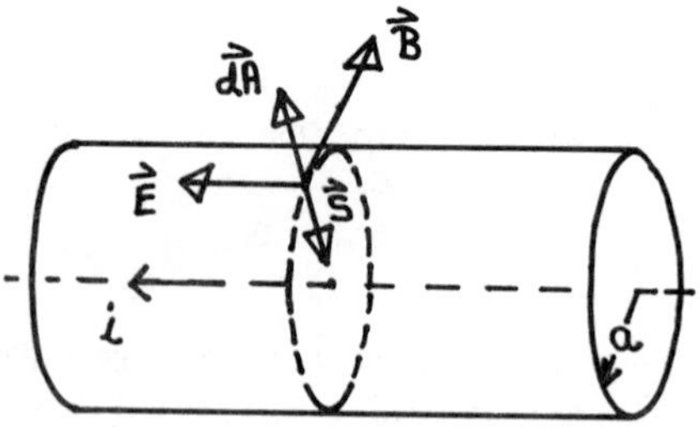

(b) Since $\vec{E}$ and $\vec{B}$ are at right angles,

$$S = \frac{EB}{\mu_0} = \frac{(V/\ell)B}{\mu_0} = \frac{i}{\mu_0 \ell}(R)B = \frac{i}{\mu_0 \ell}(\rho\ell/\pi a^2)(\mu_0 i/2\pi a) = \frac{i^2\rho}{2\pi^2 a^3}.$$

The integral desired is $\int \vec{S}\cdot\vec{dA}$ over the cylindrical surface of the wire, considering a piece of length ℓ. The angle between $\vec{S}$ and the element of area $d\vec{A}$ (called $d\vec{S}$ in previous chapters, a notation that would now conflict with that for the Poynting vector) is 180° and the magnitude of $\vec{S}$ is the same over the surface. Thus,

$$\int \vec{S}\cdot d\vec{A} = -SA = -\left(\frac{i^2\rho}{2\pi^2 a^3}\right)(2\pi a\ell) = -i^2(\rho\ell/\pi a^2) = -i^2R.$$

38-29

(a) The frequency ν is

$$\nu = \frac{c}{\lambda} = \frac{3 \times 10^8 \text{ m/s}}{3.0 \text{ m}} = 100 \text{ MHz} \quad \underline{\text{Ans.}}$$

(b) $\vec{E} \times \vec{B} = \mu_0\vec{S}$ must be in the +x-direction. By the right-hand rule, if $\vec{E}$ lies along the y-axis, $\vec{B}$ must be directed along the z-axis. Also

$$B_m = E_m/c = \frac{300 \text{ V/m}}{3 \times 10^8 \text{ m/s}} = 10^{-6} \text{ T} \quad \underline{\text{Ans.}}$$

(c) The angular frequency $\omega = 2\pi\nu = 6.28 \times 10^8$ rad/s; the wave number is $k = 2\pi/\lambda = 2.1 \text{ m}^{-1}$.

(d) The time-averaged Poynting vector may be written

$$\bar{S} = \overline{E^2}/\mu_0 c = E_m^2/2\mu_0 c = 120 \text{ W/m}^2 \quad \underline{\text{Ans.}}$$

(e) If p is the momentum delivered to the sheet of area A,

$$p = \frac{\bar{S}A}{c} = 4 \times 10^{-7} (A) \text{ kg}\cdot\text{m/s}$$

each second, so that the rate of delivery of momentum is 8×10^{-7} kg·m/s^2 or 8×10^{-7} N. The radiation pressure P is independent of the area A:

$$P = \frac{\bar{S}A/c}{A} = 4 \times 10^{-7} \text{ Pa} \quad \underline{\text{Ans.}}$$

38-32

The momentum carried off by the laser beam in time t is

$$p = \frac{U}{c},$$

where U is the energy carried by the beam that is emitted in the same time interval. If P = 10 kW is the power of the laser beam, then U = Pt and therefore

$$p = \frac{Pt}{c}$$

so that the force will be

$$F = \frac{dp}{dt} = \frac{P}{c} = ma.$$

Hence, the speed v reached in time t, assuming the spaceship started from rest is, since v = at,

$$v = \frac{Pt}{mc} = \frac{(10^4 \text{ W})(86,400 \text{ s})}{(1500 \text{ kg})(3 \times 10^8 \text{ m/s})} = 1.92 \times 10^{-3} \text{ m/s},$$

$$v = 1.92 \text{ mm/s} \quad \underline{\text{Ans}}.$$

38-34

The light being absorbed, all of its momentum p is transferred to the absorbing object, for which A is the area perpendicular to the beam. In a time t,

$$p = \frac{U}{c} = \frac{StA}{c},$$

U the energy absorbed in time t and S is the magnitude of the beam's Poynting vector (energy carried per unit area per unit time). The force F exerted on the object will be

$$F = \frac{dp}{dt} = \frac{SA}{c},$$

so that the radiation pressure becomes

$$P = \frac{F}{A} = \frac{S}{c}.$$

38-35

Let f be the fraction of the incident beam energy that is reflected. The radiation pressure due to the part of the beam energy that is absorbed is, by Problem 38-34,

$$P_a = \frac{S_a}{c} = \frac{(1 - f)S}{c},$$

S the magnitude of the Poynting vector of the incident beam. The radiation pressure due to the reflected part of the incident beam is

$$P_r = \frac{2S_r}{c} = \frac{2(fS)}{c}.$$

(The factor of two occurs because, on reflection, the momentum $\vec{p}$ of the beam is reversed, in which case $\Delta p = 2p$.) The total radiation pressure is, then,

$$P = P_a + P_r = \frac{(1 + f)S}{c}.$$

For a plane wave with energy flux S, an amount of energy SAt crosses an area A normal to the beam in time t. But in this same time the wave travels a distance ct. Hence, SAt is the energy contained in a rectangular volume of base area A and length ct, so that the energy density u in the wave is

$$u = \frac{SAt}{Act} = \frac{S}{c}.$$

In terms of the energy density u, then, the radiation pressure P found above is

$$P = (1 + f)u = u + fu.$$

The first term on the right is the energy density of the incident beam and the second is the energy density of the reflected beam. Since energy is a scalar, the total radiation energy density just outside the surface is u + fu, rather than u - fu, even though the incident and reflected beams are oppositely directed. Thus, the assertion of the problem has been established.

38-37

(a) Let r = radius, ρ = density of the particle a distance x from the sun, which has a mass M and a luminosity (rate of energy output) L. The gravitational force on the particle is

$$F_{grav} = G\frac{Mm}{x^2} = G\frac{M(4\pi r^3\rho/3)}{x^2}.$$

The force due to radiation pressure is, by Problem 38-34,

$$F_{rad} = \frac{SA}{c} = \frac{1}{c}(L/4\pi x^2)(\pi r^2),$$

since the area A perpendicular to the beam is the sphere's cross-sectional area πr^2. The critical radius $r = R_0$ occurs when these forces balance:

$$G\frac{M(4\pi R_0^3\rho/3)}{x^2} = \frac{1}{c}(L/4\pi x^2)\pi R_0^2,$$

$$R_0 = \frac{3L}{16\pi cG\rho M} \quad \underline{\text{Ans}}.$$

(b) Putting $L = 4 \times 10^{26}$ W, $\rho = 10^3$ kg/m^3 and $M = 2 \times 10^{30}$ kg gives $R_0 = 6.0 \times 10^{-7}$ m = 600 nm Ans.

(c) R_0 is independent of x since both forces, gravitation and that due to radiation pressure, are proportional to $1/x^2$; hence the x disappears when they are set equal.

38-41

Let I be the intensity of the polarized component and i the same for the unpolarized component in the incident beam; the intensity of this incident beam, then, is i + I. The transmitted intensity i_t of the unpolarized component is

$$i_t = \tfrac{1}{2}i,$$

independent of the angle θ of the polaroid sheet; the intensity I_t transmitted by the originally polarized component is

$$I_t = I\cos^2\theta.$$

Hence, the transmitted intensity is

$$i_t + I_t = \tfrac{1}{2}i + I\cos^2\theta.$$

As the polaroid sheet is rotated $\cos^2\theta$ varies between 0 and 1, and if the transmitted intensity, as a consequence, varies by a factor of five, then

$$5(\tfrac{1}{2}i) = \tfrac{1}{2}i + I,$$

$$i = \tfrac{1}{2}I.$$

The fraction of the incident beam energy that is unpolarized, then, is

$$\frac{i}{i + I} = \frac{1}{3} = 33.3\%,$$

and the relative intensity of the polarized component is 100 - 33.3 = 66.7%.

38-43

(a) As illustrated in the sketch for two sheets, use a number of sheets with a total rotation angle of 90°.

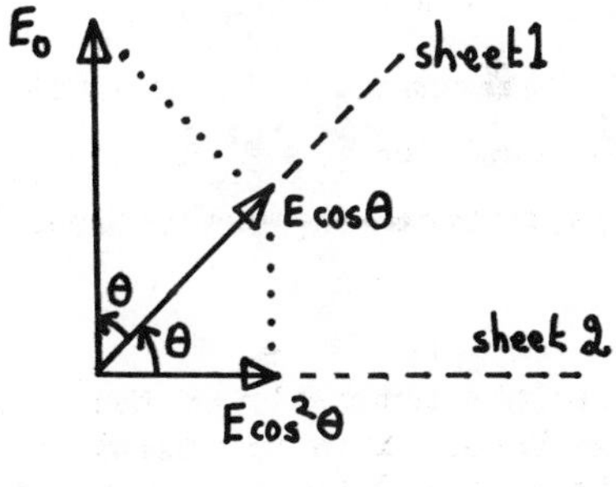

(b) For n sheets at equal angles θ between adjacent sheets, the transmitted intensity is

$$I = I_0[\cos^n(\tfrac{\pi}{2n})]^2,$$

since $n\theta = \pi/2$; I_0 is the intensity of the incident beam. For a loss in intensity of 40%, $I_0 - I = 0.4I_0$, so that $I = 0.6I_0$ giving

$$0.60I_0 = I_0[\cos^n(\frac{\pi}{2n})]^2,$$

$$0.60 = \cos^{2n}(\frac{\pi}{2n}).$$

Let $x = \pi/2n$ so that

$$0.6 = (\cos x)^{\pi/x} = (1 - \tfrac{1}{2}x^2 + \ldots)^{\pi/x}.$$

Supposing x^2 to be small,

$$\ln(0.6) \approx \ln(1 - \tfrac{1}{2}x^2)^{\pi/x} = (\frac{\pi}{x})\ln(1 - \tfrac{1}{2}x^2) \approx (\frac{\pi}{x})(-\tfrac{1}{2}x^2),$$

$$x = -\frac{2}{\pi}\ln(0.6) = \frac{\pi}{2n},$$

$$n = -\frac{\pi^2}{4\ln(0.6)} = 4.83 \approx 5 \quad \underline{\text{Ans}}.$$

38-44

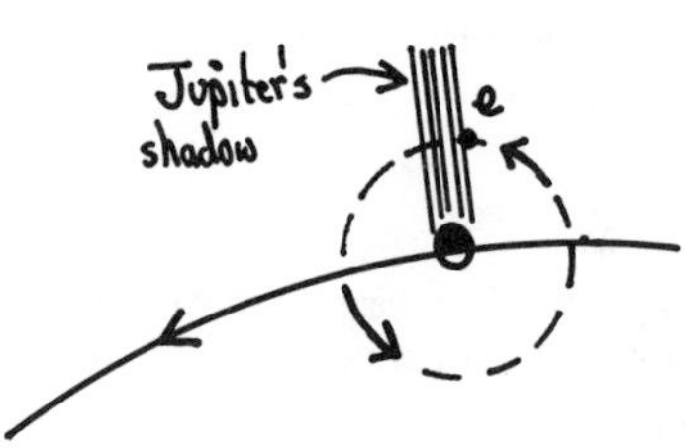

(a) Let the true period of revolution of one of Jupiter's satellites be T and the observed period T'. The latter may be determined by measuring, for example, the time between two successive passages of the satellite through point e, into eclipse. If this observation is made when the earth is near x, it is expected that T'(x) = T, since the earth, moving at a tangent to the earth-Jupiter line, does not materially alter its distance to Jupiter in the time T. However, when the earth is near y it is moving almost directly away from Jupiter, so that T'(y) will differ from T by vT/c approximately, where v is the speed of the earth in its orbit. Hence, as the earth moves from point x to y, T' will increase steadily.

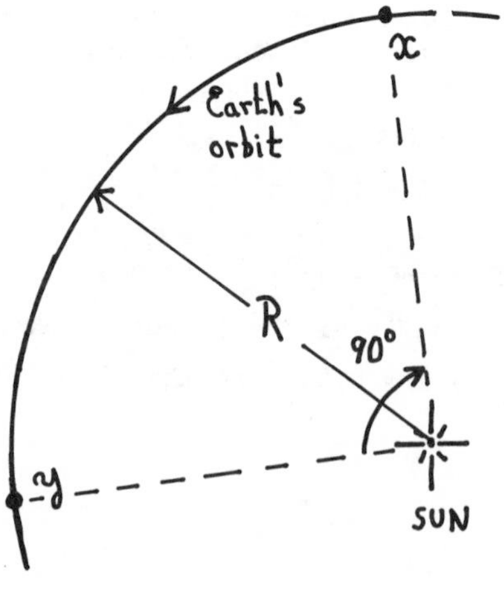

(b) With T determined (when the earth was near x) a prediction can be made, on the basis of an infinite speed of light, of the number of ingresses of the satellite into Jupiter's shadow that will be observed before the earth reaches y; also the time of the ingress that occurs closest to the moment the earth reaches y can be computed. However, this particular ingress will in fact be observed at a time R/c, approximately, later than predicted. Hence, if R (the radius of the earth's orbit) is known, c can be calculated.

38-51

(a) Let the car be approaching, say, the radar trap. Also let

ν = frequency of waves leaving radar set, in radar set's frame;

ν' = frequency of waves as received and reflected by the car, in the car's frame;

ν'' = frequency of waves as received back at the radar set, in the radar set's frame.

Then, if v is the speed of the car,

$$\nu' = \nu \frac{1 + v/c}{(1 - v^2/c^2)^{\frac{1}{2}}},$$

$$\nu'' = \nu' \frac{1 + v/c}{(1 - v^2/c^2)^{\frac{1}{2}}}.$$

Eliminating ν' and retaining only first-power terms in v/c,

$$\nu'' = \nu \frac{(1 + v/c)^2}{1 - v^2/c^2} = \nu \frac{1 + v/c}{1 - v/c} \approx \nu(1 + 2v/c),$$

$$\frac{\nu'' - \nu}{\nu} = \frac{\Delta\nu}{\nu} = 2\frac{v}{c}.$$

(b) If v = 1 mi/h = 0.447 m/s, this gives

$$\Delta\nu = 2\frac{0.447 \text{ m/s}}{3 \times 10^8 \text{ m/s}}(2.450 \times 10^9 \text{ s}^{-1}) = 7.3 \text{ Hz} \quad \underline{\text{Ans.}}$$

38-53

The Doppler formula is

$$\nu' = \nu \frac{1 \pm u/c}{(1 - u^2/c^2)^{\frac{1}{2}}}.$$

Since $\nu = c/\lambda$,

$$\frac{1}{\lambda'} = \frac{1}{\lambda} \frac{1 \pm u/c}{(1 - u^2/c^2)^{\frac{1}{2}}}.$$

Expand the denominator in a Taylor series and, for $u/c \ll 1$, retain only the first two terms:

$$\lambda = \lambda'(1 \pm u/c)(1 + \tfrac{1}{2}u^2/c^2).$$

Thus, if only linear terms are retained,

$$\lambda = \lambda'(1 \pm u/c).$$

If the source and observer are separating, the negative sign is appropriate; for this case write $\Delta\lambda = \lambda' - \lambda$. On the other hand, if the source and observer are approaching use the positive sign and let $\Delta\lambda = \lambda - \lambda'$. In either case

$$\Delta\lambda = \lambda'(\frac{u}{c}).$$

But, for $u \ll c$, $\lambda \approx \lambda'$ and therefore

$$\frac{\Delta\lambda}{\lambda} = \frac{u}{c}.$$

39-6

(a) For normal incidence there is no refraction at the water surface.

(b) By application of the law of reflection, together with the property of plane triangles that the sum of the angles is 180°, it is seen that the incident and emergent rays are parallel. The relation between the angles θ and ϕ (i.e. the law of refraction) is not required.

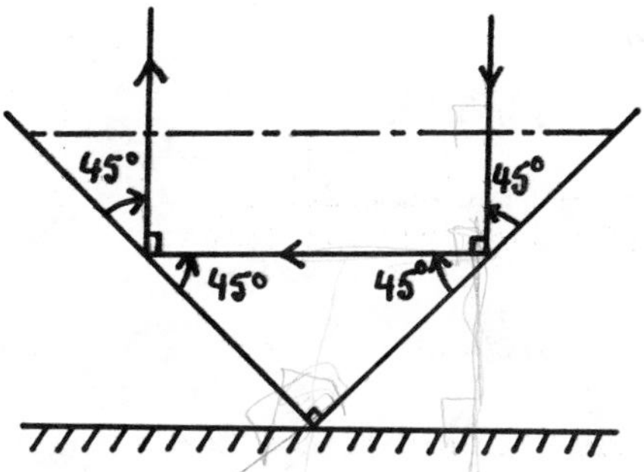

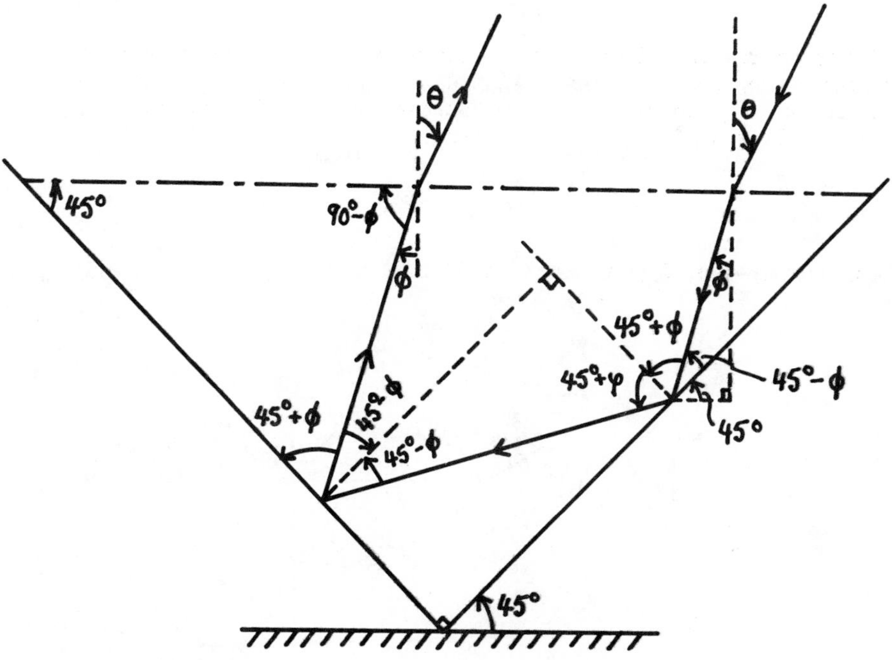

39-7

The length of the shadow is $L + \frac{1}{2}$ meters. By the law of refraction,

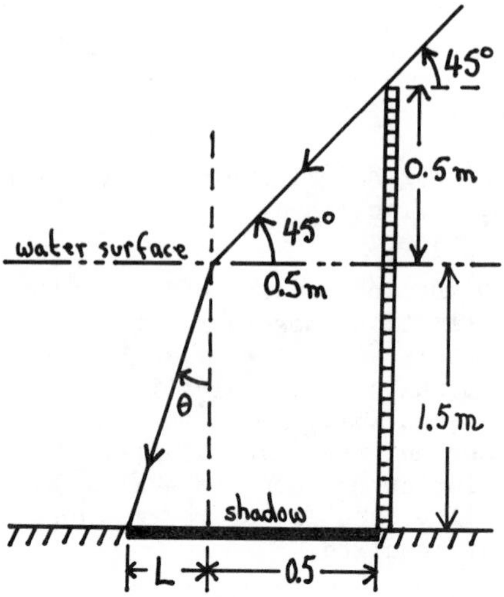

$$\sin 45° = n \cdot \sin\theta = \frac{4}{3}\sin\theta,$$

$$\sin\theta = \frac{3}{8} \cdot 2^{\frac{1}{2}}.$$

Hence

$$\tan\theta = \frac{\sin\theta}{\cos\theta} = \frac{\sin\theta}{(1 - \sin^2\theta)^{\frac{1}{2}}},$$

$$\tan\theta = 3/(23)^{\frac{1}{2}}.$$

In addition,

$$\tan\theta = \frac{L}{1.5},$$

so that

$$L = 4.5/(23)^{\frac{1}{2}},$$

and therefore the length of the shadow is

$$L + \frac{1}{2} = \frac{4.5}{(23)^{\frac{1}{2}}} + 0.5 = 1.44 \text{ m} \quad \underline{\text{Ans.}}$$

39-9

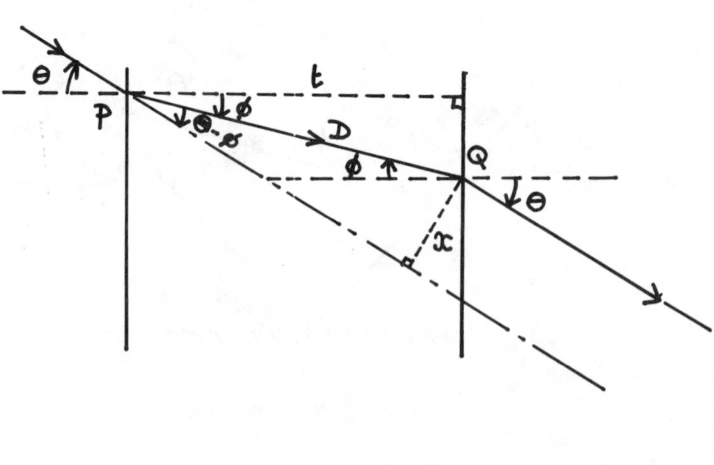

The angle of refraction ϕ at the point P of incidence equals the angle of incidence at the point Q of emergence since the normals to the two faces of the sheet are parallel, the faces themselves being parallel. Hence, the angle of emergence θ equals the angle of incidence θ at P regardless of the form of the law of refraction provided the light path is reversible. To find the deviation x, note that, if PQ = D,

$$x = D\sin(\theta - \phi),$$

and

$$D = t/\cos\phi$$

so that

$$x = t\,\frac{\sin(\theta - \phi)}{\cos\phi}.$$

If θ is small then ϕ is small also; expressing the angles in radian measure rather than degrees the law of refraction $\sin\theta = n\cdot\sin\phi$ becomes $\theta \approx n\phi$ ($\sin\alpha \approx \alpha$). In the same spirit, for $\alpha \ll 1$, $\cos\alpha \approx 1$, so that for small angles,

$$x \approx t\,\frac{\theta - \theta/n}{1} = t\theta\,\frac{n - 1}{n}.$$

39-18

(a)

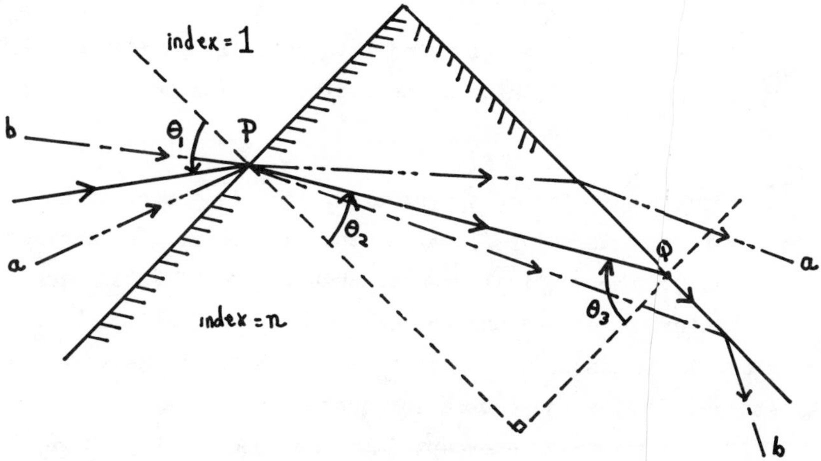

At Q,

$$n \cdot \sin\theta_3 = \sin 90^\circ = 1,$$

$$n = \frac{1}{\sin\theta_3}.$$

But $\theta_3 = 90^\circ - \theta_2$ so that

$$n = \frac{1}{\cos\theta_2}.$$

But at P,

$$\sin\theta_1 = n \cdot \sin\theta_2,$$

and therefore

$$\cos\theta_2 = (1 - \sin^2\theta_2)^{\frac{1}{2}} = [1 - (\sin\theta_1/n)^2]^{\frac{1}{2}} = \frac{1}{n}(n^2 - \sin^2\theta_1)^{\frac{1}{2}}.$$

Solving for the index of refraction,

$$n = n(n^2 - \sin^2\theta_1)^{-\frac{1}{2}},$$

$$n = (1 + \sin^2\theta_1)^{\frac{1}{2}} \quad \underline{\text{Ans}}.$$

(b) Since the maximum value of $\sin^2\theta_1 = 1$, $n_{max} = 2^{\frac{1}{2}}$.

(c) For $\theta > \theta_1$ the ray is refracted into the air, and for $\theta < \theta_1$ the ray undergoes total internal reflection at the second face.

<u>39-21</u>

(a) Under the assumption of the problem, if θ is the critical angle for total internal reflection, the rays emitted at angles with respect to the vertical greater than θ are reflected back into the water and do not escape. Only those rays emitted into directions lying inside a cone of semiangle θ about the vertical will escape. Imagine the source surrounded by a sphere of radius $R < h$. If A is the area intercepted by the cone on the sphere, the desired ratio $f = A/4\pi R^2$ since the source radiates isotropically.

But $A = 2\pi R^2(1 - \cos\theta)$ and since $n \cdot \sin\theta = 1$ (condition for onset of total internal reflection) the ratio sought is

$$f = \tfrac{1}{2}[1 - (1 - 1/n^2)^{\frac{1}{2}}] \quad \underline{\text{Ans.}}$$

(b) The index of refraction n of water is 1.33, so the ratio in this case is 0.17 Ans.

39-23

(a) Let AB represent an edge of the cube, length a. The spot S is seen by refracted light that leaves the cube. However, rays leaving S at angles greater than

$$\theta = \sin^{-1}(\tfrac{1}{n})$$

with the vertical undergo total internal reflection and, if their subsequent behaviour is ignored, do not exit the cube. Hence, rays will emerge from a circular area the radius r of which is given from

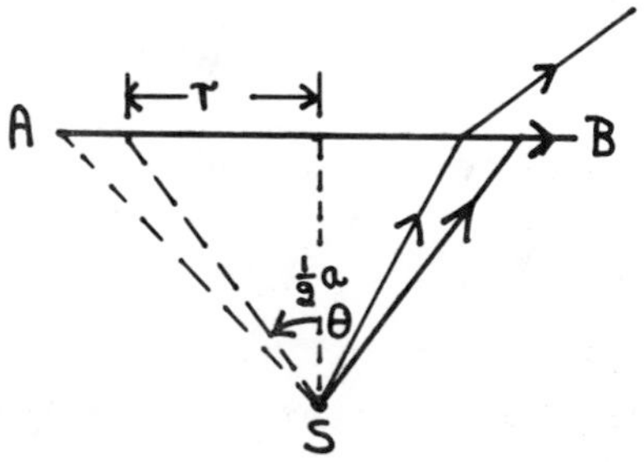

$$\tan\theta = \frac{r}{\frac{1}{2}a},$$

$$r = \tfrac{1}{2}a \cdot \tan\theta = \tfrac{1}{2}a(n^2 - 1)^{-\frac{1}{2}}.$$

Thus, if an opaque circular disc is centered on each face of the cube, the spot will not be seen if the radius of the discs equal r found above. For a = 1.0 cm and n = 1.5, the radius r = 0.45 cm.

(b) The fraction f of the cube surface covered is

$$f = \frac{6\pi r^2}{6a^2} = \frac{\pi}{4(n^2 - 1)} = 0.63 \quad \underline{\text{Ans.}}$$

39-29

In the diagram on the following page are shown the three virtual images obtained even if the object does not lie on the perpendicular bisector of the two mirrors.

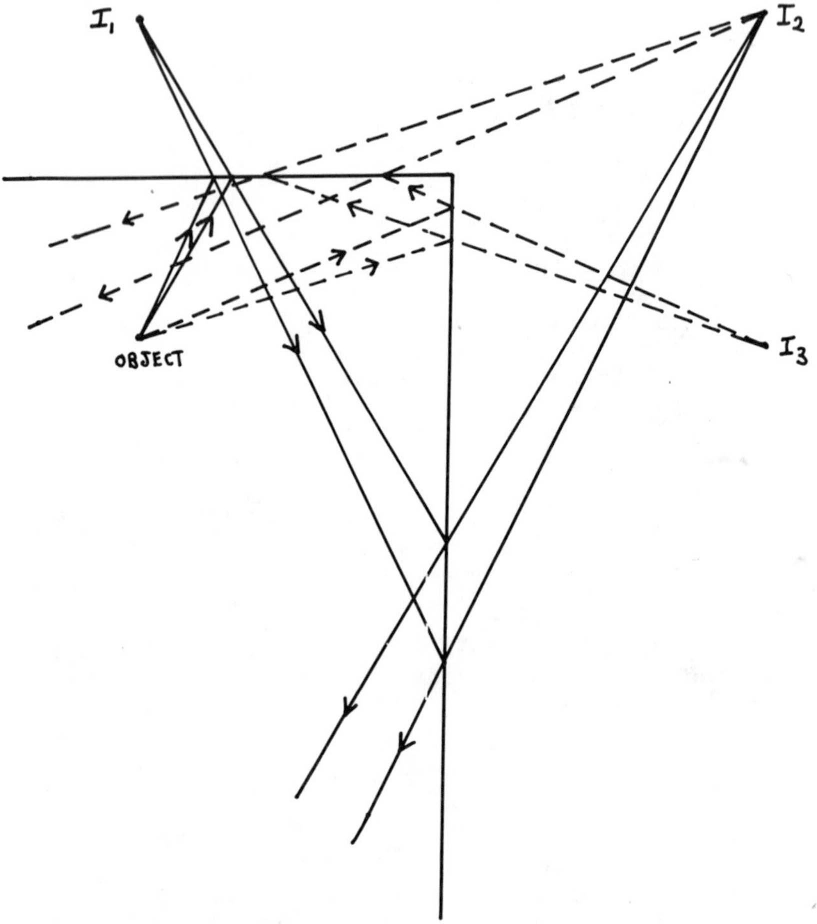

39-32

The line from image to object is perpendicular to, and bisected by, the plane mirror. If the mirror is rotated with the object held fixed, the image must move to maintain this geometrical relation. The eye will always be able to see the image (assuming perfect peripheral vision) since the situation is identical, for all practical purposes, as if the mirror be held fixed and the eye free to move about.

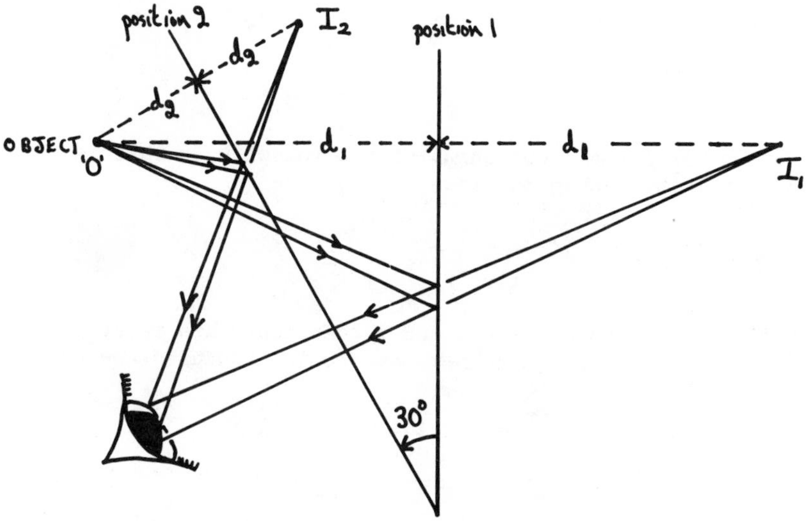
position 2
I_2
position 1
d_2
d_2
OBJECT
'O'
d_1
d_1
I_1
30°

39-34

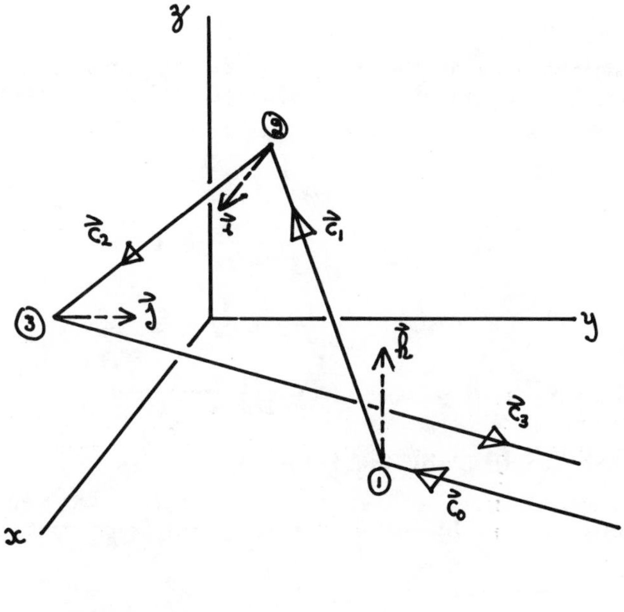
z
y
x
1
2
3
$\vec{c}_0$
$\vec{c}_1$
$\vec{c}_2$
$\vec{c}_3$
$\vec{i}$
$\vec{j}$
$\vec{k}$

Allow the mirrors to lie along the coordinate planes. Write the velocity of the incident ray as

$$\vec{c}_0 = c_x\vec{i} + c_y\vec{j} + c_z\vec{k}.$$

At reflection 1 the z-component of the velocity is reversed, the other remaining unchanged: thus,

$$\vec{c}_1 = c_x\vec{i} + c_y\vec{j} - c_z\vec{k}.$$

At reflections 2 and 3, the x and y components are reversed, respectively, leaving the other components unchanged: hence,

$$\vec{c}_2 = - c_x\vec{i} + c_y\vec{j} - c_z\vec{k},$$

$$\vec{c}_3 = - c_x\vec{i} - c_y\vec{j} - c_z\vec{k}.$$

Thus $\vec{c}_3 = -\vec{c}_0$, indicating that the ray emerges in the opposite (reversed) direction from which it entered.

39-37

(a) Let the ends of the object be at distances a, L + a from the mirror. The images of these ends are at image distances i, i' given by

$$\frac{1}{a} + \frac{1}{i} = \frac{1}{f},$$

$$i = \frac{fa}{a - f},$$

$$\frac{1}{L + a} + \frac{1}{i'} = \frac{1}{f},$$

$$i' = \frac{fL + fa}{L + a - f}.$$

Thus the length L' of the image is

$$L' = i - i' = \frac{f^2L}{(L + a - f)(a - f)}.$$

The object being short, in the denominator set $L + a \approx a \approx o$, the object distance; then

$$L' = L\left(\frac{f}{o - f}\right)^2.$$

(b) The lateral magnification $m = i/o$ (disregarding sign); from (a)

$$m = \frac{f}{o - f}.$$

Hence, if $m' = L'/L$, it is clear that $m' = m^2$.

(c) For the image of a cube to be a cube also, it is necessary that $m' = m$. Since $m' = m^2$, this will hold only if $m = m^2$, or $m = +1$. This in turn requires that $o = 2f$, placing the object at the center of curvature of the concave mirror, or symmetrically placed with respect to the center of curvature in the convex mirror.

39-43

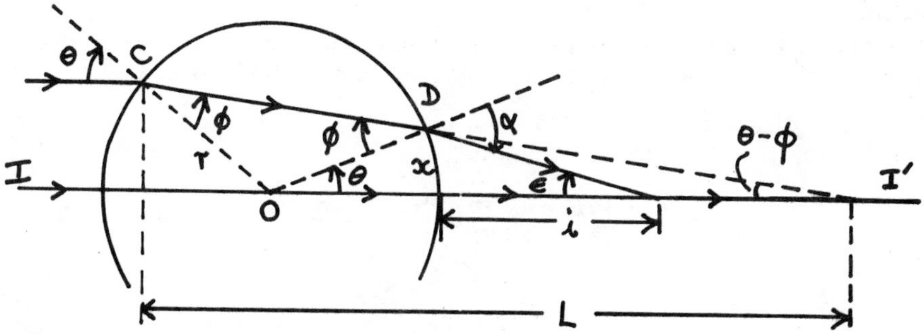

The ray II' passes undeviated through the sphere. For the other ray shown, striking the sphere at an angle θ with the normal, the first angle of refraction φ is given by

$$\sin\theta = n \cdot \sin\phi,$$

$$\phi \approx \theta/n,$$

for θ small (paraxial rays); n is the index of refraction of the glass. If no second refraction took place, an image I' would be formed a distance L behind the forward surface, with L determined from

$$\tan(\theta - \phi) = \frac{r\sin\theta}{L},$$

$$\theta - \phi \approx \frac{r\theta}{L},$$

$$L = \frac{r\theta}{\theta - \phi} = \frac{n}{n - 1} r.$$

For glass n = 1.5 approximately, so that $L \approx 3r$, and therefore I' lies outside the glass sphere. At the second refraction,

$$n \cdot \sin\phi = \sin\alpha,$$

$$\alpha = \theta.$$

The angles of triangle COD must total π rad:

$$2\phi + (\pi - \theta - \theta) = \pi,$$

$$\theta = 2\phi - \theta = \frac{2 - n}{n}\theta.$$

Hence,

$$x = r\theta = r\theta\frac{2 - n}{n}.$$

The image distance i is given by

$$i = \frac{x}{\tan\epsilon} = \frac{x}{\tan(\theta - \theta)} \approx \frac{x}{\theta - \theta} = \frac{r\theta(2 - n)/n}{\theta - (2 - n)\theta/n},$$

$$i = \frac{2 - n}{2(n - 1)} r \quad \underline{\text{Ans}}.$$

39-46

The lens maker's equation is

$$\frac{1}{f} = (n - 1)\left(\frac{1}{r'} - \frac{1}{r''}\right)$$

where n is the index of refraction of the glass $(n > 1)$; r', r'' = radius of curvature of the first and the second surface struck by the light, respectivelly. If $f > 0$ the lens is converging and if

$f < 0$ it is diverging. Now, the centers of curvature of the surfaces shown in the figures are all to the right of each lens: hence, with light assumed incident from the left of each lens, r', $r'' > 0$ since the centers of curvature lie on the R-side of each lens.

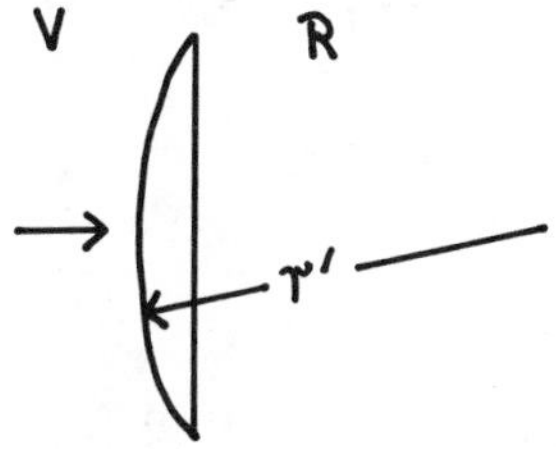

(a) $r' > 0$, $r'' = \infty$: $f > 0$, lens is converging.

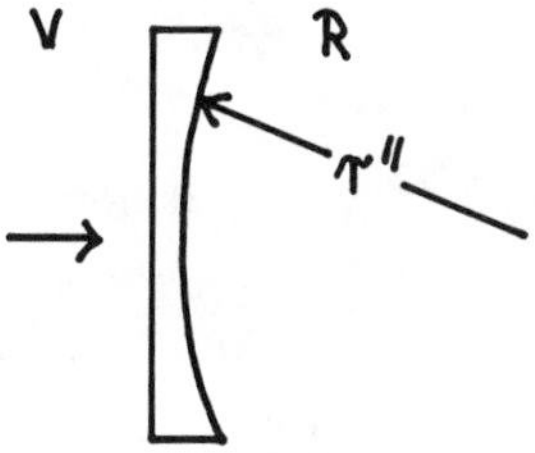

(b) $r' = \infty$, $r'' > 0$: $f < 0$, lens is diverging.

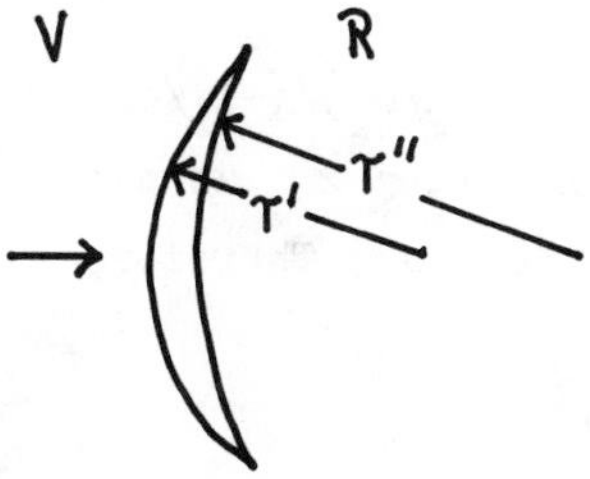

(c) $r'' > r'$: $f > 0$, lens is converging.

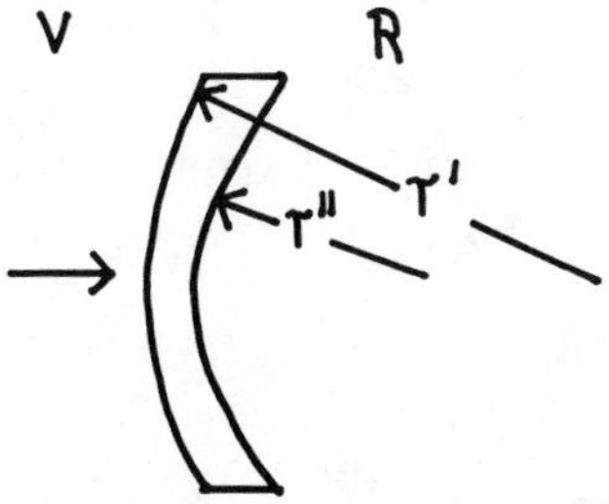

(d) $r'' < r'$: $f < 0$, lens is diverging.

These same results (diverging vs. converging) are obtained if the light is imagined incident from the right since then r' and r" are interchanged, but they also change signs as the centers of curvature will then be on the V-side of each lens.

39-49

(a) If the object distance is x, then the image distance is D - x; hence,

$$\frac{1}{o} + \frac{1}{i} = \frac{1}{f},$$

$$\frac{1}{x} + \frac{1}{D - x} = \frac{1}{f},$$

$$x^2 - Dx + Df = 0,$$

of which the solutions are

$$x_1 = \tfrac{1}{2}(D - d), \quad x_2 = \tfrac{1}{2}(D + d),$$

with

$$d = (D^2 - 4Df)^{\frac{1}{2}}.$$

Thus the separation in the two positions of the lens (object fixed) is $x_2 - x_1 = d$.

(b) The ratio of the magnifications is $m_2/m_1 = (-i_2/o_2)/(-i_1/o_1)$. But $o_1 = x_1$, $o_2 = x_2$, $i_1 = D - x_1 = x_2$, $i_2 = D - x_2 = x_1$, as can be verified from the expressions for x_1 and x_2 in (a). Therefore

$$m_2/m_1 = (x_1/x_2)^2 = \left(\frac{D - d}{D + d}\right)^2.$$

39-51

Let light be incident first on lens f_1 and assume $f_1 > 0$ (converging lens). Then,

$$\frac{1}{o_1} + \frac{1}{i_1} = \frac{1}{f_1},$$

i_1 the distance of the image formed by lens f_1. Since $f_1 > 0$ light converges toward the image which thereby forms a virtual object for lens f_2. Hence $o_2 < 0$ and $i_1 = -o_2$. But

$$\frac{1}{o_2} + \frac{1}{i_2} = \frac{1}{f_2},$$

$$-\frac{1}{i_1} + \frac{1}{i_2} = \frac{1}{f_2}.$$

By definition of f,

$$\frac{1}{o_1} + \frac{1}{i_2} = \frac{1}{f}.$$

The first and third equations yield

$$\left(\frac{1}{o_1} - \frac{1}{f_1}\right) + \frac{1}{i_2} = \frac{1}{f_2}.$$

Comparing this with the preceding equation leads to

$$\frac{1}{f} - \frac{1}{f_1} = \frac{1}{f_2}.$$

$$f = \frac{f_1 f_2}{f_1 + f_2}.$$

This holds regardless of whether f_2 be greater or less than zero. The only case not covered, then, is that of two diverging lenses ($f_1, f_2 < 0$). Here lens f_1 forms a virtual image so that $i_1 < 0$, but this forms a real object for lens f_2. Thus, although $o_2 > 0$, $i_1 = -o_2$, as before. The rest of the proof is the same as above.

39-53

The Newtonian form applies to converging lenses ($f > 0$). To derive it, start with the Gaussian form,

$$\frac{1}{o} + \frac{1}{i} = \frac{1}{f}.$$

The object and image distances in terms of x, x' are

$$o = f \pm x, \quad i = f \pm x',$$

the upper sign for a real image ($o > f$, $i > 0$), the lower sign for a virtual image ($o < f$, $i < 0$). Putting these into the first equation gives directly

$$\frac{1}{f \pm x} + \frac{1}{f \pm x'} = \frac{1}{f},$$

$$xx' = f^2,$$

in which $x, x' > 0$.

39-55

(a) The lens forms a real image at a distance i behind the lens, where i is found from the lens formula

$$\frac{1}{1} + \frac{1}{i} = \frac{1}{0.5},$$

giving i = 1.0 m. This image lies 2 - 1 = 1 m in front of the plane mirror which forms a virtual image of it 1 m behind, or 1 + 2 = 3 m behind the lens. Hence the final image is formed a distance I in front of the lens, with

$$\frac{1}{I} + \frac{1}{3} = \frac{1}{0.5},$$

$$I = 0.60 \text{ m} \quad \underline{\text{Ans}}.$$

(b) This image is real since the final object distance (3 m) is greater than the focal length (0.5 m) of the lens.

(c), (d) The magnification of the first image formed by the lens is

$$m_1 = -\frac{i}{o} = -\frac{1}{1} = -1.$$

The plane mirror retains this magnification. The second magnification by the lens is

$$m_2 = -\frac{0.6}{3.0} = -0.2.$$

Compared to the original object, the magnification is

$$m = m_1 m_2 = +0.20 \quad \underline{\text{Ans}}.$$

Since m is positive, the image must be erect.

39-62

(a) The mirror M may be considered replaced with a lens of the same focal length, the only difference introduced is that now the image is behind the lens rather than in front of the mirror. The mirror M', being flat, is of no import as far as magnification is concerned. Hence

$$m_0 = - f_{ob}/f_{eye},$$

as before, the lens also inverting the image.

(b) Considering the mirror to be spherical rather than parboloidal gives

$$\frac{1}{2000} + \frac{1}{i} = \frac{1}{16.8},$$

$$i = 16.94 \text{ m} \quad \underline{\text{Ans}}.$$

(c) The focal length f = r/2 = 5.0 m. Thus,

$$200 = 5/f_{eye},$$

$$f_{eye} = 2.5 \text{ cm} \quad \underline{\text{Ans}}.$$

40-2

For maxima, $d\sin\theta = m\lambda$, and with $m = 3$,

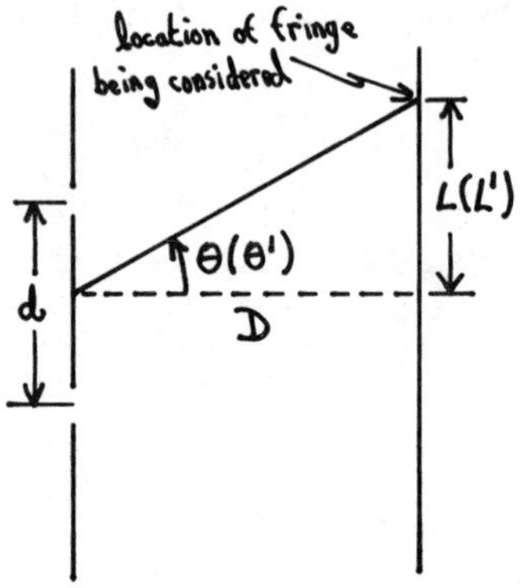

$$d\sin\theta = 3(480) = 144 \times 10^{-8} \text{ m},$$

$$d\sin\theta' = 3(600) = 18 \times 10^{-7} \text{ m}.$$

Since $d = 5 \text{ mm} = 5 \times 10^{-3}$ m,

$$\sin\theta \approx \tan\theta = \frac{144 \times 10^{-8}}{5 \times 10^{-3}},$$

$$\tan\theta = 28.8 \times 10^{-5}.$$

Similarly,

$$\sin\theta' \approx \tan\theta' = \frac{18 \times 10^{-7}}{5 \times 10^{-3}} = 3.60 \times 10^{-4}.$$

Thus, with $D = 1.0$ m,

$$L = D\tan\theta = 28.8 \times 10^{-5} \text{ m}, \quad L' = D\tan\theta' = 3.60 \times 10^{-4} \text{ m},$$

so that the separation is

$$L' - L = (3.60 - 2.88) \times 10^{-4} \text{ m} = 0.072 \text{ mm } \underline{\text{Ans}}.$$

40-9

The positions of the n th and (n + 1) st fringes are θ_n, θ_{n+1}, given by

$$d\sin\theta_n = n\lambda,$$

$$d\sin\theta_{n+1} = (n + 1)\lambda,$$

and their separation is $\Delta\theta = \theta_{n+1} - \theta_n$; therefore

$$d\sin(\theta_n + \Delta\theta) = (n + 1)\lambda,$$

$$d(\sin\theta_n\cos\Delta\theta + \cos\theta_n\sin\Delta\theta) = (n + 1)\lambda.$$

If $\Delta\theta \ll \theta_n$, then $\sin\Delta\theta \approx \Delta\theta$ and $\cos\Delta\theta \approx 1$: in this case,

$$\Delta\theta = \frac{\lambda}{d\cos\theta_n}.$$

Thus the separation between adjacent fringes depends on which pair of fringes is being examined. The problem implies that the fringe separation is the same for all fringes: this will be true for those fringes at very small values of θ, for then $\cos\theta \approx 1$ and $\Delta\theta$ is independent of θ. Thus, set $\cos\theta_n = 1$ and obtain

$$\Delta\theta = \lambda/d.$$

Hence, for two wavelengths λ_1 and λ_2 the separations are related by

$$\Delta\theta_1/\Delta\theta_2 = \lambda_1/\lambda_2,$$

$$\lambda_2 = \lambda_1\Delta\theta_2/\Delta\theta_1 = (589)(\tfrac{11}{10}\,\Delta\theta_1)/\Delta\theta_1 = 648\ \text{nm} \quad \underline{\text{Ans}}.$$

<u>40-14</u>

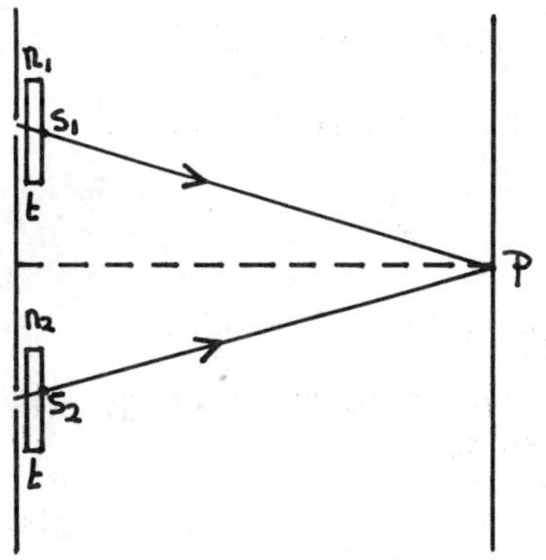

Before positioning the plates the central maximum fell at P. The geometrical path lengths from the points S_1 and S_2 at which the light emerges from the plates to P (with the plates in place) are equal. As these paths both are in air, i.e. in the same medium, no phase difference is introduced between the paths S_1P and S_2P.

The wavelengths of light in the plates are

$$\lambda_1 = \lambda/n_1,\ \lambda_2 = \lambda/n_2.$$

The number of wavelengths of light in the plates are t/λ_1 and t/λ_2 and the difference between these must be five; therefore,

$$\frac{tn_1}{\lambda} - \frac{tn_2}{\lambda} = 5,$$

$$t = \frac{5\lambda}{n_1 - n_2} = \frac{5(480)}{1.7 - 1.4} = 8.0\ \mu m \quad \underline{\text{Ans.}}$$

<u>40-19</u>

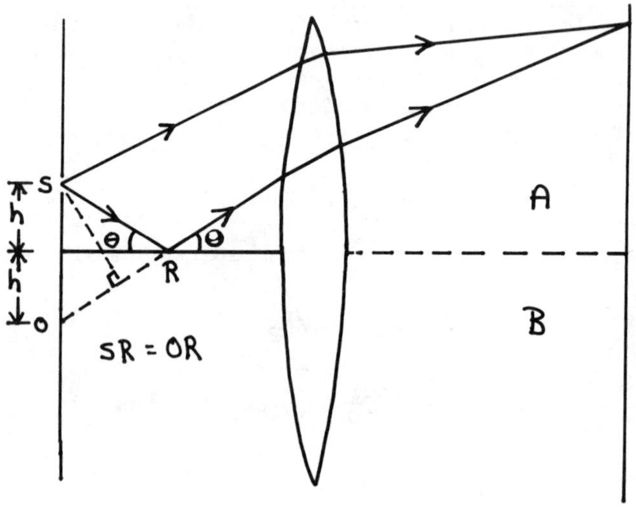

(a) As illustrated, in region A the interference may be thought of as between S and its image O except that account must be taken of the 180° phase change on reflection from the mirror; thus the usual conditions for maxima and minima are reversed $(d = 2h)$:

$$2h\sin\theta = (m + \tfrac{1}{2})\lambda, \text{ maxima}; \quad 2h\sin\theta = m\lambda, \text{ minima}.$$

(b) Clearly the mirror cannot reflect rays into region B; also, O is a virtual image. Thus, only rays directly from S can penetrate into region B and thus no interference can take place there.

40-22

The intensity I is given by

$$I = I_m \cos^2\beta,$$

with $\beta = \pi d \sin\theta/\lambda \approx \pi\theta d/\lambda$ for θ small. At a point of half-maximum intensity,

$$\tfrac{1}{2}I_m = I_m \cos^2\beta.$$

The smallest positive β satisfying this equation is $\beta = \pi/4$. Hence, the first half-intensity point occurs where θ has the value given from

$$\pi/4 = \pi\theta d/\lambda,$$

$$\theta = \lambda/4d.$$

A symmetrical half-intensity point falls at $\theta = -\lambda/4d$, with the $\theta = 0$ maximum between. Thus, the half-width of this central maximum is $2(\lambda/4d) = \lambda/2d$.

40-23

The electric field components of the two waves are

$$E_1 = E_0 \sin(\omega t + \phi),$$

$$E_2 = 2E_0 \sin(\omega t + \phi),$$

with $\phi = 2\pi d \sin\theta/\lambda$. The sum is written in the form

$$E_1 + E_2 = E = E_\theta \sin(\omega t + \beta),$$

and the intensity I of this resultant wave is proportional to E_θ^2. Applying the law of cosines to the triangle formed by E_1, E_2, E in the phasor sketch gives

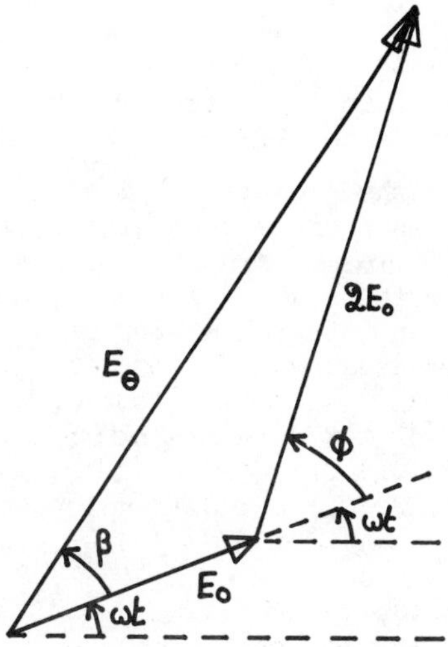

$$E_\theta^2 = E_0^2 + (2E_0)^2 - 2(E_0)(2E_0)\cos(\pi - \phi),$$

$$E_\theta^2 = E_0^2(5 + 4\cos\phi).$$

Hence,

$$I = AE_0^2(5 + 4\cos\phi),$$

with A the constant of proportionality. Now,

$$\cos\phi = \cos(\tfrac{1}{2}\phi + \tfrac{1}{2}\phi) = 2\cos^2(\tfrac{1}{2}\phi) - 1.$$

Substituting this gives

$$I = AE_0^2[1 + 8\cos^2(\tfrac{1}{2}\phi)].$$

If $\theta = 0$, $\phi = 0$ and, of course, $\tfrac{1}{2}\phi = 0$ also so that $I(\theta = 0) = 9AE_0^2$. Define I_m by $I(\theta = 0) = I_m$; then $A = I_m/9E_0^2$ and therefore in terms of I_m, the intensity of the central maximum,

$$I = \frac{1}{9} I_m[1 + 8\cos^2(\pi d\sin\theta/\lambda)] \quad \underline{\text{Ans}}.$$

<u>40-24</u>

(a) If D is the detector, the optical path difference $S_2D - S_1D$ equals the geometical path difference since the paths from both sources are in air. For maxima this path difference must equal an integral number of wavelengths:

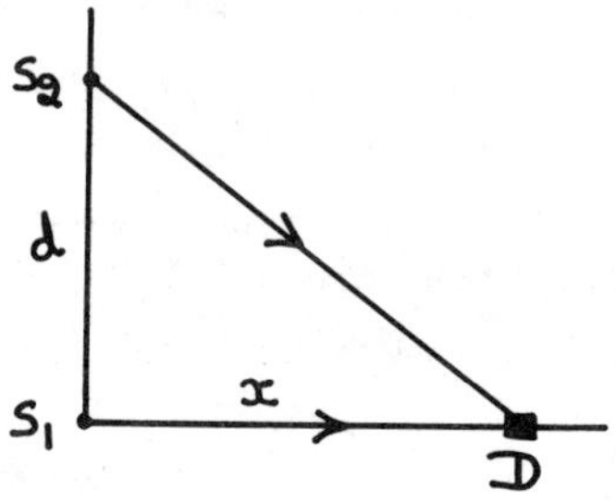

$$(d^2 + x^2)^{\frac{1}{2}} - x = n\lambda,$$

$n = 1, 2, ..$; $n = 0$ is not possible since $d \neq 0$. This gives

$$x = (d^2 - n^2\lambda^2)/2n.$$

Putting $n = 3, 2, 1$, together with $d = 4.0$ m, $\lambda = 1.0$ m yields

$$x_3 = \frac{7}{6}\text{ m},\ x_2 = 3.0\text{ m},\ x_1 = \frac{15}{2}\text{ m}\quad \underline{\text{Ans.}}$$

(b) For completely destructive interference the waves must arrive at D exactly 180° out of phase and with equal amplitudes. The first requirement can be met but the second cannot since the waves, being spherical, experience a $1/r$ fall-off in amplitude and the distances r are different for the two waves.

40-32

The phase changes on reflection for the two cases, $n < n_g$ and $n > n_g$, are shown in the figure. Thus the conditions for destructive interference in the two cases are

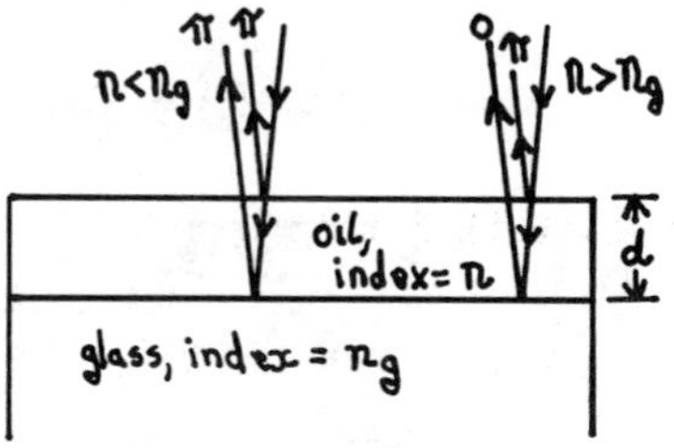

$$n > n_g:\quad 2nd = m\lambda,\quad m = 1, 2, \ldots;$$

$$n < n_g:\quad 2nd = (m + \tfrac{1}{2})\lambda,\quad m = 0, 1, 2, \ldots .$$

Now, there is destructive interference for wavelengths 500 nm and 700 nm and therefore either

$$2nd = m_5(500),\quad 2nd = m_7(700),\ n > n_g,$$

or

$$2nd = (m_5 + \tfrac{1}{2})(500),\quad 2nd = (m_7 + \tfrac{1}{2})(700),\ n < n_g.$$

In either case, since $700 > 500$, $m_7 < m_5$. In fact, there being no minimum between these wavelengths,

$$m_7 = m_5 - 1;$$

hence,

$$m_7 = \frac{5}{2},\ n > n_g;\quad m_7 = 2,\ n < n_g.$$

As m_7 must be an integer, it must be that $n < n_g$.

40-39

There is only one phase change of π; thus, for a bright B fringe and a dark D fringe, respectively,

$$2d = (m + \tfrac{1}{2})\frac{\lambda}{n} \text{ at B}; \qquad 2d = m\frac{\lambda}{n} \text{ at D.}$$

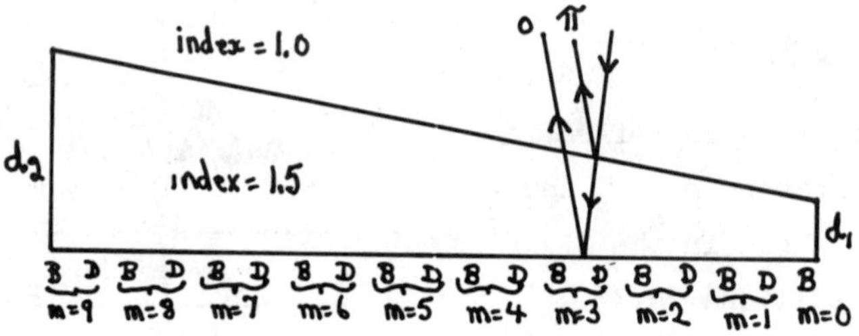

At the ends

$$2d_1 = (0 + \tfrac{1}{2})\cdot\frac{630}{1.5},$$

$$d_1 = 105 \text{ nm};$$

$$2d_2 = (9 + \tfrac{1}{2})\cdot\frac{630}{1.5},$$

$$d_2 = 1995 \text{ nm}.$$

Thus the variation in thickness is

$$d_2 - d_1 = 1890 \text{ nm} = 1.89\ \mu\text{m} \quad \underline{\text{Ans}}.$$

40-40

If the light waves are reflected from a region where the thickness of the film is y, the condition for a maximum is

$$2y = (m + \tfrac{1}{2})\lambda, \quad m = 0, 1, \ldots$$

with the $\frac{1}{2}$ compensating for the phase difference introduced in

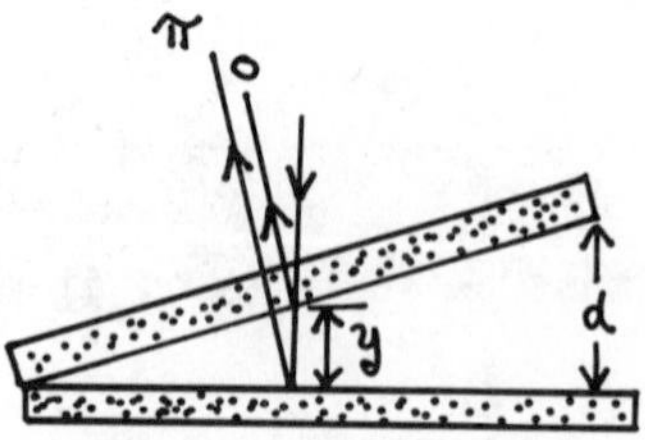

the reflection from the lower surface of the air film. If a maximum (bright fringe) appears at the end ($y = d$) it would correspond to an order number n given by

$$2d = (n + \tfrac{1}{2})\lambda,$$

$$2(48{,}000\ \text{nm}) = (n + \tfrac{1}{2})(680\ \text{nm}),$$

$$n = 140.7\,.$$

This indicates that a maximum does not lie at the end but that the maximum nearest to the end is of order $m = 140$. Since there is a bright fringe near the other end for which $m = 0$ (where the air film has a thickness $\frac{1}{4}\lambda$), there are 141 bright fringes in all.

40-41

There is a phase change of π at each reflection and therefore

$$2d = m\cdot\frac{\lambda}{n}, \text{ bright fringe,}$$

$$2d = (m + \tfrac{1}{2})\cdot\frac{\lambda}{n}, \text{ dark fringe.}$$

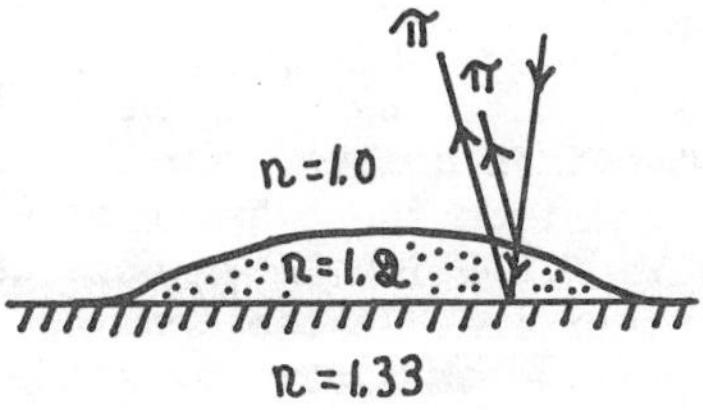

(a) At the outer regions $d \approx 0$ and with $m \geq 0$ there can be only the bright fringe $m = 0$ regardless of the wavelength of the light.

(b) For the third ($m = 3$) blue fringe,

$$2d = 3\,\frac{475\ \text{nm}}{1.2},$$

$$d = 594\ \text{nm}\quad \underline{\text{Ans}}.$$

(c) As the thicker regions of the film are examined, the fringes are seen to fall closer and closer together; eventually they cannot be distinguished by the unaided eye.

40-43

The ray from the bottom of the air film (rather than the ray from the top) undergoes a phase change of π, for it is reflected from a medium (glass) of higher refractive index than the film (air). The condition for maxima is, for perpendicular incidence,

$$2d = (m + \tfrac{1}{2})\lambda/n = (m + \tfrac{1}{2})\lambda, \quad m = 0, 1, 2, \ldots,$$

since the index of refraction of air = 1 approximately. Now,

$$d = R - (R^2 - r^2)^{\frac{1}{2}} = R - R(1 - r^2/R^2)^{\frac{1}{2}}.$$

If $r \ll R$, so that $r/R \ll 1$, then expanding the square root in a Taylor series and keeping only the first two terms reduces this to

$$d \approx R - R(1 - \tfrac{1}{2}r^2/R^2) = \tfrac{1}{2}r^2/R.$$

Combining this with the first equation to eliminate d yields

$$r = [(m + \tfrac{1}{2})\lambda R]^{\frac{1}{2}} \quad \underline{\text{Ans.}}$$

40-48

Let L be the length of the chamber. The phase difference between the two rays arises from the unequal lengths of the interferometer arms and from the passage of light in one arm through a length 2L of air of density less (once the pumping starts) than the air "in" the other arm. If the lengths of the arms are not altered during the pumping then, at any stage during the pumping process, the phase difference is introduced by the difference in air densities, and is

$$\Delta\phi = 2\pi\left[\frac{2L}{\lambda_1} - \frac{2L}{\lambda_2}\right] = 2\pi\left[\frac{2L}{\lambda/n_1} - \frac{2L}{\lambda/n_2}\right],$$

$$\Delta\phi = \frac{4\pi L}{\lambda}(n_1 - n_2).$$

Here λ is the wavelength in a vacuum and n_1, n_2 the refractive indices for the two air columns. Assuming perfect pumping, put $n_2 = 1$ and $n_1 = n$, the index for air under atmospheric pressure that remains "in" the other arm. Each fringe shift corresponds to a phase shift of 2π, so that the number N of fringes that pass across the field of view will be, since 50 mm = 5×10^7 nm,

$$N = \frac{\Delta\phi}{2\pi} = \frac{2L}{\lambda}(n - 1),$$

$$60 = \frac{2(5 \times 10^7)}{500}(n - 1),$$

$$n = 1.0003 \ \underline{\text{Ans}}.$$

40-49

If the difference between the lengths of the arms is x, the path difference between the interfering waves is 2x, and this leads to a phase difference $\Delta\phi$ given by $\Delta\phi/2\pi = 2x/\lambda$, or $\Delta\phi = 4\pi x/\lambda$. Thus the two interfering parallel waves can be written

$$E_1 = E_0 \sin\omega t, \quad E_2 = E_0 \sin(\omega t + \frac{4\pi x}{\lambda}),$$

so that

$$E = E_1 + E_2 = 2E_0 \sin(\omega t + \frac{2\pi x}{\lambda})\cos(\frac{2\pi x}{\lambda}).$$

The intensity observed is

$$I = \overline{4E_0^2 \sin^2(\omega t + \frac{2\pi x}{\lambda})\cos^2(\frac{2\pi x}{\lambda})}.$$

However, the average indicated above is

$$\frac{1}{2\pi/\omega}\int_0^{2\pi/\omega} \sin^2(\omega t + \frac{2\pi x}{\lambda})dt = \tfrac{1}{2}.$$

Therefore

$$I = I_m \cos^2(\frac{2\pi x}{\lambda}) \quad \underline{\text{Ans}},$$

with $I_m = 2E_0^2$.

CHAPTER 41

41-1

From superposition, it is anticipated that the amplitude of the wave at P due to screen A added to the amplitude at P due to screen B will give the amplitude at P due to an unscreened wave. But, with $x \gg \lambda$, P lies in the geometrical shadow of the hole: thus the last quantity is zero. Hence,

$$E_{P,A} + E_{P,B} = 0,$$

$$E_{P,A} = -E_{P,B}.$$

But the intensity I is proportional to E_P^2 and therefore $I_{P,A} = I_{P,B}$.

41-6

Assuming that parallel "rays" strike the lens, the diffraction image is formed a distance f = 70 cm behind the lens. The angular distance from the center of the pattern to the m th minimum is given by

$$\sin\theta_m = m\lambda/a.$$

But the linear distance sought is

$$x_m = f\cdot\tan\theta_m.$$

Eliminating θ_m between these equations gives

$$x_m = \frac{f}{[(a/m\lambda)^2 - 1]^{\frac{1}{2}}}.$$

Now $a/\lambda \gg 1$ so that, approximately,

$$x_m = m\lambda f/a = (0.10325 \text{ cm})m,$$

since $f = 70$ cm, $a = 0.04$ cm and $\lambda = 5.9 \times 10^{-5}$ cm.

(a) For $m = 1$, $x = 1.0325$ mm and

(b) with $m = 2$, $x = 2.065$ mm.

41-8

The conditions for minima at the two wavelengths are

$$a\sin\theta_a = m\lambda_a, \quad a\sin\theta_b = n\lambda_b, \quad m,n = 1, 2, \ldots$$

(a) Since $\theta_a = \theta_b$ when $m = 1$ and $n = 2$, it follows directly from the above that $\lambda_a = 2\lambda_b$.

(b) Assuming other coincidences (i.e., setting $\theta_a = \theta_b$), use the result from (a) above to obtain $n = 2m$; that is, every other minimum at λ_b coincides with a minimum for λ_a.

41-9

Consider a slit of given width divided into N strips of equal width so that there are N phasors. These are all parallel at the central maximum. If the slit width is doubled there will be 2N phasors, each having the same magnitude as for the original slit. Hence, the amplitude of the resultant field at the central maximum doubles and the intensity, proportional to the amplitude squared, increases by a factor of four. However, the diffraction pattern is narrower and the area under a curve of I vs. θ, equal to the rate at which energy passes through the slit, will only double.

41-10

(a) The intensity is given by

$$I = I_m \frac{\sin^2\alpha}{\alpha^2}.$$

To find the maxima and minima set $dI/d\alpha = 0$:

$$\frac{dI}{d\alpha} = 2I_m \frac{\sin\alpha}{\alpha}\left(\frac{\cos\alpha}{\alpha} - \frac{\sin\alpha}{\alpha^2}\right).$$

This will be zero if either

(i) $\sin\alpha/\alpha = 0$, giving $\alpha = m\pi$, m = 1, 2, ... These are minima since I = 0 for these values of α;

or

(ii) $\cos\alpha/\alpha - \sin\alpha/\alpha^2 = 0$, which can be arranged in the form

$$\tan\alpha = \alpha$$

and are the maxima.

(b) Clearly $\alpha = 0$ is a solution of (ii). The next solution, as shown, is close to $\alpha = 3\pi/2$. Write this second solution as $\alpha = 3\pi/2 - x$. Then, in terms of x the relation $\tan\alpha = \alpha$ becomes

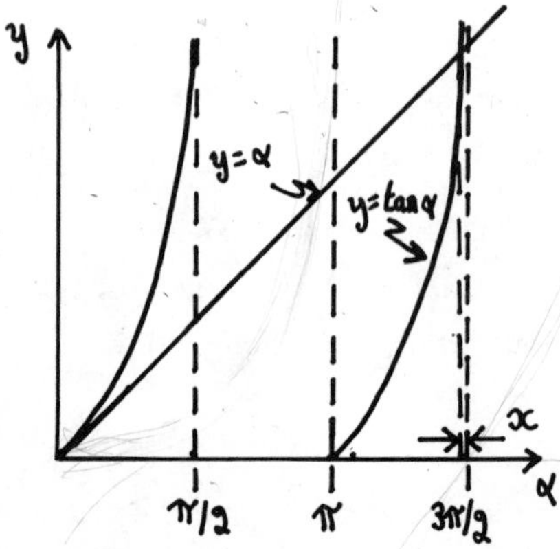

$$\frac{1}{\tan x} = \frac{3\pi}{2} - x.$$

Suppose $x \ll 1$; neglecting powers of x higher than two, this gives,

$$\frac{1}{x}\left(1 - \frac{x^2}{3}\right) = \frac{3\pi}{2} - x,$$

which can be rearranged as

$$\frac{2}{3}x^2 - \frac{3\pi}{2}x + 1 = 0,$$

of which the solution is x = 0.219 rad, corresponding to $\alpha = 4.493$ rad or $\alpha = 257°$.

(c) Write for maxima

$$\alpha = (m + \tfrac{1}{2})\pi.$$

Then, the central maximum, for which $\alpha = 0$, has $m = -\frac{1}{2}$ and the next maximum, $\alpha = 4.493$, yields m = 0.93.

41-11

Let d be the separation of the headlights a distance r away. The light rays make an angle θ with each other as they enter the eye and an angle θ' after passing through the pupil. If n be the index of refraction of the vitreous humour of the eye, then

$$\sin(\tfrac{1}{2}\theta) = n\sin(\tfrac{1}{2}\theta'),$$

by the law of refraction. With the headlights just barely resolved,

$$\sin\theta' = 1.22\,\frac{\lambda/n}{a},$$

λ the wavelength of the light in air and a the pupil diameter. If the angles are small,

$$r = \frac{d}{\theta} = \frac{d}{n\theta'} = \frac{d}{1.22\,\lambda/a} = \frac{ad}{1.22\,\lambda} = 10.4\text{ km} \quad \underline{\text{Ans}}.$$

41-17

Let θ = angle subtended at observer by the objects,
a = mirror or pupil diameter,
d = distance to objects,
D = linear separation between objects.

Then,

$$\sin\theta \approx D/d,$$

and, assuming the Rayleigh criterion,

$$a\sin\theta = 1.22\,\lambda.$$

Therefore,

$$D = \frac{1.22\lambda d}{a}.$$

Numerically, $\lambda = 550 \times 10^{-9}$ m, $d = 8 \times 10^{10}$ m, a = 0.005 m for the eye and 5.08 m for the mirror. These give

(a) $D = 1.1 \times 10^4$ km, eye; (b) D = 11 km, mirror <u>Ans</u>.

For (a) the distance D is greater than the diameter of Mars itself.

41-20

(a) The sketches illustrate the formation of a halo as the moon shines through a cloud containing suspended water droplets. The ring will appear red if blue light is absent. Since the angle θ

for the first minimum is given by $\sin\theta = 1.22\lambda/d$, blue light, as it has the shortest wavelength in the visible spectrum, will experience its first minimum closer to the moon (smallest θ gives the smallest ϕ) than any other color, giving the ring its reddish appearance.

(b) Since the rays MP and MO are virtually parallel, $\theta \approx \phi = \frac{3}{2}(\frac{1}{4}^\circ)$ = 22.5', the diameter of the moon being about $\frac{1}{2}^\circ$. Then, by (a),

$$\sin\phi \approx \sin\theta = 1.22\lambda_{blue}/d,$$

d the droplet diameter and λ_{blue} = 400 nm. These give d = 70 μm.

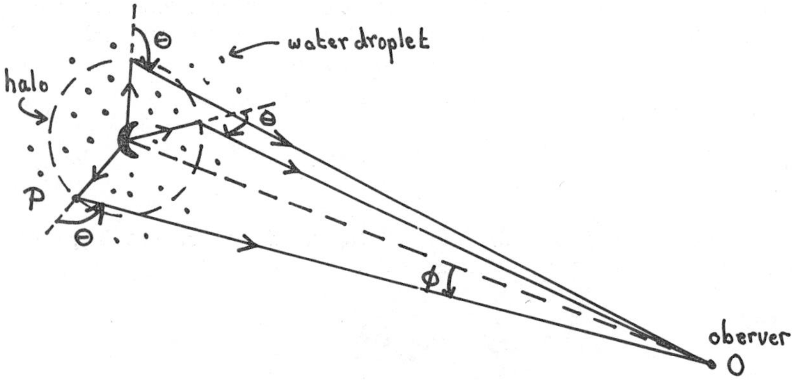

(c) A bluish ring will be seen at the scattering angle for the first minimum of red light. As $\lambda(\text{red}) \approx 700$ nm $\approx 2\lambda(\text{blue})$, the radius of the blue ring will be about twice that of the red ring or three times the lunar radius. (The intensity of the ring will be very low, however.) If water droplets of various sizes are present, various rings of these colours are present close together, giving a whitish appearance. This is, in fact, most common.

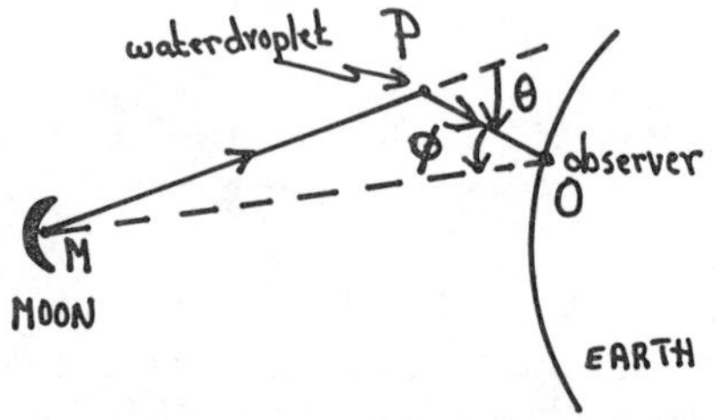

(d) These halos are a diffraction effect, a rainbow being formed by refraction.

41-25

(a) The location of the interference fringes is given by

$$d\sin\theta_i = m\lambda, \quad m = 0, 1, 2, \ldots ,$$

and the diffraction minima by

$$a\sin\theta_d = n\lambda, \quad n = 1, 2, \ldots .$$

Since $d = 5a$, the $m = 5$ interference fringe falls at the same position ($\theta_d = \theta_i$) as the $n = 1$ diffraction minimum and as a result is not seen. Inside the central diffraction envelope are the $m = 1$, 2, 3, 4 fringes on each side of the central $m = 0$ fringe, giving 9 fringes in all.

(b) The third fringe is located at an angle θ_i given by

$$\sin\theta_i = 3\lambda/d = 3\lambda/5a,$$

and has an intensity equal to that of the diffraction envelope at that location. Hence, the desired ratio is

$$\frac{I}{I_m} = \left(\frac{\sin\alpha}{\alpha}\right)^2$$

with

$$\alpha = \pi a\sin\theta_d/\lambda = \pi a\sin\theta_i/\lambda = 3\pi/5,$$

so that $I/I_m = 0.255$ Ans.

41-32

With the slit width much less than the wavelength of light, the effects associated with a diffraction envelope can be ignored. The phase difference ϕ between rays from two adjacent slits is $\phi = 2\pi d\sin\theta/\lambda$ since the path difference for these rays is $d\sin\theta$, just as for the two-slit interference "grating". The electric field components of the three waves arriving at a given point on the screen can be written as

$$E_1 = E_0\sin\omega t, \; E_2 = E_0\sin(\omega t + \phi), \; E_3 = E_0\sin(\omega t + 2\phi).$$

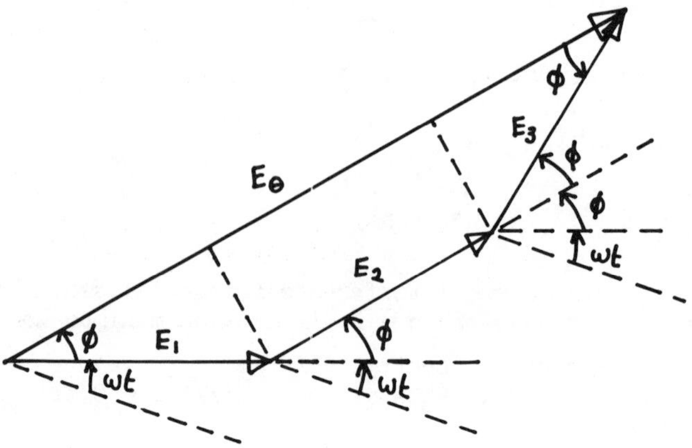

It can be seen from the phasor diagram that the amplitude of the sum of these waves is

$$E_\theta = E_0\cos\phi + E_0 + E_0\cos\phi = E_0(1 + 2\cos\phi).$$

Since the intensity is proportional to the square of this amplitude,

$$I = AE_0^2(1 + 2\cos\phi)^2.$$

At the center of the pattern $\theta = \phi = 0$ and $I(\theta = 0) = 9AE_0^2 = I_m$, the central intensity. Thus, if $A = I_m/9E_0^2$,

$$I = \frac{1}{9} I_m(1 + 4\cos\phi + 4\cos^2\phi).$$

41-33

The phase difference between waves from adjacent slits to the first minimum beyond the m th principal maximum is

$$\Delta\phi = 2\pi m + \frac{2\pi}{N},$$

which indicates a path difference ΔL given by

$$\Delta L = \frac{\Delta\phi}{2\pi}\cdot\lambda = m\lambda + \frac{\lambda}{N}.$$

But if θ_m is the position angle of the m th principal maximum, this path difference is also

$$\Delta L = d\sin(\theta_m + \Delta\theta).$$

Therefore,

$$d(\sin\theta_m\cos\Delta\theta + \cos\theta_m\sin\Delta\theta) = m\lambda + \frac{\lambda}{N}.$$

If $\Delta\theta \ll 1$, $\cos\Delta\theta \approx 1$ and $\sin\Delta\theta \approx \Delta\theta$; thus, with $d\sin\theta_m = m\lambda$,

$$m\lambda + d\cos\theta_m\Delta\theta = m\lambda + \frac{\lambda}{N},$$

$$\Delta\theta = \frac{\lambda}{Nd\cos\theta_m} \quad \underline{\text{Ans}}.$$

41-40

(a) Let the maxima in question be of order m and m + 1. Then,

$$d\sin\theta_m = m\lambda, \quad d\sin\theta_{m+1} = (m + 1)\lambda,$$

where $\sin\theta_m = 0.2$, $\sin\theta_{m+1} = 0.3$. Subtracting these equations gives

$$d(0.3 - 0.2) = \lambda = 600 \text{ nm},$$

$$d = 6.0\ \mu\text{m} \quad \underline{\text{Ans}}.$$

(b) Suppose the m = 4 maximum, which is missing, falls at the n th minimum of the diffraction envelope. These minima fall at angles ϕ given by

$$a\sin\phi = n\lambda, \quad n = 1, 2, \ldots .$$

The m = 4 maximum falls at an angle θ_4 satisfying

$$d\sin\theta_4 = 4\lambda .$$

If the two fall at the same place on the screen, $\phi = \theta_4$ so that

$$d\sin\theta_4 = 4\lambda, \quad a\sin\theta_4 = n\lambda,$$

$$a = \tfrac{1}{4}nd.$$

For the smallest a choose the smallest n, to wit n = 1. Then,

$$a_{min} = \tfrac{1}{4}\ 6000\ \text{nm} = 1.5\ \mu\text{m} \quad \underline{\text{Ans.}}$$

(c) If the m = 4 maximum is missing, the m = 8, 12, 16, ... will be also. Since d = 10λ, it is clear that the m = 10 maximum is deflected through 90° and will not fall on the screen. Hence, the visible orders are

$$m = 0,\ 1,\ 2,\ 3,\ 5,\ 6,\ 7,\ 9 \quad \underline{\text{Ans.}}$$

41-43

The dispersion D is

$$D = \frac{d\theta}{d\lambda} = \frac{m}{d\cos\theta}.$$

But $m\lambda = d\sin\theta$, so that $m = d\sin\theta/\lambda$. Putting this into the equation for D yields

$$D = \frac{d\sin\theta/\lambda}{d\cos\theta} = \frac{\sin\theta/\cos\theta}{\lambda} = \frac{\tan\theta}{\lambda}.$$

41-46

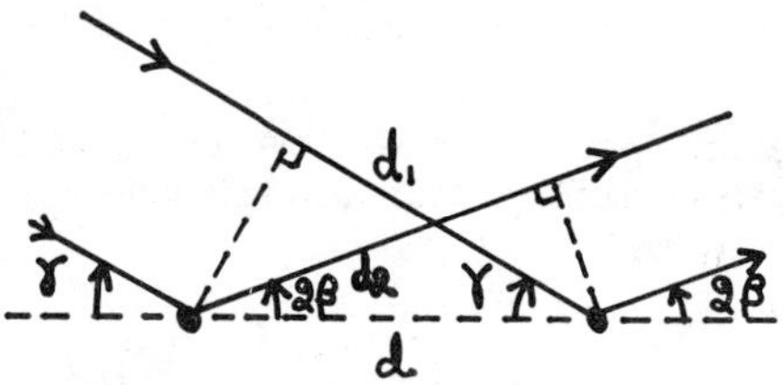

If d is the distance between adjacent scattering centers, the path difference between parallel rays striking adjacent centers is

$$d_1 - d_2 = d\cos\gamma - d\cos 2\beta.$$

For maxima this must equal an integral number m of wavelengths:

$$d(\cos\gamma - \cos 2\beta) = m\lambda.$$

With γ small, $\cos\gamma \approx 1$; since d = 1500 nm, λ = 0.5 nm and m = 1 for the first order maximum,

$$1 - \cos 2\beta \approx \frac{1}{3000};$$

evidently, then, β is a small angle also. Setting $\cos 2\beta \approx 1 - 2\beta^2$ and $\cos\gamma \approx 1 - \frac{1}{2}\gamma^2$ gives, more precisely,

$$(1 - \tfrac{1}{2}\gamma^2) - (1 - 2\beta^2) = \frac{1}{3000},$$

$$\beta = (\frac{1}{6000} + \tfrac{1}{4}\gamma^2)^{\frac{1}{2}} \quad \underline{\text{Ans}}.$$

41-51

(a) Since $\nu\lambda = c$, $\Delta\nu/\nu = \Delta\lambda/\lambda$, except for sign. Thus,

$$R = \frac{\lambda}{\Delta\lambda} = Nm = \frac{\nu}{\Delta\nu},$$

$$\Delta\nu = \frac{\nu}{Nm} = \frac{c}{Nm\lambda} \quad \underline{\text{Ans}}.$$

(b) Times of flight beyond ab are equal for the various rays. The path difference between extreme rays is

$$\Delta S = (N - 1)d\sin\theta \ .$$

Therefore,

$$\Delta t = \frac{\Delta S}{c} = \frac{(N - 1)d\sin\theta}{c} \quad \underline{\text{Ans}}.$$

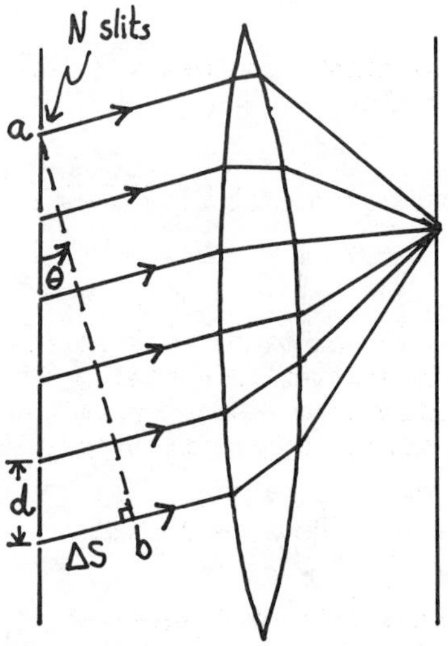

(c) Since $m\lambda = d\sin\theta$, Δt becomes

$$\Delta t = \frac{(N - 1)m\lambda}{c},$$

and

$$\Delta\nu\Delta t = \frac{c}{Nm\lambda}\,\frac{(N - 1)m\lambda}{c} = 1,$$

provided that $N \gg 1$ so that $N - 1 \approx N$ in the above.

41-53

(a)

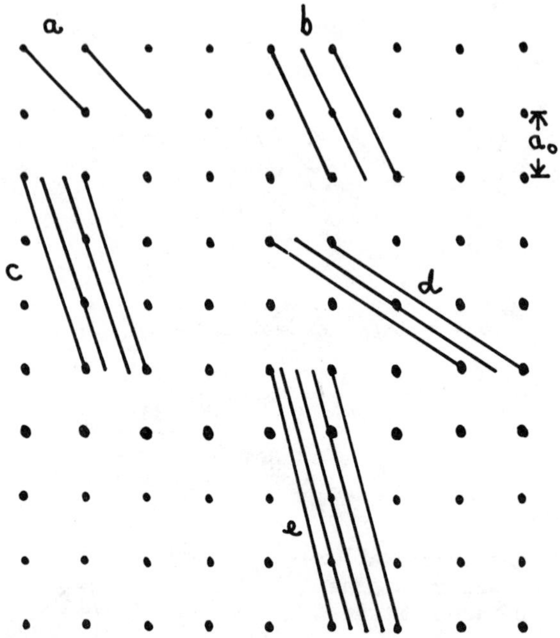

The spacings of the five sets of planes shown in the figure are

a: $a_0/(2)^{\frac{1}{2}}$;

b: $a_0/(5)^{\frac{1}{2}}$;

c: $a_0/(10)^{\frac{1}{2}}$;

d: $a_0/(13)^{\frac{1}{2}}$;

e: $a_0/(17)^{\frac{1}{2}}$.

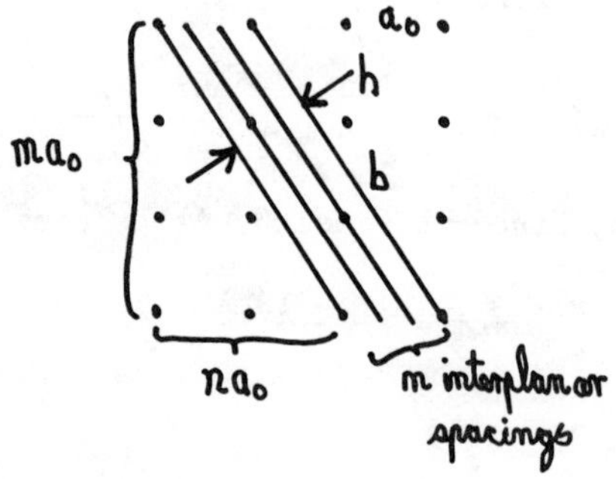

(b) For the general formula see the second sketch above. The area of the shaded parallelogram is

$$A = bh = [(ma_0)^2 + (na_0)^2]^{\frac{1}{2}}h.$$

It is also given by

$$A = (ma_0)(n + 1)a_0 - 2\cdot\tfrac{1}{2}(ma_0)(na_0) = ma_0^2.$$

Equating the two expressions for A gives

$$h = \frac{ma_0}{(m^2 + n^2)^{\frac{1}{2}}}.$$

But h = md, d the interplanar spacing and therefore

$$d = \frac{a_0}{(m^2 + n^2)^{\frac{1}{2}}}.$$

If the integers m,n contain a common factor, p say, the figure for planes with slope $\frac{m}{p}/\frac{n}{p}$ is being reproduced, yielding the same value of d: therefore, exclude these.

41-59

The radiation wavelength λ and plane spacing d constitute two unknowns; hence, observations at two orders should suffice; these observations satisfy Bragg's law

$$2d\sin\theta_1 = m_1\lambda,$$

$$2d\sin\theta_2 = m_2\lambda.$$

But upon dividing, these equations yield

$$\sin\theta_1/\sin\theta_2 = m_1/m_2,$$

the unknowns cancelling out. Only the ratio d/λ can be determined and this requires only one of these equations.

CHAPTER 42

42-4

(a) Let S = solar constant, R = radius of the earth. The rate at which energy from the sun impinges on the earth is $S\pi R^2$. Assuming a uniform distribution of this energy over the surface of the earth during a year, the average rate dE/dt at which solar energy is absorbed by a unit area of the earth's surface is given by

$$(4\pi R^2)\frac{dE}{dt} = S\pi R^2,$$

$$\frac{dE}{dt} = \tfrac{1}{4}S.$$

If the earth's average surface temperature is constant in time, the rate at which energy is reradiated back into space must, on the average, equal this; hence, the rate of reradiation is

$$\frac{dE}{dt} = \tfrac{1}{4}S = \tfrac{1}{4}(1390) = 347.5\ \text{W/m}^2 \quad \underline{\text{Ans}}.$$

(b) For a cavity radiator, with a unit area $A = 1\ \text{m}^2$,

$$\frac{dE}{dt} = \sigma AT^4,$$

$$347.5 = (5.67 \times 10^{-8})(1)T^4,$$

$$T = 280\ \text{K} = 7°\text{C} \quad \underline{\text{Ans}}.$$

42-6

Through the opening of area A the oven emits into the room energy at the rate $A\sigma T_0^4$, and receives energy from the room at the rate $A\sigma T_r^4$. Thus, converting temperatures to kelvins and the area to square meters,

$$P = A\sigma(T_0^4 - T_r^4),$$

P the net power transferred. Numerically,

$$P = (5 \times 10^{-4})(5.67 \times 10^{-8})(500^4 - 300^4) = 1.54 \text{ W} \quad \underline{\text{Ans}}.$$

42-9

The Planck law is

$$R_\lambda = \frac{2\pi c^2 h}{\lambda^5} \frac{1}{e^{hc/\lambda kT} - 1}.$$

Let $x = hc/\lambda kT$. If $x \ll 1$, then

$$e^x = (1 + x + \frac{x^2}{2!} + \ldots) \approx 1 + x,$$

so that $e^x - 1 \approx x$. The Planck law then becomes

$$R_\lambda \approx \frac{2\pi c^2 h}{\lambda^5} \frac{1}{x} = \frac{2\pi ckT}{\lambda^4}.$$

From the definition of x above, it is evident that this approximate form of Planck's law will hold for long (large) wavelengths and/or high (large) temperatures.

42-13

(a) In one second the lamp radiated 100 J of energy. The energy E of each photon is

$$E = h\nu = \frac{hc}{\lambda} = \frac{(6.626 \times 10^{-34})(3 \times 10^8)}{589 \times 10^{-9}} = 3.375 \times 10^{-19} \text{ J}.$$

Hence, the rate at which the photons are emitted is

$$\frac{100 \text{ J/s}}{3.375 \times 10^{-19} \text{ J}} = 2.96 \times 10^{20} \text{ s}^{-1} \quad \underline{\text{Ans}}.$$

(b) A flux $dn/dt = 1 \text{ photon/cm}^2\cdot\text{s} = 10^4 \text{ photons/m}^2\cdot\text{s}$ means an energy flux $I = E\cdot dn/dt = 3.375 \times 10^{-15} \text{ J/m}^2\cdot\text{s}$. The power P (100 W) of the lamp is related to the intensity I by $I = P/4\pi r^2$, the lamp radiating in all directions. Thus,

$$r = (P/4\pi I)^{\frac{1}{2}} = 4.86 \times 10^7 \text{ m} \quad \underline{\text{Ans.}}$$

(c) A rectangular column of photons of base area A normal to the path of the photons and length ct contains nAct photons, n the photon density. In the time t, the energy of the column = nActE crosses the area A since the photons move with speed c. The intensity I at A is, then,

$$I = \frac{nActE}{At} = ncE = P/4\pi r^2,$$

$$r = (P/4\pi ncE)^{\frac{1}{2}}.$$

With P = 100 W, $n = 10^6\ /m^3$ and E as found in (a), this gives r = 280.3 m <u>Ans.</u>

(d) From (b) the photon flux dn/dt is, for r = 2.0 m,

$$\frac{dn}{dt} = \frac{I}{E} = \frac{P}{4\pi r^2 E} = 5.89 \times 10^{18}\ /m^2 \cdot s \quad \underline{\text{Ans.}}$$

The photon density from (c) is

$$n = \frac{P}{4\pi r^2 cE} = \frac{dn/dt}{c} = 1.96 \times 10^{10}\ /m^3 \quad \underline{\text{Ans.}}$$

<u>42-16</u>

The electron rest energy is $E_0 = 0.511$ MeV $= 8.176 \times 10^{-14}$ J.

(a) The photon frequency is given by

$$\nu = E/h = E_0/h = \frac{8.176 \times 10^{-14}}{6.626 \times 10^{-34}} = 1.234 \times 10^{20} \text{ Hz} \quad \underline{\text{Ans.}}$$

(b) The wavelength associated with this frequency is

$$\lambda = \frac{c}{\nu} = 2.43 \times 10^{-12} \text{ m} \quad \underline{\text{Ans.}}$$

(c) The photon momentum is just

$$p = \frac{E}{c} = \frac{E_0}{c} = 2.73 \times 10^{-22} \text{ kg}\cdot\text{m/s} \quad \underline{\text{Ans.}}$$

42-21

Under the assumption that the photon gives all its energy to the initially stationary electron which, after the collision, moves off with speed v, the conservation of momentum and mass-energy gives

$$\frac{h\nu}{c} + 0 = 0 + \frac{m_0 v}{(1 - v^2/c^2)^{\frac{1}{2}}};$$

$$h\nu + m_0c^2 = 0 + \frac{m_0 c^2}{(1 - v^2/c^2)^{\frac{1}{2}}};$$

the electron, of necessity (conservation of momentum), moves away in the same direction in which the incident photon arrived. In the second equation, transposing the rest energy to the other side and dividing by c yields

$$\frac{h\nu}{c} = \frac{m_0 c}{(1 - v^2/c^2)^{\frac{1}{2}}}[1 - (1 - v^2/c^2)^{\frac{1}{2}}].$$

This agrees with the first equation if

$$c[1 - (1 - v^2/c^2)^{\frac{1}{2}}] = v,$$

which requires either $v/c = 0$ or $v/c = 1$. The former value, by the first equation, leads to $\nu = 0$ also; there is little point in further discussion of this case as the photon energy is zero. The latter "solution", $v/c = 1$, is prohibited by the Special Theory of Relativity.

42-22

(a) For the photoelectric effect,

$$h\nu = \tfrac{1}{2}mv^2 + \phi.$$

The energy $h\nu$ of the incident photon is

$$h\nu = \frac{1240}{200} = 6.2 \text{ eV}.$$

Now $\phi = \phi_0 = 4.2$ eV for the most loosely bound electron and the greatest amount of kinetic energy that a liberated electron can possess is, therefore,

$$\tfrac{1}{2}mv^2 = 6.2 - 4.2 = 2.0 \text{ eV} \quad \underline{\text{Ans.}}$$

(b) The minimum kinetic energy will be carried by an electron that has a binding energy to the surface of 6.2 eV; i.e., $K_{min} = 0$.

(c) The potential required to stop the 2.0 eV electron in (a) is, since the electron charge is one unit, 2.0 V.

(d) A photon barely able to remove the electron bound with an energy of 4.2 eV has a wavelength

$$\lambda = \frac{1240}{4.2} = 295 \text{ nm} \quad \underline{\text{Ans.}}$$

42-28

(a) The wavelength of the photon is $\lambda = 1240/E = 1240/6200 = 0.2$ nm. The Compton shift is given by

$$\Delta\lambda = \frac{h}{m_0c}(1 - \cos\theta).$$

The Compton wavelength is $\lambda_C = h/m_0c = 0.00242$ nm. Also, since $\nu\lambda = c$, it follows that $\Delta\lambda/\lambda = -\Delta\nu/\nu$. Hence, ignoring the sign,

$$\frac{\Delta\nu}{\nu} = \frac{\lambda_C}{\lambda}(1 - \cos\theta),$$

$$10^{-4} = \frac{0.00242}{0.2}(1 - \cos\theta),$$

$$\theta = 7.37^\circ \quad \underline{\text{Ans.}}$$

(b) The energy lost by the photon is given to the electron as kinetic energy. Since the energy of a photon is $E = h\nu$, the energy imparted to the electron must be

$$E = h\Delta\nu = E\cdot\frac{\Delta\nu}{\nu} = (6200 \text{ eV})(10^{-4}) = 0.62 \text{ eV} \quad \underline{\text{Ans.}}$$

42-30

(a) The Compton wavelength is $\lambda_C = h/m_0c$. For the electron put $m_0 = 9.11 \times 10^{-31}$ kg and for the proton $m_0 = 1.67 \times 10^{-27}$ kg. These give 0.00242 nm and 1.32×10^{-6} nm respectively, using 1 nm = 10^{-9}m and $h = 6.626 \times 10^{-34}$ J·s, $c = 3 \times 10^8$ m/s.

(b) In terms of its wavelength, the energy of a photon is $E = hc/\lambda$ and therefore $\lambda = hc/E$. Setting this equal to the Compton wavelength leads to

$$hc/E = h/m_0c,$$

$$E = m_0c^2,$$

the term on the right being a particle rest energy. For the electron and proton, then, the required photon energies are E = 0.511 MeV and 938 MeV, respectively.

42-40

If kinetic energy is not conserved, some of the 6.0 eV of the neutron's initial kinetic energy must be used to excite the hydrogen atom. However, to reach the first excited level (n = 2) from the ground level (n = 1) the atom must be supplied with an amount of energy E, with

$$E = 13.6 - 13.6/2^2 = 10.2 \text{ eV}.$$

Since the neutron does not possess this energy the collision will be elastic, the hydrogen atom remaining in the ground state.

42-42

A neutral helium atom has two electrons. With one electron gone only one remains, so that the singly ionized helium atom is "hydrogen-like" in that it possesses one electron only. According to the Bohr theory, the ionization energy of a one electron atom from the ground state is

$$E = \frac{Z^2me^4}{8\epsilon_0^2h^2} = 13.6 \text{ eV}\cdot Z^2 .$$

For helium Z = 2 and therefore the ionization energy is 54.4 eV. A small additional correction due to the finite mass of the nucleus is ignored.

42-45

(a) For hydrogen the energy of the allowed quantized states is proportional to m, the mass of the electron. Strictly speaking, to take account of the motion of the proton and electron relative to the center of mass of the atom, m should be the reduced mass of the system:

$$m = \frac{m_e m_p}{m_e + m_p} \approx m_e$$

since $m_e \ll m_p$. In positronium the proton is replaced with another electron so that the reduced mass is

$$m = \frac{m_e^2}{2m_e} = \tfrac{1}{2}m_e,$$

or half the value for hydrogen. But the energies of the photons emitted in transitions is the difference between the energies of the allowed states between which the transitions take place. Hence, in positronium the frequencies of photons are one-half as great as those for the corresponding transitions in hydrogen. With $\lambda = c/\nu$, the wavelengths are $1/(\frac{1}{2})$ = twice as long as in hydrogen.

(b) The radius of the relative orbit is inversely proportional to m for any n; hence, the radius of the ground state orbit in positronium is twice that for hydrogen. In hydrogen the radius of the electron orbit relative to the proton almost equals the radius of the orbit about the center of mass, since the proton is located almost exactly at the center of mass. For positronium, however, the center of mass lies midway between the electrons, and it follows that the distance of either electron from the center of mass in any state virtually equals the radius of the corresponding Bohr orbit in hydrogen.

42-47

(a) By assumption $L = I\omega = nh/2\pi$. The rotational inertia is $I = 2m(\frac{1}{2}d)^2 = \frac{1}{2}md^2$ so that

$$\tfrac{1}{2}md^2\omega = nh/2\pi,$$

$$\omega = nh/\pi md^2 \quad \underline{\text{Ans}}.$$

(b) The rotational energy is

$$E = \tfrac{1}{2}I\omega^2 = \frac{L^2}{2I} = \frac{(nh/2\pi)^2}{md^2} = n^2h^2/4\pi^2md^2 \quad \underline{\text{Ans}}.$$

42-49

The common factor $me^4/8\epsilon_0^2h^3$ cancels and the last column is just

$$100\,\frac{\nu - \nu_0}{\nu} = 100\,\frac{\dfrac{2n - 1}{(n - 1)^2n^2} - \dfrac{2}{n^3}}{\dfrac{2n - 1}{(n - 1)^2n^2}} = \frac{100}{n}\left(\frac{3n - 2}{2n - 1}\right).$$

As n becomes very large, $3n \gg 2$ and $2n \gg 1$ so that

$$\frac{3n - 2}{2n - 1} \approx \frac{3n}{2n} = \frac{3}{2},$$

and therefore

$$100\,\frac{\nu - \nu_0}{\nu} \approx \frac{150}{n}.$$

43-5

The deBroglie wavelength is $\lambda = h/p$. For nonrelativistic situations $K = \frac{1}{2}mv^2 = p^2/2m$, m the mass of the electron. In terms of the accelerating potential V, K = eV, with e the magnitude of the electron charge. Putting all this together gives

$$\lambda = \frac{h}{p} = \frac{h}{(2mK)^{\frac{1}{2}}} = \frac{h}{(2meV)^{\frac{1}{2}}} = \left(\frac{h^2}{2meV}\right)^{\frac{1}{2}}.$$

Now set $h = 6.626 \times 10^{-34}$ J·s, $m = 9.11 \times 10^{-31}$ kg, $e = 1.6 \times 10^{-19}$ C, to obtain

$$\lambda = \left(\frac{1.506 \times 10^{-18}}{V}\right)^{\frac{1}{2}}.$$

The units of the numerator are $J^2 \cdot s^2/kg \cdot C$; hence, with V in volts, the units of the quantity in parenthesis are

$$J^2 \cdot s^2/kg \cdot C \cdot V = J^2 \cdot s^2/kg \cdot J = J \cdot s^2/kg = (kg \cdot m^2/s^2) \cdot s^2/kg = m^2.$$

Since $1\ m^2 = 10^{20}\ \text{Å}^2$, this becomes

$$\lambda = \left(\frac{150.6}{V}\right)^{\frac{1}{2}}$$

yielding λ in Å if V is in volts.

43-11

The wavelength of the neutrons, which at this kinetic energy much less than their rest energy, can be calculated without resorting to relativistic mechanics, is

$$\lambda = \frac{h}{p} = \frac{h}{(2mK)^{\frac{1}{2}}}.$$

With $m = 1.67 \times 10^{-27}$ kg and $K = 4(1.6 \times 10^{-19}$ J), this gives $\lambda =$ 0.143 A. In the Bragg relation, with m = 1,

$$\lambda = 2d\sin\theta,$$

$$0.143 = 2(0.732)\sin\theta,$$

$$\theta = 5.6^\circ \quad \underline{\text{Ans.}}$$

<u>43-16</u>

(a) Consider that confining the electron in a nucleus is roughly analogous to the problem of a particle in a one-dimensional box. The energy of the electron, approximately and by nonrelativistic mechanics, would be

$$E = \frac{h^2}{8mL^2} = 3.08 \times 10^{-10}\ \text{J} = 1900\ \text{MeV} \quad \underline{\text{Ans.}}$$

(b) The nucleus cannot, with a binding energy of 2-3 MeV per nucleon, hold a 2000-MeV particle. Thus, electrons are not expected to be long-term residents in nuclei.

<u>43-18</u>

Using the result of Problem 43-17, the normalized wavefunction is

$$\psi = \left(\frac{2}{\ell}\right)^{\frac{1}{2}}\sin\left(\frac{n\pi x}{\ell}\right).$$

The desired probability P is

$$P = \int_0^{\ell/3} \psi^2 dx = \frac{2}{\ell}\int_0^{\ell/3} \sin^2\left(\frac{n\pi x}{\ell}\right)dx = \frac{1}{3} - \frac{1}{2n\pi}\sin(2n\pi/3).$$

Successively putting n = 1, 2, 3 gives (a) P = 0.20, (b) P = 0.40, (c) P = 0.33. (d) Classically, the probability of finding the particle in a region of length D is D/ℓ. Thus, with $D = \ell/3$, P = 0.33.

43-20

The motion of the electron in the ground state of the hydrogen atom is described by the wavefunction

$$\psi = (1/\pi r_B^3)^{\frac{1}{2}} e^{-R/r_B},$$

where r_B is the radius of the ground-state Bohr orbit and R is distance to the nucleus. Consider two hypothetical spherical shells centered on the nucleus, with radii R and R + dR. The probability dP of finding the electron within the infinitesimal shell bounded by these spheres is

$$dP = \psi^2(R)dV,$$

where dV is the volume between the spheres. Hence, the probability of finding the electron within a sphere of radius r centered on the nucleus is

$$P_r = \int \psi^2(R)\cdot dV = \int_0^r \psi^2(R)\cdot 4\pi R^2 dR = (1/\pi r_B^3)\int_0^r e^{-2R/r_B}\cdot 4\pi R^2 dR,$$

$$P_r = (4/r_B^3)\int_0^r e^{-2R/r_B}\cdot R^2 dR = (4/r_B^3)(\tfrac{1}{2}r_B)^3\int_0^{2r/r_B} x^2 e^{-x} dx,$$

$$P_r = 1 - e^{-2r/r_B}\left(\frac{2r^2}{r_B^2} + \frac{2r}{r_B} + 1\right).$$

(a) For r = 0, P_r = 0, and this is expected since the volume in which the electron is being sought has reduced to zero.

(b) With $r = \infty$, P_r = 1; this also is anticipated since the electron must be somewhere, so that if all space is examined it must be found.

(c) $\psi^2(r)dV$ gives only the probability of finding the electron within the distances r and r + dr, an infinitesimal region. The probability found above refers to a finite volume.

43-23

By the uncertainty principle, with $\Delta x = h/p$, the deBroglie wavelength,

$$\Delta p \Delta x \geq h,$$

$$\Delta p \cdot \frac{h}{p} \geq h,$$

$$\Delta p \geq p.$$

But $p = mv$, and therefore $\Delta p = m\Delta v$ and therefore the above gives

$$m\Delta v \geq mv,$$

$$\Delta v \geq v.$$

43-25

(a) The function ψ_n^2 is given by

$$\psi_n^2 = \psi_{max}^2 \sin^2(\frac{n\pi x}{\ell}).$$

Therefore $\psi_n^2 = 0$, the minima, when

$$\frac{n\pi x}{\ell} = m\pi, \quad m = 0, 1, 2, \ldots ;$$

$$x_m = \frac{m}{n} \cdot \ell.$$

Hence, the distance s_n between adjacent minima is

$$s_n = x_{m+1} - x_m = \frac{m + 1}{n} \cdot \ell - \frac{m}{n} \cdot \ell = \ell/n.$$

(b) Using

$$\Delta p = 2p_x = 2(2mE)^{\frac{1}{2}} = 2(2m \cdot \frac{n^2 h^2}{8m\ell^2})^{\frac{1}{2}} = nh/\ell,$$

the uncertainty principle yields

$$\Delta x \geq h/\Delta p = \ell/n = s_n.$$

(c) Since $\Delta x = \ell/n$, as n approaches infinity Δx goes to zero. To resolve the points at which the square of the wave function equals zero, the uncertainty in position Δx must be less than s_n. But the results above give the opposite, $s_n \leq \Delta x$, so that although both approach zero as n gets very large, the uncertainty in measured position is always greater than the distance between the points whose resolution is being attempted.

CHAPTER 44

44-5

For each value of n there are n possible values of ℓ ranging from 0 to n - 1, and for each ℓ there are $2\ell + 1$ possible values of m_ℓ. Hence, the number N of states at any n is

$$N = \sum_0^{n-1}(2\ell + 1) = 2\sum_0^{n-1}\ell + n = 2\left(\sum_0^{n}\ell - n\right) + n = 2\sum_1^{n}\ell - n.$$

But

$$\sum_1^{n}\ell = \tfrac{1}{2}n(n + 1),$$

since the average value of each term in the sum is $\frac{n+1}{2}$ and there are n terms. Thus

$$N = 2\cdot\tfrac{1}{2}n(n + 1) - n = n^2.$$

44-9

The desired probability P is given by

$$P = \int \psi^2 dV.$$

But, with $dr = 0.01r_B \ll r$, the wave function can be considered as approximately constant in the region of integration. Then, the probability becomes

$$P \approx \psi^2(r_B)\cdot\int dV = \psi^2(r_B)\cdot(\Delta V) \approx \psi^2(r_B)(4\pi r_B^2\cdot\Delta r).$$

Substituting the normalized ground-state wave function gives

$$P = (1/\pi r_B^3)e^{-2r_B/r_B}\cdot 4\pi r_B^2\cdot(0.01r_B) = 0.04e^{-2} = 0.0054 \quad \underline{\text{Ans.}}$$

44-12

In terms of $x = r/r_B$ the probability density is given by

$$8r_BP = x^2(2 - x)^2e^{-x} = (4x^2 - 4x^3 + x^4)e^{-x}.$$

The find the maxima, set $dP/dx = 0$ to obtain

$$8r_B dP/dx = x(8 - 16x + 8x^2 - x^3)e^{-x} = 0,$$

$$x(2 - x)(x^2 - 6x + 4)e^{-x} = 0.$$

There are five solutions. Three of these yield $P = 0$; they are $x = 0, 2, \infty$. The maxima are located by the other two, which are the roots of

$$x^2 - 6x + 4 = 0,$$

$$x = r/r_B = 3 \pm 5^{\frac{1}{2}} = 5.236;\ 0.764 \quad \underline{\text{Ans}}.$$

44-15

The quantity to be calculated is

$$\psi^2_{2p} = \frac{1}{3}(\psi^2_{211} + \psi^2_{21-1} + \psi^2_{210}) = \frac{1}{3}(2\psi^2_{211} + \psi^2_{210}),$$

$$3\psi^2_{2p} = 2[(r^2/64\pi r_B^5)e^{-r/r_B}\cdot\sin^2\theta] + [(r^2/32\pi r_B^5)e^{-r/r_B}\cdot\cos^2\theta],$$

$$3\psi^2_{2p} = (r^2/32\pi r_B^5)e^{-r/r_B}(\sin^2\theta + \cos^2\theta),$$

$$\psi^2_{2p} = (r^2/96\pi r_B^5)e^{-r/r_B},$$

a function of r only.

44-18

In the Bohr theory, the electron carrying charge $-e$ revolves in a circular orbit of radius r. If T is the period of revolution, the equivalent current i due to the circulating electron is

$$i = \frac{e}{T},$$

and the magnetic moment μ is, therefore,

$$\mu = iA = \left(\frac{e}{T}\right)(\pi r^2) = \frac{e\pi r^2}{T}.$$

This may be rewritten as

$$\mu = \tfrac{1}{2}e(2\pi r/T)r = \tfrac{1}{2}erv = \tfrac{1}{2}e\left(\frac{L}{m}\right),$$

where L is the orbital angular momentum, L = mrv. But L is a quantized quantity, with $L = n\hbar$; therefore,

$$\mu = \tfrac{1}{2}e\left(\frac{n\hbar}{m}\right) = n\left(\frac{e\hbar}{2m}\right) = n\mu_B.$$

<u>44-21</u>

The magnitude L of the orbital angular momentum is given from

$$L^2 = \ell(\ell + 1)\hbar^2.$$

Now $L^2 = L_x^2 + L_y^2 + L_z^2$ and since $L_z = m_\ell \hbar$ it follows that

$$L_x^2 + L_y^2 + m_\ell^2\hbar^2 = \ell(\ell + 1)\hbar^2,$$

$$(L_x^2 + L_y^2)^{\frac{1}{2}} = [\ell(\ell + 1) - m_\ell^2]^{\frac{1}{2}}\hbar \le [\ell(\ell + 1)]^{\frac{1}{2}}\hbar,$$

since $m_\ell^2 \ge 0$. The greatest possible value of $m_\ell = \ell$ itself, so that

$$\ell^{\frac{1}{2}}\hbar \le (L_x^2 + L_y^2)^{\frac{1}{2}}.$$

<u>44-29</u>

(a) The equivalent current i due to the circulating proton is

$$i = e/T = ev/2\pi r = erv/2\pi r^2 = eL/2\pi r^2 m = en\hbar/2\pi r^2 m.$$

Hence,

$$B = \frac{\mu_0 i}{2r} = \mu_0 enh/8\pi^2 r^3 m.$$

But

$$\frac{1}{4\pi\epsilon_0}\frac{e^2}{r^2} = \frac{mv^2}{r} = \frac{m(rv)^2}{r^3} = \frac{m(L/m)^2}{r^3} = \frac{m(n\hbar/m)^2}{r^3} = \frac{n^2\hbar^2}{mr^3},$$

so that

$$\frac{1}{r} = \frac{1}{4\pi\epsilon_0}\frac{me^2}{n^2\hbar^2} = \pi me^2/\epsilon_0 n^2h^2.$$

Substituting the cube of this into the equation for B gives

$$B = \left(\frac{\mu_0 enh}{8\pi^2 m}\right)\left(\frac{\pi me^2}{\epsilon_0 n^2h^2}\right)^3 = \frac{\mu_0 e^7\pi m^2}{8\epsilon_0^3 n^5h^5}.$$

(b) Direct numerical substitution yields B = 0.39 T for n = 2.

44-32

The energy E associated with the magnetic moment μ of the electron in a field B is

$$-E = \vec{\mu}\cdot\vec{B} = \mu B\cos\theta = (\mu\cos\theta)B = \mu_z B.$$

For parallel alignment $\theta = 0$, $\cos\theta = 1$, and for antiparallel alignment $\theta = 180°$ and $\cos\theta = -1$. Hence, the sign in the above equation can be plus or minus and therefore the difference in energy between these two arrangements is

$$\Delta E = 2\mu_z B = 2(2m_s\mu_B)B = 2\mu_B B$$

since $m_s = \frac{1}{2}$. Numerically this is

$$\Delta E = (2)(5.79 \times 10^{-5}\ \text{eV/T})(0.39\ \text{T}) = 4.52 \times 10^{-5}\ \text{eV}\quad \underline{\text{Ans.}}$$

44-35

(a) A wavelength difference $\Delta\lambda$ corresponds to a frequency interval $\Delta\nu$ given by, since $c = \nu\lambda$,

$$\Delta\nu = -\frac{\nu}{\lambda}(\Delta\lambda) = -\frac{c}{\lambda^2}(\Delta\lambda),$$

$$\Delta\nu = \frac{3 \times 10^8}{(7.672 \times 10^{-7})^2}(5.4 \times 10^{-9}) = 2.7523 \times 10^{12} \text{ Hz.}$$

The energy difference ΔE between the doublet states must be

$$\Delta E = h\Delta\nu = 1.8237 \times 10^{-21} \text{ J} = 1.14 \times 10^{-2} \text{ eV} \quad \underline{\text{Ans.}}$$

(b) Using the average wavelength, 767.2 nm, the energy is

$$E = h\nu = \frac{hc}{\lambda} = 1.62 \text{ eV} \quad \underline{\text{Ans.}}$$

44-37

The internal magnetic field B experienced by the electron is due to its orbital motion about the nucleus, for the electron "sees" the nucleus revolving about it, much as we on earth "see" the sun revolving about us. The resulting magnetic field is proportional to the electron's orbital angular momentum $\vec{L}$. The spin magnetic moment $\vec{\mu}$ is proportional to the electron spin $\vec{S}$ but opposite in direction, for the electron charge is negative. It seems likely, then, that the spin-orbit energy $E = -\vec{\mu}\cdot\vec{B}$ is proportional to $-(-\vec{S})\cdot\vec{L} = \vec{S}\cdot\vec{L}$. Then, if $\vec{S}$ is "parallel" to $\vec{L}$ ($j = \ell + \frac{1}{2} = 3/2$ for a p-state), $E > 0$ and the level is raised; on the other hand, if $\vec{S}$ and $\vec{L}$ are "antiparallel" ($j = \ell - \frac{1}{2} = \frac{1}{2}$ for a p-state), $E < 0$ and the level is lowered, i.e., the energy is more negative than it was with the spin-orbit energy ignored.

44-38

(a) If $\ell = 0$, then $j = \ell + s = s$ and $g = 2$.

(b) With $s = 0$ (spinless particle), $j = \ell + s = \ell$ and $g = 1$.

44-39

(a) A d-state has $\ell = 2$. With $s = \frac{1}{2}$ for the electron, $j = \ell \pm s = \frac{5}{2}, \frac{3}{2}$.

(b) To determine g, use the formula in Problem 44-38 with $s = \frac{1}{2}$, $\ell = 2$:

$$g = \frac{3}{2} - \frac{21/8}{j(j+1)}.$$

For $j = \frac{5}{2}, \frac{3}{2}$ this gives $g = \frac{6}{5}, \frac{4}{5}$.

(c) Since $J_z = m_j\hbar$ and $m_{j,max} = j$, it follows that the maximum values of J_z for the two states being considered are $\frac{5}{2}\hbar$ and $\frac{3}{2}\hbar$.

(d) In magnitude the total magnetic moment is $\mu_z = gm_j \mu_B$ and for the maximum value use the maximum value of m_j from (c). Hence, for $j = \frac{3}{2}$, $\mu_{j,max} = (\frac{6}{5})(\frac{5}{2})\mu_B = 3\mu_B$ and with $j = 3/2$, $\mu_{j,max} = (\frac{4}{5})(\frac{3}{2})\mu_B = \frac{6}{5}\mu_B$.

(e) In an external magnetic field, a state is generated for each value of m_j, and the number of values is $2j + 1$. Thus, with $j = \frac{5}{2}$ there are 6 states and for $j = \frac{3}{2}$ there are 4 states.

CHAPTER 45

45-4

If an isolated moving electron cannot change spontaneously into a photon, it is probably because the reaction would violate one or both of the conservation laws, of mass-energy and momentum. To see if this is so, assume the reaction to take place. In the laboratory frame the electron is moving with speed v; to conserve momentum the created photon must move off in the same direction in which the electron was moving. Writing relativistic equations, the laws of conservation of mass-energy and momentum imply that

$$E = h\nu,$$

$$p = \frac{(E^2 - E_0^2)^{\frac{1}{2}}}{c} = \frac{h}{\lambda} = \frac{h\nu}{c};$$

here E is the total energy of the electron, E_0 is its rest energy. Eliminating $h\nu$, the photon energy, gives

$$(E^2 - E_0^2)^{\frac{1}{2}} = E,$$

$$E^2 - E_0^2 = E^2,$$

$$-E_0^2 = 0 \ ?$$

clearly impossible. Hence, a third body must be present to absorb some of the momentum.(If it is "heavy", that is, with a rest mass large compared to E, it probably will not carry off much energy.) In this way the conservation laws can be satisfied and the electron to photon conversion take place.

45-8

Moseley's law with b = 1 is

$$\nu^{\frac{1}{2}} = a(Z - 1),$$

valid for K_α transitions. Writing this equation for cobalt and an unknown impurity and then dividing the two equations gives

$$\frac{\nu_c^{\frac{1}{2}}}{\nu_i^{\frac{1}{2}}} = \frac{Z_c - 1}{Z_i - 1},$$

c for cobalt and i the impurity; the unknown constant "a" in each has cancelled. With $\nu = c/\lambda$ the equation can be put in terms of the wavelengths of the K_α lines:

$$\frac{\lambda_i^{\frac{1}{2}}}{\lambda_c^{\frac{1}{2}}} = \frac{Z_c - 1}{Z_i - 1}.$$

For cobalt Z = 27, λ_c = 178.5 pm; taking λ_i = 228.5 pm for the first impurity and substituting gives

$$\left(\frac{228.5}{178.5}\right)^{\frac{1}{2}} = \frac{27 - 1}{Z_i - 1},$$

$$Z_i = 23.98 \approx 24,$$

indicating that this impurity is chromium. Turning to the other,

$$\left(\frac{153.9}{178.5}\right)^{\frac{1}{2}} = \frac{27 - 1}{Z_i - 1},$$

$$Z_i = 29,$$

which indicates copper.

<u>45-9</u>

(a) The cut-off wavelength λ_{min} is characteristic of the incident electrons, not of the material upon which they fall, and is the wavelength of a photon with an energy equal to the kinetic energy of the incident electrons, in this case 35 keV; hence,

$$h\nu = hc/\lambda_{min} = 35 \times 10^3 \times 1.6 \times 10^{-19},$$

the energy expressed in joules. Substituting $h = 6.626 \times 10^{-34}$ J·s, $c = 3 \times 10^8$ m/s yields

$$\lambda_{min} = 0.355 \times 10^{-10}\ \text{m} = 35.5\ \text{pm} \quad \underline{\text{Ans}}.$$

(b) The K_β line results from an M to K level electron jump (K to M hole transition). The energy of the emitted photon is the difference in the energies of the atom with the hole in the M and K levels. Thus,

$$hc/\lambda_\beta = \Delta E = (25.51 - 0.53)\ \text{keV},$$

$$\lambda_\beta = 49.7\ \text{pm} \quad \underline{\text{Ans}}.$$

(c) The K_α line corresponds to an L to K electron transition. The energy change is

$$\Delta E = (25.51 - 3.56) = 21.95\ \text{keV} = hc/\lambda_\alpha,$$

$$\lambda_\alpha = 56.6\ \text{pm} \quad \underline{\text{Ans}}.$$

<u>45-14</u>

(a) The n = 1 shell is filled for it can take only two electrons. The third electron must enter the n = 2 shell. With n = 2, the allowed values of ℓ are $\ell = 0, 1$ of which the $\ell = 0$ level is lower in energy. For $\ell = 0$ the only possible value of $m_\ell = 0$. The electron spin can be either "up" or "down". Thus, the quantum numbers of the third electron are $n, \ell, m_\ell, m_s, = 2, 0, 0, \pm\frac{1}{2}$.

(b) In the first excited state the electron (i.e., the one that was the "third" in (a) above) will be in the n = 2, $\ell = 1$ level, for the n = 3, $\ell = 0$ level lies higher than this. For $\ell = 1$, the values of m_ℓ = -1, 0, +1. However, if there is no external field (the hyperfine splitting is ignored here) these levels are at the same energy. Thus, the quantum numbers of the excited electron are n = 2, ℓ = 1 with m_ℓ and m_s any permitted value.

45-19

(a) There must be a node at each end if the mirrors are perfectly reflecting (in actuality, there will be a node very close to the partially-silvered "exit" end); if L is the length of the crystal,

$$N(\tfrac{1}{2}\lambda_n) = L,$$

$$N = \frac{2L}{\lambda/n} = \frac{(2)(1.75)(0.06\ m)}{(694 \times 10^{-9}\ m)} = 3.026 \times 10^5 \quad \underline{\text{Ans.}}$$

(b) In terms of the frequency,

$$N = (\frac{2Ln}{c})\nu,$$

so that

$$\Delta N = (\frac{2Ln}{c})\Delta\nu = (\frac{2Ln}{c/\nu})\cdot\frac{\Delta\nu}{\nu} = N\cdot\frac{\Delta\nu}{\nu}.$$

The frequency $\nu = (3 \times 10^8\ m/s)/(694 \times 10^{-9}\ m) = 4.323 \times 10^{14}$ Hz, and therefore, with $\Delta N = 1$,

$$\Delta\nu = \frac{\nu}{N}\cdot\Delta N = \frac{4.323 \times 10^{14}}{3.026 \times 10^5}(1) = 1.43 \times 10^9\ \text{Hz} \quad \underline{\text{Ans.}}$$

(c) From (b), with N = 1,

$$\frac{\Delta\nu}{\nu} = \frac{\Delta N}{N} = \frac{1}{N} = 3.30 \times 10^{-6} \quad \underline{\text{Ans.}}$$

45-20

The population ratio is $N_m/N_n = e^{-(E_m - E_n)/kT}$.

(a) For $E_m - E_n = 1$ eV and $T = 300$ K, $(E_m - E_n)/kT = 38.625$ giving

$$N_m/N_n = e^{-38.625} = 10^{-16.8} \quad \underline{\text{Ans}}.$$

(b) If 0.01% are in the state with energy E_m, then 99.99% must be in the other state; hence $N_m/N_n = 1/9999 \approx 10^{-4}$. Numerically, $\Delta E/k = 11587$ K. Thus,

$$10^{-4} = e^{-11587/T},$$

$$-4\cdot \ln 10 = -11587/T,$$

$$T = \frac{11587}{4\ln 10} = 1260 \text{ K} \quad \underline{\text{Ans}}.$$

CHAPTER 46

46-7

Eq.46-1 for the energy density of states is

$$n_s(E) = \left(\frac{2^{7/2}\pi m^{3/2}}{h^3}\right)E^{\frac{1}{2}} = CE^{\frac{1}{2}},$$

with

$$C = \frac{2^{7/2}\pi m^{3/2}}{h^3} = 2^{7/2}\pi \cdot \frac{(9.11 \times 10^{-31}\ kg)^{3/2}}{(6.626 \times 10^{-34}\ J\cdot s)^3},$$

$$C = 10.6238 \times 10^{55}\ kg^{3/2}/J^3\cdot s^3.$$

Examining the units,

$$kg^{3/2}\cdot J^{-3}\cdot s^{-3} = \frac{(kg/J)^{3/2}}{J^{3/2}\cdot s^3} = \frac{(s^2/m^2)^{3/2}}{J^{3/2}s^3} = m^{-3}\cdot J^{-3/2},$$

$$kg^{3/2}\cdot J^{-3}\cdot s^{-3} = m^{-3}(1.6 \times 10^{-19})^{3/2}eV^{-3/2},$$

so that

$$C = 6.80 \times 10^{27}\ eV^{-3/2}\cdot m^{-3}.$$

If E = 5 eV,

$$n_s(E) = (6.80 \times 10^{27})(5)^{\frac{1}{2}} = 1.52 \times 10^{28}\ eV^{-1}\cdot m^{-3}$$ Ans.

46-13

At T = 0 K,

$$p(E) = \begin{matrix} 1, & 0 \le E \le E_F, \\ 0, & E_F < E. \end{matrix}$$

Hence, the total number N of particles is

$$N = \int_0^{\infty} n_s(E)p(E)dE = \int_0^{E_F} CE^{\frac{1}{2}}dE = \frac{2}{3}CE_F^{3/2},$$

$$C = \frac{3}{2}NE_F^{-3/2}.$$

The average energy per conduction electron is, then,

$$\overline{E} = \frac{1}{N}\int_0^{\infty} En_s(E)p(E)dE = \frac{1}{N}\int_0^{E_F}(E)(CE^{\frac{1}{2}})dE = \frac{C}{N}\cdot\frac{2}{5}E_F^{5/2}:$$

Substituting the expression for C found above gives

$$\overline{E} = (\frac{3}{2}E_F^{-3/2})(\frac{2}{5}E_F^{5/2}) = \frac{3}{5}\cdot E_F.$$

<u>46-15</u>

In Problem 46-13 it is shown that if

$$n_s(E) = CE^{\frac{1}{2}},$$

then

$$C = \frac{3}{2}(n_0)E_F^{-3/2},$$

n_0 the total number of particles (called N in Problem 46-13). Thus,

$$n_s = \frac{3}{2}\cdot n_0 E_F^{-3/2}E^{\frac{1}{2}}.$$

The state distribution function n_s is independent of material because E_F itself depends implicitly on the electron density n_0 (different for different substances), such that $n_0 E_F^{-3/2} = \frac{2}{3}C$, where

$$C = 2^{7/2}\pi m^{3/2}/h^3,$$

which does not depend on the material considered, m being the electron mass.

46-17

The number # of excited electrons is

$$\# = \int_{E_F}^{\infty} n_s(E)p(E)dE.$$

Now, at temperatures above absolute zero p(E) differs from its T = 0 value only within a range of energies of about kT, centered at the Fermi energy E_F; i.e., p(E) drops from 1 to 0 within an energy range kT about E_F. Hence,

$$\# = n_s(E_F)\int_{E_F-kT}^{E_F+kT} (\tfrac{1}{2})dE = (kT)n_s(E_F),$$

approximately, using as p(E) about E_F the average of its value (1) for energies much less than E_F and its value (0) at energies significantly greater than E_F. Thus, the fraction of excited electrons is

$$\#/n_0 = \frac{(kT)n_s(E_F)}{n_0}.$$

46-19

(a) By the kinetic theory of gases,

$$v_{rms} = (\frac{3kT}{m})^{\frac{1}{2}}.$$

The electron mass is $m = 9.11 \times 10^{-31}$ kg; assuming room temperature $T \approx 300$ K,

$$v_{rms} = 1.17 \times 10^5 \text{ m/s} \quad \underline{\text{Ans}}.$$

(b) The Fermi speed is defined by $E_F = \frac{1}{2}mv_F^2$. For copper the Fermi energy $E_F = 7.05$ eV $= 1.128 \times 10^{-18}$ J and therefore

$$v_F = \left[\frac{(2)(1.128 \times 10^{-18})}{(9.11 \times 10^{-31})}\right]^{\frac{1}{2}} = 1.57 \times 10^6 \text{ m/s} \quad \underline{\text{Ans}}.$$

(c) The ratio is

$$v_F/v_{rms} = 1.57/0.117 = 13.4 \quad \underline{\text{Ans}}.$$

46-22

For $E \gg E_F$, $e^{(E - E_F)/kT} \approx e^{E/kT} \gg 1$ so that

$$n_p(E) = C\cdot\frac{E^{\frac{1}{2}}}{e^{(E - E_F)/kT} + 1} \approx C\cdot\frac{E^{\frac{1}{2}}}{e^{E/kT}} = CE^{\frac{1}{2}}e^{-E/kT}.$$

In energy dependence this is the same as for the classical Boltzmann distribution (see Problems 46-20, 21); that is, at high enough temperatures quantum effects disappear.

CHAPTER 47

47-2

(a) The nuclear force is extremely short-ranged, effective over distances considerably smaller than the diameters of nuclei. This means that any one nucleon can form nuclear bonds only with its very nearest neighbours and therefore the number N of bonds formed by any one nucleon is independent of the number A of nucleons in the nucleus, depending solely on the nuclear density, which is roughly the same for all nuclides. Hence, the total number of nuclear bonds in a given nucleus is equal to AN, that is, directly proportional to A; it follows also that the nuclear energy, itself proportional to the total number of bonds, likewise is proportional to A.

(b) In contrast to the nuclear force, the coulomb interaction is long-ranged, each proton interacting electrically with all of the charged nucleons (i.e., protons) in the nucleus. Thus, the number of coulomb bonds present in a given nucleus is equal to the number of distinct pairs of protons that can be counted; for Z protons, this is $\frac{1}{2}Z(Z - 1)$, so that the coulomb energy is proportional to $Z(Z - 1)$.

(c) Although, as larger nuclei are examined Z does not increase as fast as A (due to the neutron excess), Z^2 increases much faster than A, so that the coulomb energy becomes more and more important.

47-6

(a) As the kinetic energy $K = 200$ MeV is much greater than the rest energy $E_0 = 0.511$ MeV, the extreme relativistic approximation can be used, as follows:

$$E^2 = p^2c^2 + E_0^2,$$

$$(E_0 + K)^2 = p^2c^2 + E_0^2,$$

$$2E_0K + K^2 = p^2c^2,$$

$$K^2 \approx p^2c^2,$$

$$p = \frac{K}{c} = \frac{200\text{ MeV}}{c},$$

since $E_0K << K^2$. The deBroglie wavelength is

$$\lambda = \frac{h}{p} = \frac{4.136 \times 10^{-21}\ \text{MeV}\cdot\text{s}}{200\ \text{MeV}/(3 \times 10^8\ \text{m/s})} = 6.2 \times 10^{-15}\ \text{m},$$

$$\lambda = 6.2\ \text{fm}\quad \underline{\text{Ans.}}$$

(b) By comparison, the diameter of a copper nucleus is about 8.6 fm, only a little larger than the deBroglie wavelength of the electron in (a). For sensitive probing, the wavelength/diameter ratio should be as small as possible. Thus, 200 MeV seems to be about the minimum energy for electrons to be useful tools in disciminating the details of nuclear structure in medium-sized nuclei.

47-8

The uncertainty in momentum will be $\Delta p \approx h/R$. Adopting the "hint", set $\Delta p \approx p$ to obtain

$$p \approx \frac{h}{R} = \frac{h}{R_0A^{1/3}}.$$

For $A = 100$, $R_0 = 1.1 \times 10^{-15}$ m; this gives $p = 1.3 \times 10^{-19}$ kg·m/s or,

$$p = \left(\frac{1.30 \times 10^{-19}\ \text{kg}\cdot\text{m/s}}{c}\right)\cdot\frac{(3 \times 10^8\ \text{m/s})}{(1.6 \times 10^{-13}\ \text{J/MeV})},$$

$$p = 244\ \text{MeV}/c.$$

The rest energy E_0 of a neutron or proton is about 940 MeV. Hence,

$$E^2 = p^2c^2 + E_0^2 = (244)^2 + (940)^2,$$

$$E = 971\ \text{MeV},$$

so that the kinetic energy $K = E - E_0 \approx 31$ MeV. This is on the same order as the binding energy per nucleon, indicating that protons and neutrons can remain bound in the nucleus. (The opposite conclusion would be reached for electrons, with $E_0 = 0.511$ MeV.)

47-16

A nucleus contains Z protons and N neutrons. The binding energy of the nucleus is

$$E = (Zm_p + Nm_n - m)c^2,$$

where m_p, m_n, m are the atomic masses of the proton, neutron and the original nucleus. In terms of the mass excesses,

$$\Delta_p = (m_p - 1)c^2, \quad m_p c^2 = \Delta_p + c^2;$$

$$\Delta_n = (m_n - 1)c^2, \quad m_n c^2 = \Delta_n + c^2;$$

$$\Delta = (m - A)c^2, \quad mc^2 = \Delta + Ac^2.$$

Hence,

$$E = (Z\Delta_p + N\Delta_n - \Delta) + (Z + N - A)c^2,$$

$$E = Z\Delta_p + N\Delta_n - \Delta,$$

since A = N + Z. For $^{197}_{79}Au$, Z = 79, N = 197 - 79 = 118. Therefore

$$E = (79)(7.29) + (118)(8.07) - (-31.2) = 1559.37 \text{ MeV}.$$

The average binding energy per nucleon for this gold nucleus is

$$\bar{E}_B = \frac{E}{A} = \frac{1559.37 \text{ MeV}}{197} = 7.92 \text{ MeV} \quad \underline{\text{Ans}}.$$

This compares with 7.91 MeV listed in the table.

47-18

(a) Since the nucleus is "free" (no orbital electrons and therefore no electronic angular momentum), the total angular momentum equals the spin. The number N of ground-state levels present due to space quantization is

$$N = 2I + 1 = 4 \quad \underline{\text{Ans}}.$$

(b) The energy E of any level in a magnetic field B is

$$E = -\vec{\mu}\cdot\vec{B} = -\mu B\cos\theta = -(\mu\cos\theta)B = -\mu_z B,$$

assuming $\vec{B}$ to lie along the z-axis. Suppose $\vec{\mu}$ is parallel to the nuclear spin. The maximum projected value of the latter is $I\hbar = \frac{3}{2}\hbar$ and in that configuration the angle between the nuclear spin vector, magnitude $[I(I+1)]^{\frac{1}{2}}\hbar$, and the magnetic field is given by

$$\cos\theta = \frac{(3/2)}{[\frac{3}{2}(\frac{3}{2}+1)]^{\frac{1}{2}}} = 3(15)^{-\frac{1}{2}}.$$

Under the assumption above, the maximum projected value of $\vec{\mu}$ is $\mu\cos\theta = (3)(2.27)(15)^{-\frac{1}{2}}\mu_N$. This corresponds to an energy

$$E = \mu_z B = (3)(2.27)(15)^{-\frac{1}{2}}\mu_N B = 11.04 \times 10^{-8} \text{ eV},$$

in, say, the parallel configuration. In the antiparallel situation, $E = -11.04 \times 10^{-8}$ eV. The energy difference between these two states is 22.08×10^{-8} eV. Since there are 4 levels in all,

$$\Delta E = \frac{1}{3}(22.08 \times 10^{-8}) = 7.36 \times 10^{-8} \text{ eV} \quad \underline{\text{Ans}}.$$

(c) The wavelength is

$$\lambda = \frac{hc}{\Delta E} = 17 \text{ m} \quad \underline{\text{Ans}}.$$

(d) A photon with this wavelength lies in the radio region of the spectrum, between the TV-FM band and the standard broadcast band.

<u>47-22</u>

Let N_0 be the number of ^{64}Cu atoms present initially. The number N present after time t is $N = N_0e^{-\lambda t}$. Hence, the fraction that decay between times t' and t" is

$$\frac{\Delta N}{N_0} = e^{-\lambda t''} - e^{-\lambda t'}.$$

Now,

$$\lambda = \frac{\ln 2}{t_{\frac{1}{2}}} = 0.05458\ h^{-1}.$$

Hence $\lambda t'' = 0.8733$ and $\lambda t' = 0.7641$, and the fraction becomes

$$\frac{\Delta N}{N_0} = -0.0482 \quad \underline{\text{Ans}},$$

the negative sign indicating that the number of radioactive atoms decreases during the period.

47-25

In 1.00 g of elemental samarium there are 0.151 g of ^{147}Sm. The number of atoms in this 0.151 g is

$$N = \frac{0.151}{147}(6.023 \times 10^{23}) = 6.187 \times 10^{20}.$$

The decay rate is

$$\frac{dN}{dt} = \lambda N,$$

$$120\ s^{-1} = \lambda(6.187 \times 10^{20}),$$

$$\lambda = 1.940 \times 10^{-19}\ s^{-1}.$$

This means a half-life of

$$t_{\frac{1}{2}} = \frac{\ln 2}{\lambda} = 3.573 \times 10^{18}\ s = 1.13 \times 10^{11}\ y \quad \underline{\text{Ans}}.$$

47-26

Let the initial number of ^{32}P atoms (half-life $t_2 = 14.3$ d) be N_{02} and the initial number of ^{33}P atoms (half-life $t_3 = 25.3$ d) be N_{03}. Initially,

$$dN_2/dt = 9(dN_3/dt),$$

$$\lambda_2 N_{02} = 9\lambda_3 N_{03}.$$

Let T be the time that must elapse until the conditions are reversed, at which time

$$9(\lambda_2 N_2) = \lambda_3 N_3.$$

But $N_2 = N_{02}e^{-\lambda_2 T}$, $N_3 = N_{03}e^{-\lambda_3 T}$; the above then yields

$$9\lambda_2 N_{02}e^{-\lambda_2 T} = \lambda_3 N_{03}e^{-\lambda_3 T}.$$

Divide this equation by the second equation to obtain

$$9e^{-\lambda_2 T} = \frac{1}{9}\cdot e^{-\lambda_3 T},$$

$$\ln(81) - \lambda_2 T = -\lambda_3 T,$$

$$T = \frac{\ln(81)}{\lambda_2 - \lambda_3} = \frac{(\ln 81)}{(\ln 2)}\cdot\frac{t_2 t_3}{t_3 - t_2} = 209\ \text{d}\quad \underline{\text{Ans}}.$$

<u>47-29</u>

Let N be the number of atoms of the radionuclide present at time t. Then,

$$\frac{dN}{dt} = +R - \lambda N,$$

the first term on the right due to production, the second to radioactive decay. To integrate, write

$$\frac{dN}{R - \lambda N} = dt,$$

$$-\frac{1}{\lambda}\cdot\ln(R - \lambda N) + C = t,$$

where C is the constant of integration. Suppose that N = 0 at t = 0; then,

$$C = +\frac{1}{\lambda}\cdot\ln R.$$

Substituting this into the previous equation gives

$$-\frac{1}{\lambda}\cdot\ln(R - \lambda N) + \frac{1}{\lambda}\cdot\ln R = t,$$

$$\ln\left(\frac{R - \lambda N}{R}\right) = -\lambda t,$$

$$N = \frac{R}{\lambda}(1 - e^{-\lambda t}).$$

After a time $t \gg 1/\lambda$, $e^{-\lambda t} \ll 1$ and

$$N \approx R/\lambda.$$

47-30

(a) Refer to Problem 47-29, where N = number of ^{56}Mn atoms present a time t after the bombardment starts. Since $t_{\frac{1}{2}} = 2.58$ h = 9288 s, $\lambda = \ln 2/t_{\frac{1}{2}} = 7.4628 \times 10^{-5}\ s^{-1}$. The activity is λN so that, with the conditions for secular equilibrium established,

$$R = \lambda N = (2.4)(3.7 \times 10^{10}),$$

$$R = 8.88 \times 10^{10}\ s^{-1} \quad \underline{\text{Ans}}.$$

(b) During the bombardment the decay rate is $\lambda N = R(1 - e^{-\lambda t})$; numerically,

$$\text{decay rate} = (8.88 \times 10^{10}\ s^{-1})[1 - e^{-(7.4628 \times 10^{-5})t}],$$

with t in seconds.

(c) At the end of the bombardment,

$$\lambda N = 2.4\ \text{Ci},$$

$$(7.4628 \times 10^{-5})N = (2.4)(3.7 \times 10^{10}),$$

$$N = 1.190 \times 10^{15} \quad \underline{\text{Ans}}.$$

(d) The mass of these ^{56}Mn atoms is

$$m = \frac{1.190 \times 10^{15}}{6.023 \times 10^{23}}(56) = 0.11\ \mu g \quad \underline{\text{Ans}}.$$

47-32

The Q for the reaction is the sum of the kinetic energies of the α-particle and the ^{234}Th nucleus (the ^{238}U nucleus is presumed to have been at rest). By conservation of momentum, their momenta must be equal and oppositely directed. The rest energy of an α-particle is about (4 u)(931.481 MeV/u) = 3726 MeV which is much greater than the α-particle's kinetic energy. Using, then, classical expressions

$$p_{Th} = p_\alpha = (2mK_\alpha)^{\frac{1}{2}} = (2MK_{Th})^{\frac{1}{2}},$$

$$K_{Th} = \frac{m}{M}\cdot K_\alpha = (\frac{4}{234})(4.196) = 0.0717 \text{ MeV}.$$

Hence,

$$Q = 4.197 + 0.072 = 4.269 \text{ MeV} \quad \underline{\text{Ans}}.$$

47-36

Substituting the given numbers,

$$(16)(\frac{E}{U_0})(1 - \frac{E}{U_0}) = (16)(\frac{4.00}{15.0})(\frac{11}{15}) = (16)\frac{44}{225} = 3.129.$$

Turning to the exponential term

$$2U_0(1 - \frac{E}{U_0}) = (2)(15)(\frac{11}{15}) = 22 \text{ MeV}.$$

Now,

$$m = 6.7 \times 10^{-27} \text{ kg} = 6.7 \times 10^{-27} \text{ (kg)}c^2/c^2 = \frac{60.3 \times 10^{-11} \text{ J}}{c^2},$$

$$m = 3769 \text{ MeV}/c^2 .$$

Since $h = 4.136 \times 10^{-21}$ MeV·s, it follows that

$$k = \frac{437.445}{c}\ s^{-1} = 1.458 \times 10^{15}\ m^{-1}.$$

Hence, with $\ell = 20$ fm $= 2 \times 10^{-14}$ m, $2k\ell = 58.32$ and

$$T = 3.13e^{-58.32} = 3.13 \times 10^{-25.3} \quad \underline{\text{Ans.}}$$

47-39

Since the electron is emitted with maximum kinetic energy K_{max} = 1.71 MeV, assume that no neutrino is emitted. Conservation of momentum requires that, with the ^{32}P nucleus originally at rest, the ^{32}S nucleus and the electron move away along the same straight line in opposite directions and with momenta of equal magnitudes. For the electron

$$p_e = \frac{1}{c}(E^2 - E_0^2)^{\frac{1}{2}} = \frac{1}{c}(K^2 + 2E_0K)^{\frac{1}{2}} = \frac{2.1614\ \text{MeV}}{c},$$

since $E_0 = 0.511$ MeV. Using nonrelativistic expressions for the ^{32}S nucleus,

$$p_n = (2MK)^{\frac{1}{2}} = (2AuK)^{\frac{1}{2}} = \frac{(2Auc^2K)^{\frac{1}{2}}}{c},$$

$$K = \frac{c^2p_n^2}{2A(uc^2)} = \frac{(cp_e)^2}{2A(uc^2)} = \frac{(2.1614)^2}{(2)(32)(931.481)} = 7.84 \times 10^{-5}\ \text{MeV},$$

$$K = 78.4\ \text{eV} \quad \underline{\text{Ans,}}$$

using $u = 931.481$ MeV/c^2 as the energy equivalent of the atomic mass unit u.

47-44

The free-neutron decay scheme is

$$n = p + e^- + \nu.$$

For the maximum β-energy, assume that no neutrino is emitted. Then,

$$Q_{max} = (m_n - m_p - m_e)c^2 = m_n c^2 - (m_p + m_e)c^2 = (m_n - m_H)c^2.$$

Using $m_n - m_H = 840 \times 10^{-6}$ u, and 1 u = 931.481 MeV/c^2,

$$Q_{max} = (931.481)(840 \times 10^{-6}) = 0.782 \text{ MeV} \quad \underline{\text{Ans}}.$$

47-49

(a) In any reaction the charge and mass number must be conserved; hence, for the first reaction,

$$^{19}_{9}F + ^{1}_{1}p = ^{4}_{2}\alpha + ^{A}_{Z}Y_1 + Q_1,$$

$$19 + 1 = 4 + A,$$

$$A = 16;$$

$$9 + 1 = 2 + Z,$$

$$Z = 8.$$

Thus, $^{A}_{Z}Y_1 = ^{16}_{8}O$. Similarly,

$$^{16}_{8}O + ^{2}_{1}d = ^{1}_{1}p + ^{A}_{Z}Y_2 + Q_2,$$

giving A = 17, Z = 8 so that $^{A}_{Z}Y_2 = ^{17}_{8}O$.

(b) Clearly $Q_3 = -Q_1 - Q_3$. Treating the reaction equations above as actual algebraic equations,

$$-Q_1 = ^{4}_{2}\alpha + ^{16}_{8}O - ^{19}_{9}F - ^{1}_{1}p,$$

$$-Q_2 = -\,^{2}_{1}d - ^{16}_{8}O + ^{17}_{8}O + ^{1}_{1}p.$$

Adding these equations yields

$$Q_3 = ^{4}_{2}\alpha - ^{2}_{1}d + ^{17}_{8}O - ^{19}_{9}F,$$

$$^{17}_{8}O + ^{4}_{2}\alpha = ^{19}_{9}F + ^{2}_{1}d + Q_3,$$

or $^{17}O(\alpha, d)^{19}F$ as the reaction sought.

47-50

(a) Assuming that classical mechanics holds, the velocity V of the center of mass is found from

$$(m_X + m_a)V = m_X(0) + m_a v_a,$$

$$V = \frac{m_a}{m_X + m_a} \cdot v_a.$$

In the reaction, momentum and not velocity is conserved. Only if no mass is lost (or gained), i.e., Q = 0, will the velocity of the center of mass be unchanged.

(b) The velocities of X and a before reaction are -V and v_a - V in the center of mass system. Hence,

$$K_{cm} = \tfrac{1}{2}m_X(-V)^2 + \tfrac{1}{2}m_a(v_a - V)^2 = \tfrac{1}{2}(m_X + m_a)V^2 - m_a v_a V + \tfrac{1}{2}m_a v_a^2;$$

invoking (a),

$$K_{cm} = \tfrac{1}{2} \cdot \frac{m_a^2 v_a^2}{(m_X + m_a)} - \frac{m_a^2 v_a^2}{(m_X + m_a)} + \tfrac{1}{2}m_a v_a^2 = \tfrac{1}{2}m_a v_a^2 \left(1 - \frac{m_a}{m_a + m_X}\right),$$

$$K_{cm} = K_{lab} \cdot \frac{m_X}{m_a + m_X}.$$

The kinetic energy will change in the reaction unless it is elastic (Q = 0).

(c) For the deuteron,

$$K_{lab} = (2.1 \times 10^6)(1.6 \times 10^{-19}) = \tfrac{1}{2}(2)(1.66 \times 10^{-27})v_d^2,$$

$$v_d = 1.42 \times 10^7 \text{ m/s} \quad \underline{\text{Ans}}.$$

Then,

$$V = v_d \frac{2}{2 + 27} = 9.81 \times 10^5 \text{ m/s} \quad \underline{\text{Ans.}}$$

Finally,

$$K_{cm} = (2.1)\frac{27}{2 + 27} = 1.96 \text{ MeV} \quad \underline{\text{Ans.}}$$

<u>47-51</u>

(a) In the center of mass system, by definition,

$$Q = (K_b' + K_Y') - (K_a' + K_X'),$$

K' = center of mass system kinetic energy. For the threshold energy, set $K_b' = K_Y' = 0$ so that

$$-Q = K_a' + K_X' = K_{cm} = K_{lab}\left(\frac{m_X}{m_X + m_a}\right),$$

by Problem 47-50. Hence, the threshold kinetic energy is

$$K_{lab} = K_{th} = \left(\frac{m_X + m_a}{m_X}\right)(-Q).$$

Since Q is negative, $K_{th} > 0$.

(b) It is expected that $K_{th} > -Q$ because the rest-mass deficiency must be supplied by the incoming particle's kinetic energy before the reaction can proceed at all, and more than this as kinetic energy must be imparted to the products.

<u>47-60</u>

Let the nuclear temperature be T, defined as in the kinetic theory of gases by

$$\overline{\tfrac{1}{2}mv^2} = \bar{K} = \tfrac{3}{2}kT.$$

As the average kinetic energy, use the "typical" value quoted of

5 MeV = $(5 \times 10^6)(1.6 \times 10^{-19}) = 8 \times 10^{-13}$ J. As k = 1.38×10^{-23} J/K, this gives

$$T = \frac{2\bar{K}}{3k} = 3.86 \times 10^{10} \text{ K} \quad \underline{\text{Ans}}.$$

47-68

For strontium Z = 38; as this is even, the protons couple to zero angular momentum. The number of neutrons is N = 87 - 38 = 49. Now 50 is a magic number, with an angular momentum of zero. Hence, the angular momentum of ^{87}Sr plus the angular momentum of a 50th neutron sums to zero. This implies that the angular momentum of ^{87}Sr = angular momentum of the missing 50th neutron. This neutron would go into the $5g_{9/2}$ level, with j = 9/2. Therefore, I = 9/2.

CHAPTER 48

48-16

Let N be the number of free neutrons that will initiate fission that are present in the reactor at any moment. Since about 200 MeV is released in each fission, the power output P is

$$P = \frac{\text{energy}}{\text{time}} = \frac{N(200\ \text{MeV})}{t_{gen}}.$$

If at $t = 0$, say, there are N' free neutrons, then $P' = 200N'/t_{gen}$ MeV/s. After a time t_{gen} there are kN' neutrons present; a time t_{gen} after this there are $k(kN') = k^2N'$ present. Hence, at a time nt_{gen} after $t = 0$ there are

$$N = k^nN'$$

free neutrons present, so that at this time the power output P is

$$P = \frac{200N}{t_{gen}} = \frac{200(k^nN')}{t_{gen}} = k^n\left(\frac{200N'}{t_{gen}}\right) = k^nP'.$$

For $P = 2P'$, then,

$$k^n = 2,$$

$$n\cdot\ln(k) = \ln(2),$$

$$n = \frac{\ln(2)}{\ln(k)} = \frac{\ln(2)}{\ln(1.0005)} = 1387,$$

or 1387 neutron generations have elapsed. This corresponds to an actual time $t = nt_{gen} = (1387)(10^{-3}\ \text{s}) = 1.387$ s for the power output to double.

48-18

In one gram of ^{238}Pu there are N nuclei of ^{238}Pu where

$$N = \frac{1}{238}(6.023 \times 10^{23}) = 2.53 \times 10^{21},$$

6.023×10^{23} being Avogadro's number. The rate of decay is

$$\frac{dN}{dt} = \lambda N = \frac{\ln 2}{t_{\frac{1}{2}}}(N).$$

Using $t_{\frac{1}{2}} = (87.7)(3.16 \times 10^7) = 2.77 \times 10^9$ s and N from above,

$$\frac{dN}{dt} = 6.33 \times 10^{11}\ s^{-1}.$$

Each decay yields 5.5 MeV = $5.5 \times 10^6 \times 1.6 \times 10^{-19} = 8.8 \times 10^{-13}$ J so that the power generated is

$$P = (8.8 \times 10^{-13}\ J)(6.33 \times 10^{11}\ s^{-1}) = 0.557\ W \quad \underline{\text{Ans}},$$

per gram of pure ^{238}Pu.

48-30

(a) In one second the sun generates $E = 3.9 \times 10^{26}$ J of energy. The mass m destroyed in this time must be

$$m = \frac{E}{c^2} = \frac{3.9 \times 10^{26}\ J}{(3 \times 10^8\ m/s)^2} = 4.33 \times 10^9\ kg \quad \underline{\text{Ans}}.$$

(b) In 4.5×10^9 y = $(4.5 \times 10^9)(3.16 \times 10^7) = 1.422 \times 10^{17}$ s, the sun converts

$$\Delta M = (4.33 \times 10^9\ kg/s)(1.422 \times 10^{17}\ s) = 6.157 \times 10^{26}\ kg$$

of its mass to energy. The fraction f of the original mass M_0 that has been converted up to now is, therefore,

$$f = \frac{\Delta M}{M_0} \approx \frac{\Delta M}{M} = \frac{6.157 \times 10^{26}}{2 \times 10^{30}} = 3.1 \times 10^{-4} \quad \underline{\text{Ans}}.$$

48-32

(a) The heat of combustion is 3.3×10^4 J/g of atomic carbon. In one gram of carbon there are

$$\frac{1}{12}(6.023 \times 10^{23}) = 5.02 \times 10^{22}$$

atoms of carbon, 12 being the atomic mass of a carbon atom. Hence, the heat of combustion per carbon atom is

$$\frac{3.3 \times 10^4}{5.02 \times 10^{22}} = 6.57 \times 10^{-19} \text{ J/atom} \quad \underline{\text{Ans}}.$$

(b) Two oxygen atoms are required to combine with each carbon atom. The atomic mass of oxygen being 16, it appears that the mass of reactants involved in the liberation of 6.57×10^{-19} J of energy is

$$12 \text{ u} + 2(16 \text{ u}) = 44 \text{ u} = 44(1.66 \times 10^{-24} \text{ g}) = 7.30 \times 10^{-23} \text{ g}.$$

Thus, the energy released per gram of reactants is

$$\frac{6.57 \times 10^{-19} \text{ J}}{7.30 \times 10^{-23} \text{ g}} = 9000 \text{ J/g},$$

or 9×10^6 J/kg <u>Ans</u>.

(c) At the current rate of radiation of energy = 3.9×10^{26} J/s, the sun, if made of carbon and oxygen, would "burn"

$$\frac{3.9 \times 10^{26} \text{ J/s}}{9 \times 10^6 \text{ J/kg}} = 4.33 \times 10^{19} \text{ kg/s},$$

and would be converted entirely to CO_2 in a time

$$\frac{2.00 \times 10^{30} \text{ kg}}{4.33 \times 10^{19} \text{ kg/s}} = 4.62 \times 10^{10} \text{ s} = 1460 \text{ y} \quad \underline{\text{Ans}}.$$

48-38

(a) From Example 6, the particle density n in the compressed pellet is $n = 6 \times 10^{26}\ cm^{-3}$. The volume V_c of the compressed fuel pellet in terms of the volume V_0 of the pellet before compression is

$$V_c = 10^{-4}\ V_0$$

since the density (= m/V) is increased by a factor of 10^4 by the laser-induced compression. Hence, the number N of particles (of deuterium and tritium) in the compressed fuel pellet is

$$N = (10^{-4}\ V_0)n = (10^{-4})[\tfrac{4\pi}{3}(0.1\ cm)^3](6 \times 10^{26}\ cm^{-3}),$$

$$N = 2.513 \times 10^{20}.$$

Since only 10% of these particles participate in the fusion in this case and as each fusion requires two particles, one each of tritium and deuterium, then 1.26×10^{19} fusions take place within each pellet. With $Q = 17.59$ MeV $= 2.814 \times 10^{-12}$ J, the energy released in each microexplosion is

$$E = (1.26 \times 10^{19})(2.814 \times 10^{-12}) = 3.55 \times 10^7\ J \quad \underline{Ans}.$$

(b) Since 1 cal = 4.186 J, $E = 8.47 \times 10^6$ cal $= 8.47 \times 10^3$ kcal. As the heat of combustion of TNT is 500 kcal/lb, each microexplosion is "equivalent" to the detonation of

$$\frac{8470}{500} = 16.9\ lb \quad \underline{Ans},$$

of TNT.

(c) The power generated in 100 pellet microexplosions per second is

$$P = \frac{(100)(3.55 \times 10^7\ J)}{1\ s} = 3.55 \times 10^9\ W \quad \underline{Ans}.$$

SUPPLEMENTARY TOPIC

SPECIAL RELATIVITY

SR-7

(a) Assuming that the lifetime in flight as measured from the earth is $t' = 2.2 \times 10^{-6}$ s, the measured distance L' traveled would be

$$L' = vt' = (0.99)(3 \times 10^{8}\ \text{m/s})(2.2 \times 10^{-6}\ \text{s}),$$

$$L' = 6.534 \times 10^{2}\ \text{m} = 0.6534\ \text{km} \quad \underline{\text{Ans}}.$$

(b) The lifetime measured from the earth actually is

$$t' = \frac{t}{(1 - \beta^2)^{\frac{1}{2}}} = \frac{2.2 \times 10^{-6}}{(1 - 0.99^2)^{\frac{1}{2}}} = 15.595 \times 10^{-6}\ \text{s}.$$

Hence, the distance traveled in this time as seen from the earth is

$$L' = vt' = (0.99)(3 \times 10^{8}\ \text{m/s})(15.595 \times 10^{-6}\ \text{s}),$$

$$L' = 46.32 \times 10^{2}\ \text{m} = 4.632\ \text{km} \quad \underline{\text{Ans}}.$$

(c) For an observer traveling "on" the muon, the muon's lifetime is $t = 2.2 \times 10^{-6}$ s, since such an observer sees the muon at rest. But this observer sees the earth approaching at 0.99c. Hence, the distance this observer sees the muon move relative to the earth's atmosphere, for him (her) a moving frame, is

$$L' = \frac{L}{(1 - \beta^2)^{\frac{1}{2}}} = \frac{vt}{(1 - \beta^2)^{\frac{1}{2}}} = \frac{(0.99c)(2.2 \times 10^{-6}\ \text{s})}{(1 - 0.99^2)^{\frac{1}{2}}},$$

$$L' = 4.632\ \text{km} \quad \underline{\text{Ans}},$$

in agreement with (b).

SR-14

Direct application of Equation 7 gives

$$\nu = \nu'\left(\frac{c - v}{c + v}\right)^{\frac{1}{2}} = \nu'\left(\frac{1 - \beta}{1 + \beta}\right)^{\frac{1}{2}} = (10^8)\left(\frac{1 - 12/13}{1 + 12/13}\right)^{\frac{1}{2}},$$

$$\nu = (10^8)\left(\frac{1}{25}\right)^{\frac{1}{2}} = 2 \times 10^7 \text{ Hz} \quad \underline{\text{Ans}}.$$

SR-15

Classically there is no Doppler shift since the distance between source and observer is not changing. However, the relativistic transverse Doppler shift is observed. By Equation 8,

$$\nu = \nu'(1 - \beta^2)^{\frac{1}{2}},$$

$$\frac{c}{\lambda} = \frac{c}{\lambda'}(1 - \beta^2)^{\frac{1}{2}},$$

$$\lambda = \lambda'(1 - \beta^2)^{-\frac{1}{2}} = (5890)(1 - 0.1^2)^{-\frac{1}{2}},$$

$$\lambda = 5920 \text{ Å}.$$

Hence, the wavelength shift is

$$\Delta\lambda = \lambda - \lambda' = 30 \text{ Å} \quad \underline{\text{Ans}}.$$

SR-18

(a) The final speed v of the electron, accelerated from rest, is related to the accelerating potential V by

$$\tfrac{1}{2}m_0v^2 = eV.$$

With $v = c = 3 \times 10^8$ m/s, $m_0 = 9.11 \times 10^{-31}$ kg,

$$\tfrac{1}{2}m_0c^2 = 4.0995 \times 10^{-14} \text{ J} = 0.2562 \text{ MeV}.$$

Since the charge e on the electron is one quantum unit,

$$V = 0.2562 \times 10^6 \text{ volts} \quad \underline{\text{Ans}}.$$

(b) Relativistically $K = mc^2 - m_0c^2 = mc^2 - 0.511$ MeV, so that

$$K = eV,$$

$$mc^2 - 0.511 = 0.2562,$$

$$mc^2 = 0.7672 \text{ MeV}.$$

But $m = m_0(1 - \beta^2)^{-\frac{1}{2}}$ and therefore

$$(m_0c^2)(1 - \beta^2)^{-\frac{1}{2}} = 0.7672,$$

$$(0.511)(1 - \beta^2)^{-\frac{1}{2}} = 0.7672,$$

$$(1 - \beta^2)^{-\frac{1}{2}} = 1.50,$$

$$\beta = 0.745,$$

$$v = \beta c = 2.24 \times 10^8 \text{ m/s} \quad \underline{\text{Ans}}.$$

(c) At this speed the electron's mass is

$$m = m_0(1 - \beta^2)^{-\frac{1}{2}} = 1.5m_0 = 1.37 \times 10^{-30} \text{ kg}$$

and its kinetic energy is $K = eV = 0.2562$ MeV <u>Ans</u>.

<u>SR-23</u>

In terms of the wavelength, the "effective mass" m is

$$m = \frac{E}{c^2} = \frac{h\nu}{c^2} = \frac{h}{\lambda c},$$

since $\nu = c/\lambda$. But

$$\frac{h}{c} = \frac{6.626 \times 10^{-34} \text{ J·s}}{3 \times 10^8 \text{ m/s}} = 2.2087 \times 10^{-42} \text{ kg·m},$$

or

$$m = \frac{0.02424\ m_e \cdot \text{Å}}{\lambda},$$

where m_e is the rest mass of the electron and λ is in Angstroms.

For the visible light photon (λ = 5000 Å), this gives for the "photon mass" $m = 4.85 \times 10^{-6}\ m_e$, and for the X-ray photon (λ = 1Å) $m = 0.02424\ m_e$ Ans.